Julian Trappe

The influence of medieval building activity on relief development within the Spessart uplands, Germany

WÜRZBURGER GEOGRAPHISCHE ARBEITEN

Herausgegeben vom Institut für Geographie und Geologie der Universität Würzburg in Verbindung mit der Geographischen Gesellschaft Würzburg

Band 125

Die Schriftenreihe Würzburger Geographische Arbeiten wird vom Institut für Geographie und Geologie zusammen mit der Geographischen Gesellschaft herausgegeben. Die Beiträge umfassen mit wirtschafts-, sozial- und naturwissenschaftlichen Forschungsperspektiven die gesamte thematische Bandbreite der Geographie. Der erste Band der Reihe erschien 1953.

Julian Trappe

The influence of medieval building activity on relief development within the Spessart uplands, Germany

A sedimentological, geophysical and GIS-based approach at different castle and mining sites

Würzburg
University Press

Dissertation, Julius-Maximilians-Universität Würzburg
Philosophische Fakultät, 2021
Gutachter: Prof. Dr. Roland Baumhauer, Prof. Dr. Julia Meister

Impressum

Julius-Maximilians-Universität Würzburg
Würzburg University Press
Universitätsbibliothek Würzburg
Am Hubland
D-97074 Würzburg
www.wup.uni-wuerzburg.de

Print on Demand

ISSN 0510-9833 (print)
ISSN 2194-3656 (online)
ISBN 978-3-95826-184-6 (print)
ISBN 978-3-95826-185-3 (online)
DOI 10.25972/WUP-978-3-95826-185-3
URN urn:nbn:de:bvb:20-opus-261499

Acknowledgments

I would like to say thank you to a vast number of people. Without them, this work would not have been possible.

First, I want to express my gratitude to Prof. Dr. Roland Baumhauer, the head of the chair of Physical Geography in Würzburg, for giving me the opportunity to do this work. He was my academic supervisor and my main mentor during this work. Second, thank you to my two other mentors, Prof. Dr. Christof Kneisel for teaching me several of the applied methods and allowing me to practice them in these studies, and Prof. Dr. Stefan Winkler for our academic exchange and his language and structural support of this work. Moreover, I would like to thank Jun.-Prof. Dr. Julia Meister for supervising this work and helping me finish it despite the deadline pressure in such troubled times.

I offer my thanks to my colleagues from the Chair of Physical Geography in Würzburg: Dr. Christian Büdel, Dr. Adrian Emmert, Julius Kunz, Dr. Angela Tintrup gen. Suntrup and Dr. Tobias Ullmann for our scientific exchange, their ideas in the field and in the office, their improvement of this work and, of course, their physical strength during fieldwork.

Additionally, I would like to thank all members of the "Archäologisches Spessartprojekt" (ASP): first and foremost, Harald Rosmanitz, Sabrina Bachmann and Michael Geißlinger for their archaeological support during excavation and beyond, but also Herrmann Bender, Claus Bergmann, Wolfgang and Angelika Beyer, Rainer Geschwinder, Harald Rosenberg and many more active and non-active members of the ASP for logistical support, hot meals, helpful input whenever something went wrong or whenever a Pürckhauer had to be pulled out of a stone wall. I also want to thank Joachim Lorenz for video-calling me and discussing the iron deposits at Eisenberg, and Dr. Nils Ostermeier for his help during the work with fluxgate magnetic surveying.

Not to be forgotten, many others who helped me in the field or otherwise: Felix Camenzuli, Anika Donner, Philipp Garbe, Georgius Giantidis, Patrick Horst, Nils Karges, Jenny Klemm, Alina Plotzki, David Scheer, Alexander Schreck, Jenny Weiß and Tim Wiegand, and all participants of the "invasive methods" course in 2018 and the students of several 'project courses' that took place in these study sites, all of whom helped me with measurements.

Further, thank you to my crisis-proof, lockdown office maid (and girlfriend) Nicole Liebers, with whom I shared an eight-square-meter office for the last ten months of preparing this work with no major crises.

Finally, I want to thank all my friends and family for supporting me during my entire studies.

Julian Trappe, December 2021

Contents

Contents VII
List of Figures X
List of tables XV
Abbreviations XVI
Abstract XVIII
Zusammenfassung XX

1 Introduction 1
1.1 Settlement history and research history of the relief development in the Spessart 1
1.2 Objectives 5
1.2.1 Objectives (O)1: Relief development and medieval building activity 5
1.2.2 Objectives (O)2: Suitability and transferability of the methodical approach 6

2 Study site 8
2.1 Overview 8
2.2 Waldaschaff (W) 11
2.3 Hauenstein (H) 15
2.4 Mömbris (M) 18
2.5 Alzenau (A) 20
2.6 Eisenberg (E) 22

3 Methods 26
3.1 GIS-based relief analysis and morphometric description 29
3.2 Geophysical methods 30
3.2.1 Electrical resistivity tomography 32
3.2.2 Ground-penetrating radar 42
3.2.3 Seismic refraction tomography 45
3.2.4 Magnetic surveying 47
3.3 Ground-truth 49
3.3.1 Invasive methods 50
3.3.2 Outcrop 51
3.4 Positioning 51
3.4.1 Global positioning system 51
3.4.2 Tachymeter 52
3.4.3 Manual positioning 52
3.5 Quantification of the medieval anthropogenic influence 52

4 Results and interpretation 55
4.1 Waldaschaff 55
4.1.1 Morphometric analyses for the creation of site-specific questions 55
4.1.2 W1 – *Origin of the rough relief*: results 58
4.1.3 W1 – *Origin of the rough relief*: interpretation 75
4.1.4 W2 – *Neck ditch*: results 78
4.1.5 W2 – *Neck ditch*: interpretation 83
4.1.6 W3 – *Fault system*: results 85
4.1.7 W3 – *Fault system*: interpretation 95
4.1.8 W4.1 – *Building structures on-site*: results 98
4.1.9 W4.1 – *Building structures on-site*: interpretation 103
4.1.10 W4.2 – *Outer bailey*: results 104
4.1.11 W4.2 – *Outer bailey:* interpretation 107
4.1.12 Quantification of the anthropogenic, medieval influence in Waldaschaff 109
4.2 Hauenstein 110
4.2.1 Morphometric analysis for the creation of site-specific questions 111
4.2.2 H2 – *Development of the talweg:* results 113
4.2.3 H2 – *Development of the talweg:* interpretation 121
4.2.4 H3 – *Center of the castle hill:* results 123
4.2.5 H3 – *Center of the castle hill*: interpretation 126
4.2.6 H4.1 – *Building structures on-site:* results 127
4.2.7 H4.1 – *Building structures on-site*: interpretation 132
4.2.8 H4.2 – *Outer bailey*: results 134
4.2.9 H4.2 – *Outer bailey:* interpretation 136
4.2.10 Quantification of the medieval, anthropogenic influence in Hauenstein 137
4.3 Mömbris 138
4.3.1 Morphometric analysis for the creation of site-specific questions 139
4.3.2 M4.1 – *Building structures:* results 140
4.3.3 M4.1 – *Building structures:* interpretation 148
4.3.4 Quantification of the medieval, anthropogenic influence in Mömbris 151
4.4 Alzenau 151
4.4.1 Morphometric analysis for the creation of site-specific questions 151
4.4.2 A1 – *Loess coverage:* results 153
4.4.3 A1 – *Loess coverage*: interpretation 161
4.4.4 A4.1 – *Building structures on-site:* results 162
4.4.5 A4.1 – *Building structures on-site:* interpretation 173
4.4.6 Quantification of the medieval, anthropogenic influence in Alzenau 176

4.5 Eisenberg ... 177
4.5.1 Morphometric analyses for the creation of site-specific questions . 177
4.5.2 E2 – *Collapsed shafts:* results ... 179
4.5.3 E2 – *Collapsed shafts:* interpretation ... 186
4.5.4 E3 – *Hydrological situation:* results ... 190
4.5.5 E3 – *Hydrological situation:* interpretation ... 195
4.5.6 Quantification of the medieval, anthropogenic influence at the Eisenberg ... 197

5 Discussion and outlook ... 198
5.1 Site-specific ... 198
5.1.1 Waldaschaff ... 199
5.1.2 Hauenstein ... 203
5.1.3 Mömbris ... 205
5.1.4 Alzenau ... 206
5.1.5 Eisenberg ... 207
5.2 Transferability of the results ... 209
5.3 Applied methods ... 213

6 Conclusions and key findings ... 224
6.1 Site-specific key findings ... 224
6.1.1 Waldaschaff ... 224
6.1.2 Hauenstein ... 225
6.1.3 Mömbris ... 225
6.1.4 Alzenau ... 226
6.1.5 Eisenberg ... 226
6.2 Key findings for the Spessart and at a larger scale ... 227
6.3 Methodological key findings ... 228
6.4 Synthesis ... 229

7 References ... 231

Appendix ... 245

List of Figures

Figure 1: Medieval castle sites within and near the Spessart uplands and landscape units on DEM ... 2
Figure 2: Overview of the study sites ... 9
Figure 3: Climate graphs ... 12
Figure 4: Study site – Waldaschaff ... 13
Figure 5: Field pictures – Waldaschaff ... 14
Figure 6: Study site – Hauenstein ... 16
Figure 7: Field pictures – Hauenstein ... 17
Figure 8: Study site – Mömbris ... 18
Figure 9: Field pictures – Mömbris ... 19
Figure 10: Study site – Alzenau ... 20
Figure 11: Field pictures – Alzenau ... 21
Figure 12: Study site – Eisenberg ... 23
Figure 13: Field pictures – Eisenberg ... 24
Figure 14: Model of GIS-based relief analyses ... 27
Figure 15: Model of the workflow ... 28
Figure 16: Homogeneous halfspace ... 33
Figure 17: Electrode configuration, geometric value and calculation of apparent resistivity ... 34
Figure 18: Flowchart – boundary analysis ... 38
Figure 19: Flowchart – guidelines for ERT description ... 40
Figure 20: Guidelines for ERT description – overview ... 41
Figure 21: Setup of GPR ... 43
Figure 22: Simplified setup of SRT ... 46
Figure 23: Field picture – Breaking-off edge Waldaschaff ... 56
Figure 24: Morphometric analyses – Waldaschaff ... 56
Figure 25: Location of measurements – W1 ... 59
Figure 26: Electrical resistivity tomography – W1_ERT_a ... 59
Figure 27: Boundary analysis – W1_ERT_a ... 60
Figure 28: Electrical resistivity tomography – W1_ERT_b ... 61
Figure 29: Boundary analysis – W1_ERT_b ... 61
Figure 30: Electrical resistivity tomography – W1_ERT_c ... 62
Figure 31: Boundary analysis – W1_ERT_c ... 63
Figure 32: Electrical resistivity tomography – W1_ERT_d ... 64
Figure 33: Boundary analysis – W1_ERT_d ... 64
Figure 34: Electrical resistivity tomography – W1_ERT_e ... 65
Figure 35: Boundary analysis – W1_ERT_e ... 66
Figure 36: Electrical resistivity tomography – W1_ERT_f ... 66
Figure 37: Boundary analysis – W1_ERT_f ... 67
Figure 38: Electrical resistivity tomography – W1_ERT_g ... 68
Figure 39: Boundary analysis – W1_ERT_g ... 68
Figure 40: Ground-penetrating radar measurement – W1_GPR_a ... 69

Figure 41: Seismic refraction tomography: W1_SRT_a 70
Figure 42: Boreholes – W1 71
Figure 43: Pürckhauer soundings – W1 73
Figure 44: Edelman soundings – W1 74
Figure 45: Field pictures – Outcrop W1_OC_a 75
Figure 46: Conceptual models – W1 77
Figure 47: Location of measurements – W2 79
Figure 48: Electrical resistivity tomography – W2_ERT_a 79
Figure 49: Boundary analysis – W2_ERT_a 80
Figure 50: Electrical resistivity tomography – W2_ERT_b 81
Figure 51: Boundary analysis – W2_ERT_b 81
Figure 52: Electrical resistivity tomography – W2_ERT_c 82
Figure 53: Boundary analysis – W2_ERT_c 83
Figure 54: Location of measurements – W3 85
Figure 55: Electrical resistivity tomography – W3_ERT_a 86
Figure 56: Boundary analysis – W3_ERT_a 86
Figure 57: Electrical resistivity tomography – W3_ERT_b 87
Figure 58: Boundary analysis – W3_ERT_b 88
Figure 59: Ground-penetrating radar measurement – W3_GPR_a 89
Figure 60: Ground-penetrating radar measurement – W3_GPR_a 89
Figure 61: Ground-penetrating radar measurement – W3_GPR_c 90
Figure 62: Ground-penetrating radar measurement – W3_GPR_d 91
Figure 63: Ground-penetrating radar measurement – W3_GPR_e 91
Figure 64: Ground-penetrating radar measurement – W3_GPR_f 92
Figure 65: Ground-penetrating radar measurement – W3_GPR_g 92
Figure 66: Seismic refraction tomography – W3_SRT_a 93
Figure 67: Seismic refraction tomography – W3_SRT_b 94
Figure 68: Seismic refraction tomography – W3_SRT_c 94
Figure 69: Boreholes – W3 95
Figure 70: Conceptual models – W3 97
Figure 71: Location of measurement – W4.1 98
Figure 72: Electrical resistivity tomography – W4.1_ERT_a 99
Figure 73: Electrical resistivity tomography – W4.1_ERT_b 99
Figure 74: Boundary analysis – W4.1_ERT_a 100
Figure 75: Boundary analysis – W4.1_ERT_b 100
Figure 76: 3D electrical resistivity tomography – W4.1_3DERT_a 101
Figure 77: 3D ground-penetrating radar – W4.1_3DGPR_a 102
Figure 78: Magnetic survey – W4.1_MAG_a 102
Figure 79: Interpretation – W4.1 105
Figure 80: Location of measurements – W4.2 105
Figure 81: Magnetic survey – W4.2_MAG_a 106
Figure 82: Magnetic surveys – W4.2 106
Figure 83 Outcrop – W4.2_OC_a 107
Figure 84: Interpretation – northern part of W4.2 108
Figure 85: Intensity and type of medieval, anthropogenic influence – Waldaschaff 110

Figure 86: Morphometric analyses – Hauenstein 111
Figure 87: Location of measurements – H2 113
Figure 88: Electrical resistivity tomography – H2_ERT_a 114
Figure 89: Boundary analysis – H2_ERT_a 114
Figure 90: Electrical resistivity tomography – H2_ERT_b 115
Figure 91: Boundary analysis – H2_ERT_b 116
Figure 92: Boreholes and 14C-ages – H2 117
Figure 93: Pürckhauer soundings – H2 119
Figure 94: Field pictures – outcrops H2 121
Figure 95: Conceptual model – H2 (south-north orientated) 123
Figure 96: Location of measurements – H3 124
Figure 97: Electrical resistivity tomography – H3_ERT_a 124
Figure 98: Boundary analysis – H3_ERT_a 125
Figure 99: Borehole – H3_BH_a 126
Figure 100: Conceptual model – H3 127
Figure 101: Location of measurements – H4.1 128
Figure 102: Electrical resistivity tomography – H4.1_ERT_a 128
Figure 103: Boundary analysis – H4.1_ERT_a 129
Figure 104: Electric resistivity tomography – H4.1_ERT_b 129
Figure 105: Boundary analysis – H4.1_ERT_b 130
Figure 106: Electrical resistivity tomography – H4.1_ERT_c 131
Figure 107: Boundary analysis – H4.1_ERT_c 131
Figure 108: Interpretation – H4.1 133
Figure 109: Location of measurements – H4.2 134
Figure 110: 3D ground-penetrating radar measurement – H4.2_3DGPR_a 135
Figure 111: Magnetic survey – H4.2_MAG_a 136
Figure 112: Borehole – H4.2_BH_a 136
Figure 113: Intensity and type of medieval, anthropogenic influence – Hauenstein 138
Figure 114: DEM of the castle site in Mömbris 139
Figure 115: Location of measurements – M4.1 140
Figure 116: 3D electrical resistivity tomography – M4.1_ERT3D_a 141
Figure 117: 3D model – M4.1_ERT3D_a 141
Figure 118: 3D ground-penetrating radar measurement – M4.1_3DGPR_a 142
Figure 119: Lines of 3D ground-penetrating radar – M4.1 143
Figure 120: 3D ground-penetrating radar 144
Figure 121: Lines of 3D ground-penetrating radar 145
Figure 122: 3D mode –: M4.1_3DGPR_a and -b 146
Figure 123: Magnetic survey – M4.1_MAG_a 146
Figure 124: Borehole – M4.1_BH_a 147
Figure 125: Interpretation – M4.1_3DERT_a 149
Figure 126: Interpretation – M4.1_3DGPR_a and -b 150
Figure 127: Morphometric analyses – Alzenau 152
Figure 128: Location of measurements – A1 153
Figure 129: Electrical resistivity tomography – A1_ERT_a 154

Figure 130: Boundary analysis – A1_ERT_a 154
Figure 131: Electrical resistivity tomography – A1_ERT_b 155
Figure 132: Boundary analysis – A1_ERT_b 156
Figure 133: Electrical resistivity tomography – A1_ERT_c 157
Figure 134: Boundary analysis – A1_ERT_c 157
Figure 135: Electrical resistivity tomography – A1_ERT_d 158
Figure 136: Boundary analysis – A1_ERT_d 159
Figure 137: Boreholes – A1 160
Figure 138: Outcrops – A1_OC_a 160
Figure 139: Conceptual model – A1 162
Figure 140: Location of measurements – A4.1 163
Figure 141: Electrical resistivity tomography – A4.1_ERT_a 163
Figure 142: Boundary analysis – A4.1_ERT_a 164
Figure 143: Electrical resistivity tomography – A4.1_ERT_b 164
Figure 144: Boundary analysis – A4.1_ERT_b 165
Figure 145: Electrical resistivity tomography – A4.1_ERT_c 166
Figure 146: Boundary analysis – A4.1_ERT_c 166
Figure 147: Ground-penetrating radar measurement – A4.1_GPR_a 167
Figure 148: Ground-penetrating radar measurement – A4.1_GPR_b 168
Figure 149: Ground-penetrating radar measurement – A4.1_GPR_c 169
Figure 150: Ground-penetrating radar measurement – A4.1_GPR_d 169
Figure 151: 3D ground-penetrating radar measurement – A4.1_3DGPR_a 170
Figure 152: 3D ground-penetrating radar measurement – A4.1_3DGPR_b 171
Figure 153: 3D ground-penetrating radar measurement – A4.1_3DGPR_c 172
Figure 154: Magnetic surveys – A4.1_MAG 173
Figure 155: Field picture – Outcrop A4.1_OC_e 174
Figure 156: Intensity and type of medieval, anthropogenic influence – Alzenau 177
Figure 157: Morphometric analyses – Eisenberg 178
Figure 158: Location of measurements – E2 179
Figure 159: Electrical resistivity tomography – E2_ERT_a 180
Figure 160: Boundary analysis – E2_ERT_a 180
Figure 161: Electrical resistivity tomography – E2_ERT_b 181
Figure 162: Boundary analysis – E2_ERT_b 182
Figure 163: Ground-penetrating radar measurement – E2_GPR_a 183
Figure 164: Ground-penetrating radar measurement – E2_GPR_b 184
Figure 165: Outcrops – E2 185
Figure 166: Field pictures – Outcrops E2 185
Figure 167: Conceptual model – E2 188
Figure 168: Geological profile 188
Figure 169: Location of measurements – E3 189
Figure 170: Ground-penetrating radar measurement – E3_GPR_a 190
Figure 171: Ground-penetrating radar measurement – E3_GPR_b 191
Figure 172: Ground-penetrating radar measurement – E3_GPR_c 192
Figure 173: Ground-penetrating radar measurement – E3_GPR_d 193

Figure 174: Ground-penetrating radar measurement – E3_GPR_e 193
Figure 175: Magnetic survey – E3_MAG_a 194
Figure 176: Edelman and Pürckhauer soundings – E3 194
Figure 177: Intensity and type of medieval, anthropogenic influence – Eisenberg 197
Figure 178: Comparison of the lithological units after geological map and after the results 200
Figure 179: Model of the landscape development – Waldaschaff 201
Figure 180: Models of the distribution of different topographic features and sediment bodies 204
Figure 181: Field picture: iron ore at the Eisenkopf 208
Figure 182: Model of micro-relief development in an area of medieval usage 211
Figure 183: Comparison of different models and inversion processes for the boundary analysis 220

List of tables

Table 1: Guidelines for ERT description – values....42
Table 2: Classified anthropogenic influences on the relief during the medieval period....54
Table 3: Abbreviation code: study site, question type and method....55
Table 4: Site-specific questions in Waldaschaff....58
Table 5: ERT guidelines – W1_ERT_a....60
Table 6: ERT guidelines – W1_ERT_b....62
Table 7: ERT guidelines – W1_ERT_c....63
Table 8: ERT guidelines – W1_ERT_d....65
Table 9: ERT guidelines – W1_ERT_e....66
Table 10: ERT guidelines – W1_ERT_f....67
Table 11: ERT guidelines – W1_ERT_g....69
Table 12: ERT guidelines – W2_ERT_a....80
Table 13: ERT guidelines – W2_ERT_b....81
Table 14: ERT guidelines – W2_ERT_c....83
Table 15: ERT guidelines – W3_ERT_a....87
Table 16: ERT guidelines – W3_ERT_b....88
Table 17: ERT guidelines – W4.2_ERT_a and -b....100
Table 18: Site-specific questions in Hauenstein....112
Table 19: ERT guidelines – H2_ERT_a....115
Table 20: ERT guidelines – H2_ERT_b....116
Table 21: ERT guidelines – H3_ERT_a....125
Table 22: ERT guidelines – H4.1_ERT_a....129
Table 23: ERT guidelines – H4.1_ERT_b....131
Table 24: ERT guidelines – H4.1_ERT_c....132
Table 25: Grain-size distribution – M4.1_BH_a pipetting....148
Table 26: Site-specific questions in Alzenau....153
Table 27: ERT guidelines – A1_ERT_a....155
Table 28: ERT guidelines – A1_ERT_b....156
Table 29: ERT guidelines – A1_ERT_c....158
Table 30: ERT guidelines – A1_ERT_d....159
Table 31: ERT guidelines – A4.1_ERT_a....164
Table 32: ERT guidelines – A4.1_ERT_b....165
Table 33: ERT guidelines – A4.1_ERT_c....166
Table 34: Site-specific questions at the Eisenberg....179
Table 35: ERT guidelines – E2_ERT_a....181
Table 36: ERT guidelines – E2_ERT_b....182
Table 37: Guideline profile with a generalized, relative resistivity distribution in Waldaschaff....217

Abbreviations

A	Alzenau (study site)
AMS	Accelerator mass spectrometry
a.s.l.	Above sea level
ASP	Archäologisches Spessartprojekt: the archeology project of the Spessart region
BCE	Before the common era
BH	Borehole
BP	Before present
CE	Common era
DEM	Digital elevation model
E	Eisenberg (study site)
EDE	Edelman sounding
ERT	Electrical resistivity tomography
GIS	Geoinformation system
GPR	Ground-penetrating radar
GPS	Global positioning system
H	Hauenstein (study site)
M	Mömbris (study site)
MAG	Fluxgate magnetic survey
OC	Outcrop
PCB	Periglacial coverbeds
PUE	Pürckhauer sounding
RTK	Real-time kinematic
SRT	Seismic refraction tomography
W	Waldaschaff (study site)

Die Umwelt-Mensch-Beziehungen haben sich seit der letzten Eiszeit verkehrt:
An der Pleistozän/Holozän-Wende wurden kulturelle Entwicklungen
durch Umwelt-Veränderungen beeinflusst […].
Heute erleben wir eine umgekehrte Entwicklung, nämlich eine anthropogene
Veränderung der Geosysteme […]. Dadurch entstehen globale Umweltprobleme;
sie sind die größten Herausforderungen an die Menschheit [….] Heine (1994)

Contrary to the transition between the Pleistocene and the Holocene, when
cultural development was impacted by the environment; today, the development
of culture and environment seems to be upside down. Cultural development
is strongly influencing the development of the environment and the environment
itself. This situation causes problems for different parts of the environment,
even on a global scale. (free translated after Heine 1994)

Abstract

Mostly taking place in the Spessart, a low mountain range in central Germany, a feud between the counts of Rieneck and the bishops in Mainz escalated during the medieval period. This feud led to many castles being built at both sites, as well as other building activities. Many of these buildings and castles were located at specific and unique relief positions. This study analyzes the mutual influence of (paleo-) relief development and medieval building activity by utilizing a multi-method approach.

For this purpose, GIS-based terrain analyses and geophysical measurements were conducted and combined with sedimentological information from invasive methods and outcrops to assess information on the topographic relief at four medieval castle sites and a medieval mining area. From the recent relief, the influence of medieval building activity was revealed and the paleo-relief was derived. The study sites were chosen in conjunction with the Archaeology Project of the Spessart Region (Archäologisches Spessartprojekt), which performed or is planning to perform excavations at these sites.

With the help of GIS-based terrain analyses, the main objective was split into a number of site-specific questions. Spatial data were collected with the help of electrical resistivity tomography, ground-penetrating radar, seismic refraction tomography and fluxgate magnetic surveying to answer these questions. The ambiguous, spatial geophysical results were proven using Pürckhauer and Edelman soundings and with the help of boreholes and outcrops. More precise information about the selective soundings was extrapolated using geophysical data. Additionally, with these established methods, a new approach for the description of the relative resistivity distribution within the subsurface was developed. In addition to a quantitative description of the data, additional parameters were derived with the help of a constructed ArcGIS-tool.

By interpreting all these data, most of the site-specific questions were answered. These questions are of variable complexity and focus on different geomorphological, geological, geoarchaeological and archaeological topics. For example, the genesis of the rough topography surrounding *Wahlmich Castle* near Waldaschaff was deciphered. Additionally, mainly boreholes and AMS-^{14}C-datings revealed a connection between the medieval usage of a castle with a mill, including mill ponds, and the sedimentation history in a talweg at *Hauenstein Castle*. With the help of a geophysical prospection, the location of *Mömbris Castle* was precisely predicted, which resulted in a targeted excavation off-target the initially suspected areas. At *Rannenburg Castle* near Alzenau, massive, anthropogenically used loess loam deposits were prospected. Furthermore, at *Eisenberg* near Steinau an der Straße, important information was derived based on the position and form of iron deposits and collapsed shafts.

In addition to all these key contributions for the archaeological and/or geomorphological interpretation of individual study sites, quantification of the anthropogenic influence on the relief was conducted. A generalized model that focuses on the mutual influence of relief development and medieval building activity was derived.

Furthermore, new research approaches were developed that focus not only on positive location factors for a specific use, but also on the negative location factors for another type of use. The newly developed approach for describing the relative resistivity distribution can also be transferred to other studies to increase objectivity. Moreover, it is possible to create generalized datasets for individual study sites.

Thus, this study links geomorphological/geological and (geo-)archaeological investigations at a medieval site and delivers important information on the mutual influence between the research items of all these disciplines within the Spessart and beyond.

Zusammenfassung

Nachdem im Mittelalter eine Fehde zwischen den Grafen von Rieneck und den Mainzer Bischöfen eskalierte, wurden im Spessart, einem deutschen Mittelgebirge, häufig in sehr spezifischen und ungewöhnlichen Reliefpositionen von beiden Streitparteien diverse Burgen errichtet sowie andere Bautätigkeiten vorgenommen. Die vorliegende Arbeit widmet sich dieser Landschaft, indem sie den wechselseitigen Einfluss von (Paläo-)Reliefentwicklung und mittelalterlicher Bauaktivität mit Hilfe eines Multimethodenansatzes analysiert.

Im Rahmen dieser Arbeit wurden an vier mittelalterlichen Burgenstandorten und einem mittelalterlichen Bergbaugebiet GIS-basierte Reliefanalysen und geophysikalische Messungen durchgeführt sowie sedimentologische Informationen aus invasiven Methoden und Aufschlüssen aufgenommen, um das rezente Relief zu erfassen, den Einfluss der mittelalterlichen Bauaktivität darzustellen und das Paläorelief abzuleiten. Die Untersuchungsgebiete wurden mithilfe des *Archäologischen Spessartprojektes* ausgesucht, welches an den jeweiligen Standorten Ausgrabungen durchgeführt hat oder plant.

Mit Hilfe einer GIS-basierten Reliefanalyse wurde die Hauptfragestellung in eine Anzahl standortspezifischer Fragestellungen untergliedert. Für die Beantwortung dieser Fragestellungen wurden mit geoelektrischer Widerstandstomographie, Bodenradar, Refraktionsseismik und Fluxgatemagnetik geophysikalische Daten erhoben. Die mehrdeutigen aber flächenhaften geophysikalischen Daten wurden mit punktuellen Pürckhauer-, Edelman- und Rammkernsondagen genauer definiert; diese präziseren Informationen der punktuellen Sondagen konnten wiederum mit Hilfe der geophysikalischen Daten extrapoliert werden. Zusätzlich zu diesen etablierten Methoden wurde ein neuer Ansatz zur Beschreibung von der relativen elektrischen Widerstandsverteilung im Untergrund entwickelt und neben einer quantitativen Beschreibung der Daten auch zusätzliche Parameter mit Hilfe eines konstruierten ArcGIS-Tools abgeleitet.

Durch die übergreifende Interpretation der Daten konnten die aufgeworfenen standortspezifischen Fragestellungen beantwortet werden. Nachdem diese Fragestellungen in ihrer Komplexität variierten und zudem verschiedene geomorphologische, geologische, geoarchäologische und archäologische Themen fokussierten, konnte eine große Bandbreite an Befunden generiert werden. So konnte beispielsweise die Genese einer markanten und rauen Topographie in der Umgebung des Burgstalls Wahlmichs in Waldaschaff entschlüsselt werden. Die Sedimentation in einer Bachaue bei Hauenstein konnte mit Hilfe von Bohrungen und ^{14}C-Datierungen klar zu der mittelalterlichen Nutzung eines Burgstalls mit Mühle und entsprechenden Mühlteichen zugeordnet werden. In Mömbris konnte die Lage der eigentlichen Burganlage mit Hilfe einer geophysikalischen Prospektion präzisiert werden, was zu einer gezielten Ausgrabung außerhalb der eigentlichen Verdachtsflächen geführt hat. An der Rannenburg in der Nähe von Alzenau konnten mächtige, anthropogen genutzte Lössauflagen prospektiert werden und am Eisenberg in der Nähe von Steinau an der Straße konnten wichtige Informationen zur Lage und Form von Erzlagerstätten und Pingen gewonnen werden.

Neben all diesen wichtigen Beiträgen zur archäologischen und geomorphologischen Interpretation der einzelnen Standorte konnte der anthropogene Einfluss an jedem Standort quantifiziert werden und ein generalisiertes Modell zur wechselseitigen Einflussnahme von Reliefentwicklung und mittelalterlicher Bebauung abgeleitet werden. Des Weiteren konnten neue Forschungsansätze entwickelt werden, welche nicht nur die Standortvorteile für bestimmte Nutzungen einzelner Standorte fokussieren, sondern auch die Standortnachteile für andere Nutzungsformen betrachten. Außerdem kann der neu entwickelte Ansatz zur Beschreibung der relativen Widerstandsverteilung im Untergrund auch auf zukünftige Fragestellungen übertragen werden, um die Objektivität bei der Betrachtung von verschiedenen Datensätzen zu erhöhen und gegebenenfalls auch generalisierte Datensätze für einzelne Standorte zu erstellen.

Somit bildet diese Arbeit ein Bindeglied zwischen der geomorphologischen beziehungsweise geologischen und (geo-)archäologischen Erforschung eines mittelalterlichen Standortes und liefert wichtige Informationen über die wechselseitige Einflussnahme dieser Forschungsgegenstände im Spessart und darüber hinaus.

1 Introduction

Certainly, not many people would contradict the abovementioned statements of Heine (1994). However, is the influence between cultural and environmental development only one way, or do they always influence each other, with changing intensity?

The influence of human activities on the processes, structures and functions of landscapes over decades and centuries has been known since antiquity (Strabo, 1836 after Karl Kärcher) and has been researched in various investigations (e.g. Hooke & Martín-Duque, 2012; Ellis *et al.*, 2013). Humans should be considered as a factor for landscape development since there is no place on earth not affected by them (Schütt, 2006a). Hence, the influence of land use on the function and development of landscapes in past centuries must be considered as a major issue for modern landscape development (Bork, 1998).

The relevance, motivation and historical background for analyzing medieval building activity's influence on relief development within the Spessart uplands are provided in the following section.

1.1 Settlement history and research history of the relief development in the Spessart

The Spessart, an upland region in central Germany (Figure 1), was settled during prehistoric times, at least in the river valleys, as the Altenburg between Sulzbach and Leinenbach reveals for the period between about 500 and 400 BCE (Lagies, 2005; Rosmanitz, 2012). However, the mountain ridges and other parts of the Spessart were not settled in 2000 BP, even though many parts of lower Franconia were settled at this time (Kolb & Krenig, 1989). Although the use of the forest is documented within vast parts of the Spessart in the 8th century CE (Cramer, 1952), until the High Middle Ages, settlements were rare in the Spessart uplands (Kolb & Krenig, 1992; see also for the prehistory of the region Marquart, 2017).

It seems likely that the High Middle Ages were characterized by the pressure of population growth. Before the great plague in central Europe and Scandinavia in the middle of the 14th century CE, an increase in this area's population from four million in 1000 CE up to 11.5 million in 1340 CE has postulated by various authors (e.g. Bork, 1998). Due to the position of the Spessart within different agglomeration areas, this increase may have affected the region. Additionally, parts of the area were often granted as donations to dioceses or other religious corporations. Therefore, the Spessart uplands were divided between several authorities, such as Mainz, Fulda and Würzburg, but also the county of Rieneck (Kolb & Krenig, 1992).

Figure 1: Medieval castle sites within and near the Spessart uplands and landscape units on DEM

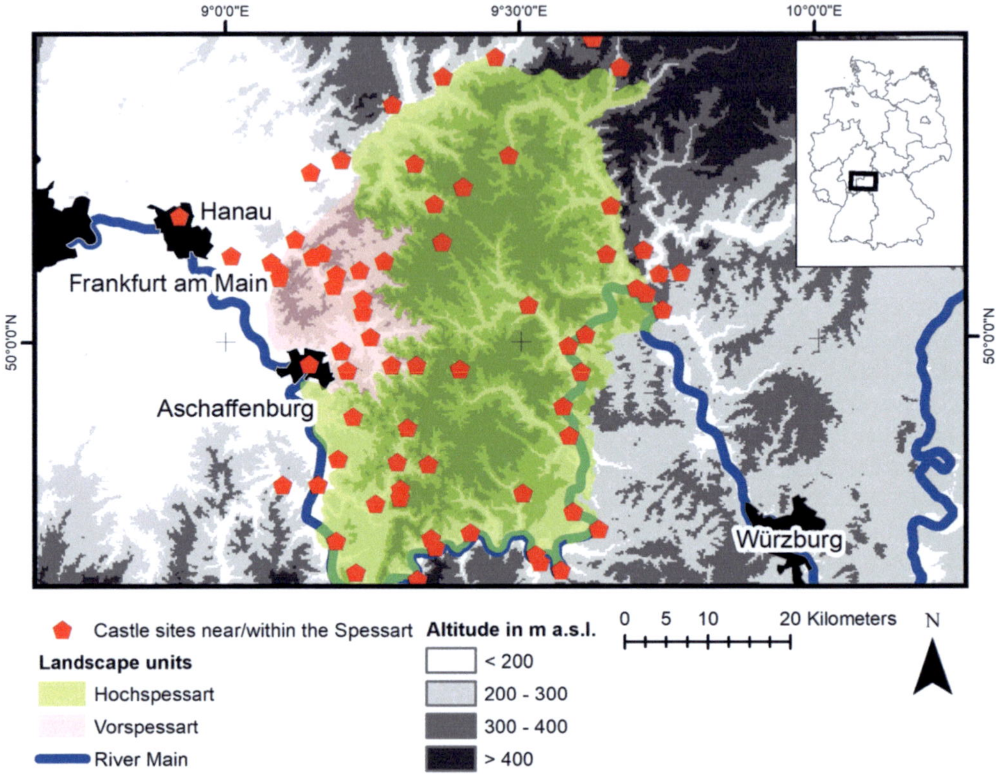

Sources: Jung, 2006; Hessian Administration for Land Management and Geoinformation, 2007; Bavarian Environment Agency, 2020a; Hessian Agency for Nature Conservation, Environment and Geology, 2020
Map creation: J. Trappe

Within the Spessart uplands, the soils were poor for agriculture, as in other forest-rich, low mountain ranges in central Europe outside of loess deposits (Bork, 1998). The small-sized, nutrient-poor soils made of Buntsandstein have particularly low fertility; therefore, even today the border between crystalline and Buntsandstein is mainly the border between field crops and forest (Kampfmann & Krimm, 1988). Especially in the medieval period, the forests were economically important (Büdel *et al.*, 2021). Pollen diagrams of the Wiesbüttmoor reveal an increase of copper beech since 1950 BCE (Lagies, 2005). The potential natural vegetation is in line with these diagrams (Zerbe, 1997). Beginning in the 11th century CE, the beeches were felled in small clearings. From 1200 to 1250 CE, larger clearings without the existence of beeches are reported in the diagrams (Lagies, 2005). These beech-rich forests were needed for glass manufacturing or, at least, for the production of charcoal (Lagies, 2005). Since the 8th century CE, in Franconian territory, glass could be smelted using ash instead of soda. At that time, glass production was located in present-day France, but from the 12th century CE, glassworks came into production in different parts of present-day Germany, too. Either ferns or beeches could be used for the production of the necessary ash (Wedepohl, 2003). The enormous consumption of beeches for glass production was mentioned for the Spessart (Ermischer, 2019). Due to urban growth, especially in the area of the Lower Rhine in the 12th and 13th centuries CE, glass manufactories became increasingly crucial (Wamser, 1979; Boss & Wamser, 1984). Early Gothic construction especially had a greater need for glass than earlier epochs (Wedepohl, 2003). One document attests to a guild running glasswork in the Spessart uplands in 1406 CE, while the first authentic mention is dated to 1349 CE (Schäfer, 1991). Most of the excavated glass manufactories are of later periods, up to the 18th century CE (Wamser, 1982). However, there was at least one found in Schöllkrippen dating to the 13th century CE (Boss & Wamser, 1984), while the earliest findings of glass production within the Spessart can be dated to the 11th century CE (Wamser, 1979).

In several parts of the Spessart uplands, quarrying was economically important (Preuschoff, 1996). In the 19th century CE, which was the prime period for mining in the Spessart, the industry was still limited compared with other areas (Okrusch *et al.*, 2011). However, in a few locations, for example, at Bieber (Freymann, 1991), archival sources provide the first evidence of mining in the 15th century CE.

Not only were the natural resources used in the Spessart, but the relief and the geological setting of the Buntsandstein were also economically important. Old pathways and roads tended to be built through the Spessart. Buntsandstein provides a dry and stable foundation for roads and pathways, and the relief is characterized by elongated ridges without huge differences in gradient. Additional dense, clay-rich layers, with a large amount of stagnant water and many springs providing necessary drinking water for draft animals, allowed pathways to run through the area without it being settled first. Different archival sources suggest that pathways have crossed the Spessart since the 8th century CE (Landau, 1958; Kampfmann & Krimm, 1988), and while definitive proof has been found for the 13th century CE (Kampfmann & Krimm, 1988; Himmelsbach, 2014). The most important supra-regional pathways seem to be the *Eselsweg* in a north-south direction and the *Birkenhainer Landstraße* in an east-west direction (Himmelsbach, 2014). These pathways and the River Main waterway were not only essential for the collection of tolls, but also allowed the transportation of important economic goods such as wood, charcoal and cattle (Kampfmann & Krimm, 1988; Ermischer, 2008; Himmelsbach, 2014).

In combination with political factors, all these economic and strategic reasons help explain the decision of the counts of Rieneck to extend their territory in the Spessart uplands during the 13th century CE. Subsequently, "von Eppstein" placed the bishop (Ruf, 2019) and used this powerful position to defend the growth of the counts, whom they traditionally regarded as their enemies. Therefore, both parties built castles and other settlements in the area of investigation (cf. Figure 1). Even though the position of the bishop as chancellor was far more powerful compared with the counts, the counts have been described as economically advanced, although the reason for this is as yet unknown (Ruf, 2019). The conflict escalated in 1260 CE and resulted in the destruction or abandonment of many of these castles (Kolb &

Krenig, 1992). Several contracts dated between July 21, 1260 CE and June 25, 1271 CE describe the destruction of all the castles and clearings of Rieneck in the territory of Mainz. Additionally, some reparations had to be made, and Rieneck was forbidden to build castles again on Mainz ground or to provoke hostilities with Mainz and their allies (Schecher, 1969; Ruf, 1984). A document dated March 17, 1266 CE concerns the destruction of several castles within a period of 14 days (Ruf, 1984, 2019).

Other authorities were involved in this feud as well, for example, "von Hanau" and "von Kälberau" (e.g. Ruf, 1984; Schäfer, 2000; Ruf, 2019), mostly in the bishop's party. However, due to imprecise archival sources, the extent of involvement and party affiliations are often unclear; these party affiliations could also be changed during the feud (Ruf, 2019). Therefore, an exact assignment of castles to authorities is not always possible. Names of castles, collected from written sources, cannot always be matched precisely to respective castle sites (e.g. Ruf, 2019). Some of the castles were occupied by the lower nobility seeking social advancement. However, in Aschaffenburg, the number of members of the nobility increased after the conflict. On the one hand, there was no need for such castles within the uplands anymore. On the other hand, it could be more socially desirable to live in the city than in a castle in the uplands (Ermischer, 2006). Other members of the lower nobility began assaulting escorts. Therefore, the cities of Gelnhausen and Frankfurt removed these so-called robber knights and their castles at the beginning of the 15th century CE in the so-called "Wetterauer feud" (e.g. Griebel, 1982).

Many of the described castles and other medieval buildings have been known and catalogued since early modern times (Gröber & Karlinger, 1916). However, systematic research on them started in 1998 with the founding of the ASP ("Archäologisches Spessartprojekt"; the Archeology Project of the Spessart Region) and annual excavations of these buildings and structures (e.g. Rosmanitz & Gresse, 2011).

In recent times, Larsen *et al.* (2013; 2016) described periodic erosion processes and gully forming within the area of investigation. Periglacial processes of the Pleistocene were studied in recent works by Mueller (2011) and Mueller and Thiemeyer (2014). Their pedological research focuses on lessivation and podsolisation in the periglacial cover beds (PCBs) of the region. In addition, Jung (2004) describes the geomorphological importance of PCBs and explains the reworking of Tertiary weathering horizons within them. Also, the relief development in the Spessart in the context of the tectonic influence is described by Jung (2006).

As mentioned above, the High Middle Ages were the first significant period of settlement founding, building construction and land use within the Spessart uplands. Nevertheless, studies that have focused on the Holocene relief development and that have systematically linked it with medieval building activity are rare for this area. The influence on flora and fauna is often discussed in the context of medieval activity within the area of investigation (e.g. Lagies, 2005; Dannath *et al.*, 2013). Additionally, the anthropogenic effect of the medieval period on the relief is mentioned in various archaeological studies because its description is necessary for every archaeological excavation in Bavaria (e.g. Bavarian State Office for Monument

Preservation, 2020). However, the quantification, classification and systematization of this anthropogenic impact have not yet been performed in a geomorphological work for the Spessart.

1.2 Objectives

The High Middle Ages' importance as the first significant period of settlement-founding and castle construction within the Spessart uplands is well-known and has been investigated. Although the castles' topographic positions are generally known and have been described (e.g. Scherer-Hall, 1998; Biller, 2016), the effect on the relief development and how the specific micro-relief impacted castle construction have not been investigated from a geomorphological perspective. There remains somehow a disconnection between relief development and medieval building activity in the research literature. Therefore, this work reconstructs, classifies and quantifies the paleo-relief, predominantly on the microscale (Leser, 1995), including the remnants of natural processes and the anthropogenic effect on these processes and the relief. This is performed by analyzing the recent surface and subsurface geomorphological setting and the anthropogenic structures and earthworks within or on it. For this reason, this work tries to link geomorphological research with archaeological knowledge to answer the following questions:

> *How do relief development and medieval building activity in the Spessart uplands influence each other?*
>
> *How can this influence be detected, measured and classified?*

These main questions raise several detailed further questions, which can be split into two "bunches." The first bunch focuses on relief development and medieval building activity and can be divided into two groups as well. The first group focuses on site-specific questions and relevance. The first group's spatially limited information is transferred and generalized to a wider area within the second group.

The second bunch focuses on the methodological approach used in this study.

1.2.1 Objectives (O)1: Relief development and medieval building activity

Group 1: Regional objectives

- *O1.1.1 Is it possible to decide which parts of the present relief are predominantly of natural origin and which have been intensely anthropogenically influenced at the individual study sites?*

- *O1.1.2 What specific interactions of relief development and medieval building activity exist at the single study sites? How can these be detected, measured and quantified to answer the site-specific questions according to the following categories?*

  1. *Surficial, natural features*
  2. *Surficial, anthropogenic features*
  3. *Buried, natural features*
  4. *Buried, anthropogenic features*
     4.1. *On-site*
     4.2. *Off-site*

- *O1.1.3 What general interactions can be derived from the site-specific questions to reveal the mutual influence of the paleo-surface and medieval building activity at each study site?*

Group 2: Transferability of the regional objectives

- *O1.2.1 Is the number and representativity of the chosen site-specific questions and study site high enough?*
- *O1.2.2 Which general assumptions can be derived for the interaction of medieval building activity and relief development in the entire Spessart region or even on a larger scale?*

The answers to such questions improve our knowledge of the general importance of topographic factors for medieval building activity and castle construction, in addition to military and representative factors (e.g. Biller, 2016). Additionally, the man-made topographic changes during the medieval period are revealed, and this helps transfer the meaning of the results to other study sites.

Such variable geomorphological questions in the context of archaeological excavations are generally addressed using a multi-method approach (Fischer *et al.*, 2016; Lange-Athinodorou *et al.*, 2019). In this study, a multi-method approach of geophysical, invasive and GIS-based methods was the most promising way to answer such variable, complex and interactive questions for all the different study sites. These methods combine the spatial benefits of geophysical and GIS-based data with the clarity of invasive methods. In addition to answering the main question, a more objective way of describing electrical resistivity tomography (ERT) data was developed. Using this method, the first steps for the creation of generalized resistivity guideline profiles of archaeologically important geomorphological sites were taken. Therefore, another bunch of questions can be derived from this approach, with a focus on the methodology, and possible innovations within it are questioned.

1.2.2 Objectives (O)2: Suitability and transferability of the methodical approach

Methodological objectives:

- *O2.1 Is the chosen multi-method approach suitable for answering the main research-question in such variable relief?*

- *O2.2 What are the most important factors for improving the quality of geophysical prospection in the context of archaeological excavations?*
- *O2.3 How transferable are the guidelines for ERT description for other study sites?*
- *O2.4 Is it possible to create a guideline profile with a generalized description of the subsurface's relative resistivity distribution for a single study site?*

All these further questions allow one to approach the main question from different points of view in order to answer it.

2 Study sites

First, an overview of the Spessart is provided, and then the individual study sites are described in detail. Archaeological relevance determined the choice of individual study sites. They have either already been described (Gröber & Karlinger, 1916) or are known for medieval usage. The site description is considered as Step 1 (S1) of the methodological approach of the workflow (Chapter 3).

Most of the information within the overview chapter was generated via a map description of a vast number of digital topographical, geological and soil maps at various scales (digital topographical maps in BayernAltas: Bavarian Ministry of Finance and Regional Identity 2020; geological and soil maps in Umweltaltas: Bavarian Environment Agency 2020b). The interpreted maps of the individual study sites are mentioned and cited in detail in the chapter on each study site. The interpretations vary in detail in line with the complexity of each study site's subsurface. However, detailed analyses of the relief are part of the GIS-based relief analyses and, therefore, described in detail within the results (Chapter 4).

2.1 Overview

The Spessart is divided into the Vorspessart, with uncovered crystalline bedrock, and the Hochspessart, where Triassic rocks of Buntsandstein (Early Triassic) cover the crystalline bedrock. Locally, thin strata of the Permian Rotliegend (early Lopingian and Guadalupian) and Zechstein (late Lopingian) sediments prevail and separate the crystalline bedrock from the Buntsandstein cover (Meynen & Schmithüsen, 1955; STD, 2016). As illustrated in Figure 2, all the study sites are situated within the western areas of the Spessart. Only the Eisenberg is in the northern edge of the uplands. The study areas of Alzenau, Hauenstein and Mömbris are entirely located within the Vorspessart, with Waldaschaff almost on the edge, between the Vorspessart and the Hochspessart. In contrast, the Eisenberg study site is fully situated within the Hochspessart. However, the Vogelsberg vulcanism plays a significant role in the landscape development of parts of the Spessart. Therefore, its location within the Hochspessart, by definition, needs to be approached with caution.

The division into the crystalline Vorspessart and the Hochspessart is visible in the topography of the two landscape units. Within the Hochspessart, huge ridges with no single peaks are common, as are numerous valleys and valley systems (Meynen & Schmithüsen, 1955). However, also within the Vorspessart, huge ridges without precise peaks are common, such as the Hahnenkamm (cf. Figure 2).

Geology

The lithology of the Spessart, as mentioned above, is divided into the crystalline Vorspessart and the Hochspessart, mainly consisting of Buntsandstein capping on top. Locally, thin strata of the Permian Rotliegend (early Lopingian and Guadalu-

Figure 2: Overview of the study sites: a) location of the Spessart and the study sites on DEM 200 and landscape units; b) close up of the Vorspessart, including the study sites

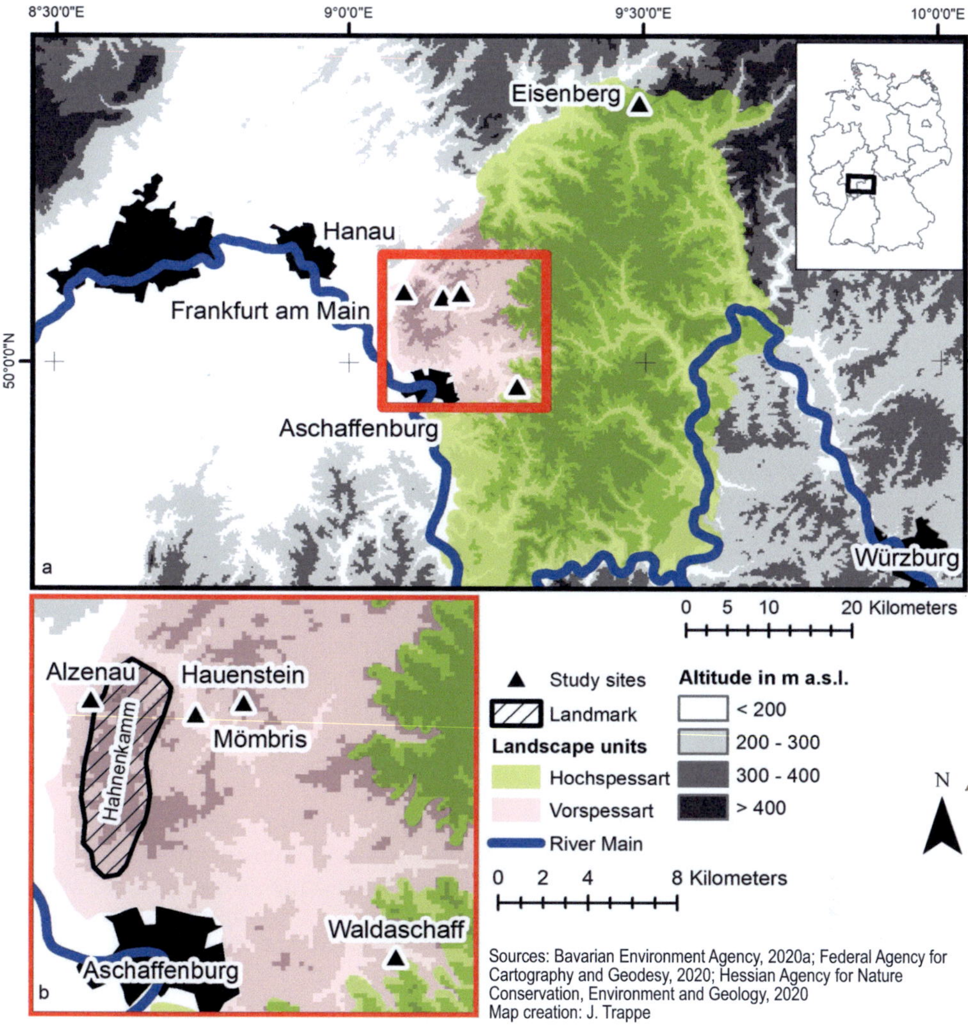

pian) and Zechstein (late Lopingian) sediments prevail and separate the crystalline bedrock from the Buntsandstein cover (Okrusch *et al.*, 2011), and Tertiary volcanism (Stegemann, 1964) affects the northern part of the Spessart. Both play a significant role in landscape development within parts of the study area.

The crystalline bedrock consists of sediments accumulated within the Neoproterozoic's timespan to the Silurian/Devonian turnaround, mixed with extrusive rocks. During the late phase of accumulation, an intrusive rock melt diffused into this mixture. All these rocks underwent metamorphic processes during the Variscan orogeny. Within these metamorphic rocks, a melt diffused that crystallized into granites, granodiorites and quartz-diorites (Okrusch *et al.*, 2011).

The accumulation of the sediments of the burden cap started during the Rotliegend. While most of these terrestrial rocks were eroded, a small edge of the marine sediment of the Zechstein occurs in a band between the crystalline and the following massive layers of Buntsandstein, which were produced in a hot and dry climate within the braided river systems of the early Triassic (Scheinpflug, 1992; Okrusch *et al.*, 2011).

Pedology

The formation of soils is basically divided by the geological setting of the two units of the crystalline Vorspessart and the Hochspessart (Meynen & Schmithüsen, 1955; Sabel, 1996). However, in the uplands within the Pleistocene's periglacial, the accumulation of loess or loess loam is more important for soil development (Semmel, 1964; Mueller, 2011).

Within the crystalline Vorspessart, the accumulated loess allows the development of luvisols, which are rich in nutrient saturation and are fertile. These soils are more often used for agriculture than those of the Hochspessart. The lower orographic position with highs around 300 m above sea level (a.s.l.) reduces the erosion of the PCBs. Nevertheless, depending on the topographic position of the soils, parts of or the whole profile of these luvisols can be eroded. On steeper slopes, the higher amount of cryoplastic weathered gneiss saprolite within the thinner loess layers developed less fertile luvisols. Additionally, the dense central layer can lead to stagnic conditions, decreasing fertility. Some of the loess-rich eroded material is accumulated more downslope as colluvium or within the floodplains of the valleys, which can be used for agriculture. The Hahnenkamm is a ridge mainly composed of quartzite with a maximum altitude of 420 m a.s.l., providing an acid material for soil development. In exposed positions, where most of the loess is eroded, eluvial leaching of iron and manganese ions starts, and the developed cambisols become podsolized or even become podzols (Sabel, 1996).

Soils developed on Buntsandstein are comparable to the soils on the Hahnenkamm, including a low nutrient saturation and high acid concentration. Therefore, the ridges are dominated by podsolic cambisols or even podzols themselves, whereas in the valleys and slopes, some loess deposits allow for development of cambisols or luvisol. On the small band of Zechstein-dolomite, carbonate-rich leptosols are developed. The claystones of the Bröckelschiefer deliver clay-rich material on which regosols and cambisols have developed (Sabel, 1996).

Water network

The Main is the most important river for and within the Spessart. Due to its U-shaped course, it not only passes the uplands in the southern part but also surrounds the uplands to the east and west. Therefore, the entire drainage system of the Spessart ends up in the River Main and River Rhine in the North Sea. In conjunction with the drainage system of the Spessart, the Kinzig and Sinn must also be named as tributaries. While the Sinn drains the northeastern part of the Spessart, flowing southeastward, and discharges into the Main in Gemünden, the Kinzig flows westward, draining the northwestern part and discharging into the Main outside the

Spessart. In addition, the Lohr and Hafenlohr drain the Hochspessart eastward, the Hasslochbach southward and the Elsava westward, to name but the most important. However, several small creeks and vast areas drain directly into the Main due to its position within and surrounding the area of investigation. The Vorspessart is almost entirely drained westward via the Kahl and Aschaff or directly into the Main; only a small part in the northern Vorspessart drains via the Kinzig and later into the Main. The watershed within the Spessart is comparable to the location of the Eselsweg (Ruf, 2019).

There are no natural bodies of standing water in the Spessart. Wetlands outside the floodplain are rare, too. The most important bog or fen is the Wiesbüttmoor in the northern part of the Spessart. Even though springs characterize the area, the fen itself is manmade by the impounding of the Wiesbüttsee in the 18th century CE. Small, natural wetlands are only located in the area of the Lohr.

Climate

The climate graphs (Figure 3) of Steinfeld (WGS 84: 9.674617 E, 49.946854 N; 300 m a.s.l.), Großostheim (WGS 84: 9.052198 E, 49.930197 N; 129 m a.s.l.) and Heppdiel (WGS 84: 9.358056 E, 49.665 N; 335 m a.s.l.) display a Cfb-climate after Köppen and Geiger. To create comparable datasets for all stations, the data from November 2002 to October 2018 were calculated. The average temperature within Steinfeld and Heppdiel was 9.7 °C, while the temperature in Großostheim was around 11.2 °C. Due to each study site's altitude, an average annual temperature of around 10 °C can be suggested. The average annual precipitation was around 700 mm/a at all stations. Caused by Luv and Lee effects, including west being the main wind direction and the western exposure of the study sites within the uplands, a precipitation value higher than the one displayed is suggested, around 800 mm/a. All stations display the temperature and precipitation maximum for the summer months.

All the climate stations surround the study area. However, they are still several kilometers away from each individual study site. The data from these stations offer some understanding of the climate within the Spessart uplands. Meynen and Schmithüsen (1955) have assumed climate parameters that differ from those here, at least for some parts of the investigation area. For the windward side of the Spessart, an oceanic climate is suggested with annual precipitation values of 900–1,100 mm, and in areas with lower altitudes (e.g. the Main Valley), smaller values of around 600 mm are proposed. Even though this is an older source, there seems no need to contradict it for such exposed areas.

2.2 Waldaschaff (W)

Wahlmich Castle was excavated in 2016 and 2018 by the *Archäologisches Spessartprojekt*. Almost no written sources existed about the castle. Even though the *Kunstdenkmäler des Königreichs Bayern* (art monuments of the Kingdom of Bavaria; Gröber and

Figure 3: Climate graphs of Steinfeld, Großostheim and Heppdiel

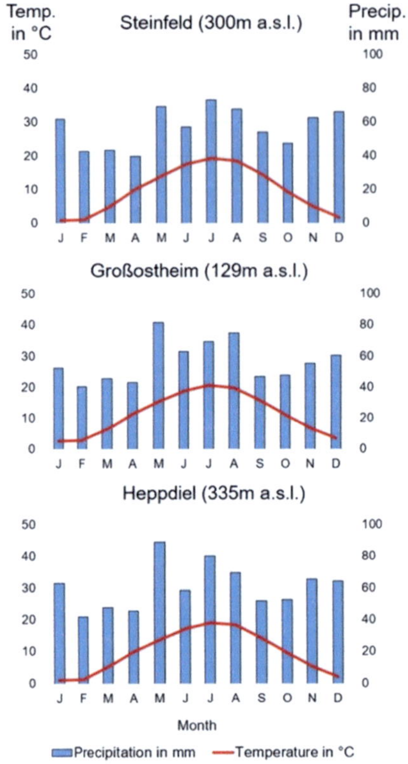

	P in mm Steinfeld	T in °C Steinfeld	P in mm Groß-ostheim	T in °C Groß-ostheim	P in mm Heppdiel	T in °C Heppdiel
J	62.0	0.8	52.7	2.6	63.2	1.0
F	43.0	1.0	40.5	2.8	42.0	1.2
M	43.6	5.0	45.9	6.5	47.8	5.1
A	39.9	10.1	43.3	11.2	46.0	10.0
M	69.6	14	81.7	15.4	89.0	13.7
J	57.2	17.4	63.2	18.8	58.5	17.2
J	73.5	19.3	69.5	20.6	80.5	19.0
A	67.8	18.4	75.0	19.7	69.7	18.4
S	54.1	14.4	47.1	15.7	52.2	14.4
O	47.5	9.4	48.0	11.0	53.0	9.6
N	62.8	5.1	55.5	6.9	65.7	5.3
D	66.3	1.8	60.7	3.4	64.4	2.0
Σ/ Average	687.3	9.7	683.1	11.2	732	9.7
Climate Köppen/ Geiger	Cfb		Cfb		Cfb	

Time of recording: November 2002 – October 2018

Source: Bavarian State Research Center for Agriculture, 2019
Figure creation: J. Trappe

Karlinger, 1916) mention the church and the so-called *Weiler* in Waldaschaff, nothing is written about the castle (Rosmanitz & Bachmann, 2017; Rosmanitz *et al.*, 2019b).

The castle was constructed around 1220 CE as a stony weir facility with a few outbuildings on top of an isolated hard-rock hill. The castle can be most likely classified as a spur castle. Its position on a road toward the important Eselsweg was significant for strategic reasons. After the construction of the outer wall, a palas was constructed around 1250 CE. However, it was destroyed in the late 1260s CE, most likely sometime after the defeat of the counts of Rieneck in March 1266 CE. Subsequently, as mentioned above, the counts were contractually bound to the destruction of their castle in Mainz territory (Rosmanitz & Bachmann, 2017; Rosmanitz *et al.*, 2019b).

Wahlmich Castle is situated south of Waldaschaff, separated from the present-day town by the A3 motorway. The study site is located on a north-exposed (Figure 5a) slope of the V-shaped Aschaff Valley (cf. Figure 4a).

Today, the castle hill and its direct surroundings are covered with forest (Figures 5b, c, d); meadows are located south, east and west of the forest (Figure 5e).

Figure 4: Study site – Waldaschaff: a) altitude and hillshade combined with the soil map; b) geological map

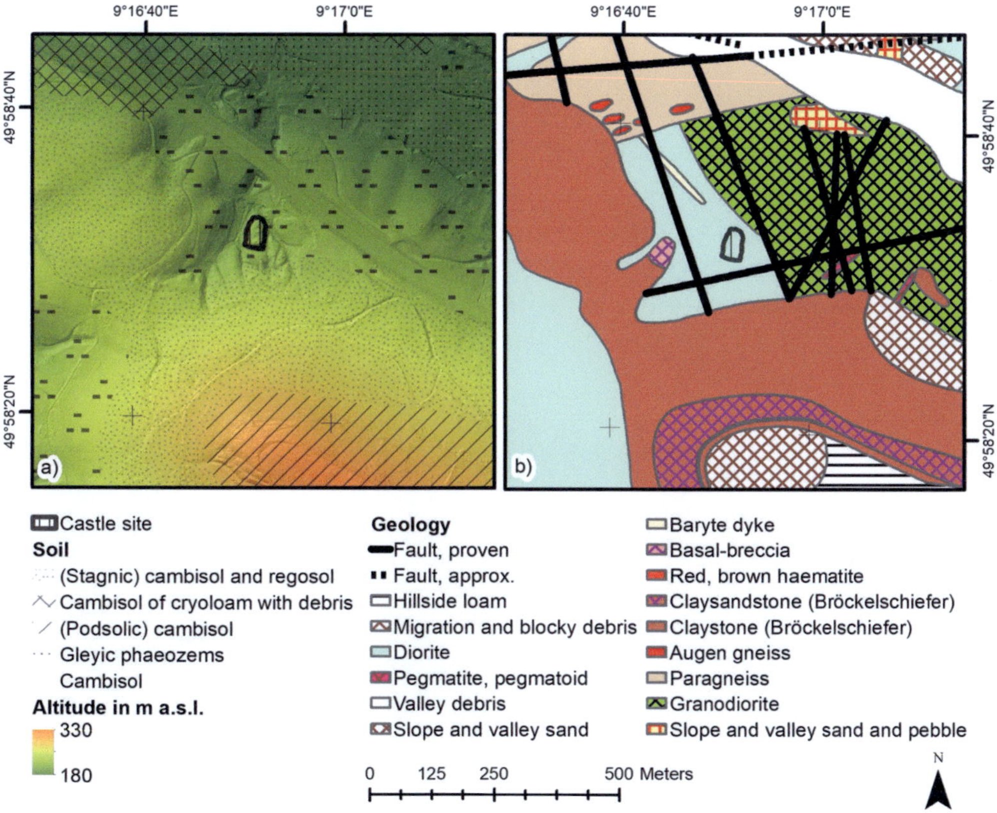

Sources: Weinelt, 1962; Bavarian Agency for Digitalisation, High-Speed Internet and Surveying, 2014; Bavarian Environment Agency, 2017c
Map creation: J. Trappe

Geology

The study area's geological map (Figure 4b) illustrates magmatic diorites and granodiorites situated below sedimentary claystones. The diorites and granodiorites of the research site belong to the described sole magmatic rocks of the Vorspessart (Lorenz *et al.*, 2010). The granodiorite has a higher content of quartz and orthoclase than the diorite, even though orthoclase-rich areas occur diffusely within the diorite (Lorenz *et al.*, 2010; Okrusch *et al.*, 2011). Therefore, a clear border between these two rock types is usually not described (Weinelt, 1962). The diorite-granodiorite complex is dated, on average, to 329.6±0.8 Ma using the $^{207}Pb/^{206}Pb$ ratio of zircons (Anthes & Reischmann, 2001).

In the southern section of the study area, Bröckelschiefer formations of the Fulda sequence, with a maximum thickness of over 20 m (Käding, 1978), mark the Permo-Triassic boundary. These sediments accumulated within the margin of the retreating Zechstein Sea. The Bröckelschiefer incorporates red claystone layers and,

Figure 5: Field pictures – Waldaschaff: a) overview; b) the largest of the small ridges south(east) of the castle, with the castle hill in the background; c) smaller ridges and striking step; d) neck ditch and ridges; e) meadows, southeast of the castle

Sources: Bavarian Agency for Digitalisation, High-Speed Internet and Surveying, 2014; own pictures, 2020
Map creation: J. Trappe

in part, gray claystone, siltstone and sandstone. At the bottom of the Bröckelschiefer, the so-called "Basalbrekzie," basal-breccia is recorded (Okrusch *et al.*, 2011). This breccia is described as having a groundmass of general Bröckelschiefer rocks mixed with the underlying crystalline layer, with no sharp border (Scheinpflug, 1992).

The faults in the broader study area mainly strike in a southeasterly direction, following the general Hercynian trend, which predominantly structures the sedimentary rock cover (Scheinpflug, 1992; Jung, 2006; Okrusch *et al.*, 2011). The development of the Hercynian trend is linked to the Younger Variscan Orogeny but was active in different phases, especially within the Tertiary (Murawski, 1965). Baryte dykes observable in the study area are associated with these tectonic activities (Weinelt, 1962). In addition to these now-inactive striking zones, a north-northeast-striking Rhenish trend exists, as well as the characteristic northeast-striking older Variscan trend, which defines the anticlinal structures of the Spessart's underlying crystalline bedrock (Okrusch *et al.*, 2011) and is also essential for the study area (cf. Figure 4b). The Rhenish trend is associated with the tectonic activity of the Lower Rhine Graben (Jung, 2006; Okrusch *et al.*, 2011) and has been active since the Tertiary (Meschede, 2018).

Quaternary sediments cover the Bröckelschiefer and (grano-)diorites of the bedrock in the study area. In addition to the formation of terraces and floodplains associated with the local drainage pattern's genesis, solifluction deposits and PCBs (Figure 4b) have developed and now play a significant role in determining soil quality in the area (Jung, 2004).

Remarkably, the border between the crystalline and Rotliegend/Zechstein is the border between two of the important landscape units of southern and central Germany, the *Mitteldeutsche Kristallinschwelle* zone (Brinkman, 1948) and the southwestern German cuesta landscape (Dongus, 2000; Meschede, 2018).

Pedology

The diorites and granodiorites affect soil distribution (e.g. Figure 4a) in the area of investigation. On the grus of these rocks, cambisols have developed. In contrast, in areas with Bröckelschiefer as bedrock, the soils consist of clay-rich material. Here, not only have regosols developed, but cambisols also, and both tend to stagnic conditions. Furthermore, PCBs play a significant role within the area of investigation as the substrate for soil development (e.g. Jung, 2004). Upslope, the Buntsandstein bedrock creates mainly podzolic cambisols.

2.3 Hauenstein (H)

In contrast to *Wahlmich Castle, Hauenstein Castle* is often mentioned in archival records (Reimer, 1897; Gröber & Karlinger, 1916). Furthermore, an archaeological excavation was conducted in 2017.

Hauenstein Castle was most likely constructed during the last third of the 13th century CE. Due to its conversion and expansion in about 1375 CE, this building phase is hardly archaeologically detectable. During this phase, the castle's role changed from a pure weir castle on a feeder road to the Birkenhainer Landstraße to an economic and representative center, including a mill in the lower parts of the castle ground.

However, the castle was destroyed in the early 15th century CE during the *Wetterauer feud* (Wetterauer Fehde; Griebel, 1982 Rosmanitz *et al.*, 2019a). At this time, several castles associated with robber knights needed to be destroyed. Indeed, *Hauenstein Castle* is mentioned as one of these (Rosmanitz *et al.*, 2019a).

The castle hill of the former *Hauenstein Castle* is located at the junction of two small creeks (Figure 6a) in the bottom of a valley southwest of the present-day Court Hauenstein, between Krombach and Mömbris.

Figure 6: Study site – Hauenstein: a) altitude and hillshade combined with the soil map; b) geological map

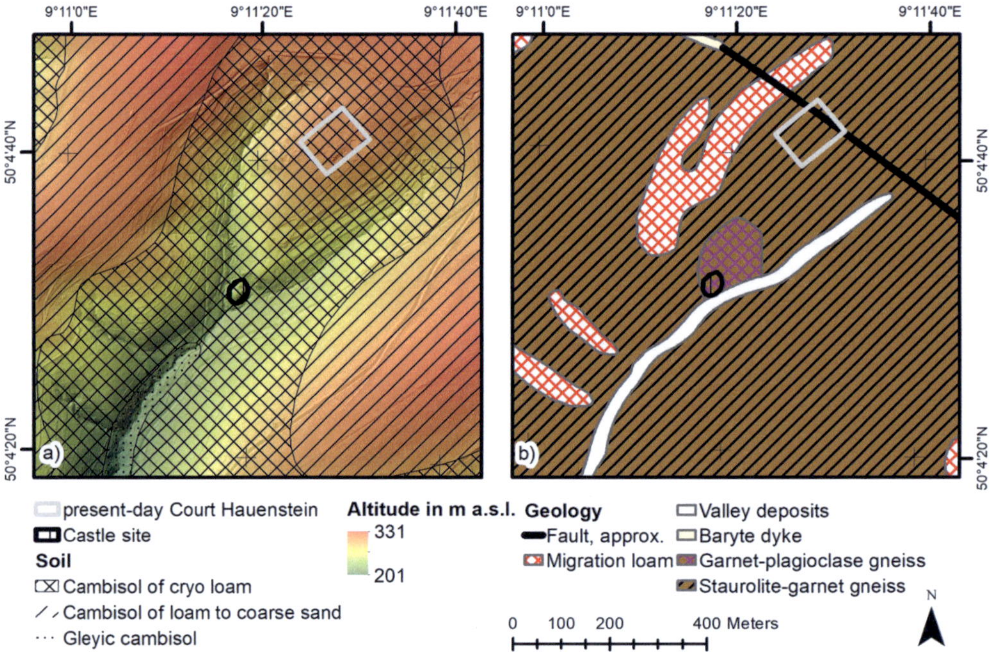

Sources: Okrusch & Weinelt, 1965; Bavarian Agency for Digitalisation, High-Speed Internet and Surveying, 2014; Bavarian Environment Agency, 2017b
Map creation: J. Trappe

Today, the valleys (Figure 7b) surrounding the castle hill itself (Figure 7c) and also parts of the slopes (Figure 7d) are covered with forest. Only on the southern slope of the southern creek and the northeastern hill, near Court Hauenstein, are meadows located (Figure 7e).

Geology

On the geological map (Figure 6b) of the *Hauenstein Castle* area, staurolite-garnet-plagioclase gneiss is illustrated. The gneiss is part of the Elterhof-Formation, which is mainly made of gneiss and can is associated with the Alzenau-Formation. The Elterhof-Formation is dated to the Cambrian. The gneiss is deeply weathered and folded within the investigation area, revealing a laminated matrix of the eponymous minerals, added to by gneiss's quartz and mica (Okrusch & Weinelt, 1965). Marble is embedded within the gneiss, as is characteristic of this formation (Okrusch *et al.*, 2011). Valley deposits are only mapped in a small section of the valley.

Figure 7: Field pictures – Hauenstein: a) overview; b) northern creek; c) castle hill and northern creek; d) neck ditch; e) meadows near the present-day court

Sources: Bavarian Agency for Digitalisation, High-Speed Internet and Surveying, 2014; own pictures, 2020
Map creation: J. Trappe

Pedology

The soil map (Figure 6a) in the area surrounding *Hauenstein Castle* describes the soils as cambisols made of the gneiss's weathering products, partly mixed with loess loam of the PCBs.

2.4 Mömbris (M)

Whether the castle in Mömbris is the *Womburg Castle* mentioned in the archives has been intensively discussed in the literature (e.g. Friedel, 2000; Huggenberger, 2015; Rosmanitz *et al.*, 2020). In the following, the castle is named *Mömbris Castle*. It was excavated in 2019.

Mömbris Castle was founded in the mid-13th century CE and destroyed in 1405 CE, apparently as part of the *Wetterauer feud*, and it can be classified as a spur castle

Figure 8: Study site – Mömbris: a) altitude (in contrast to the other study sites calculated with DGM 50) combined with the soil map; b) geological map

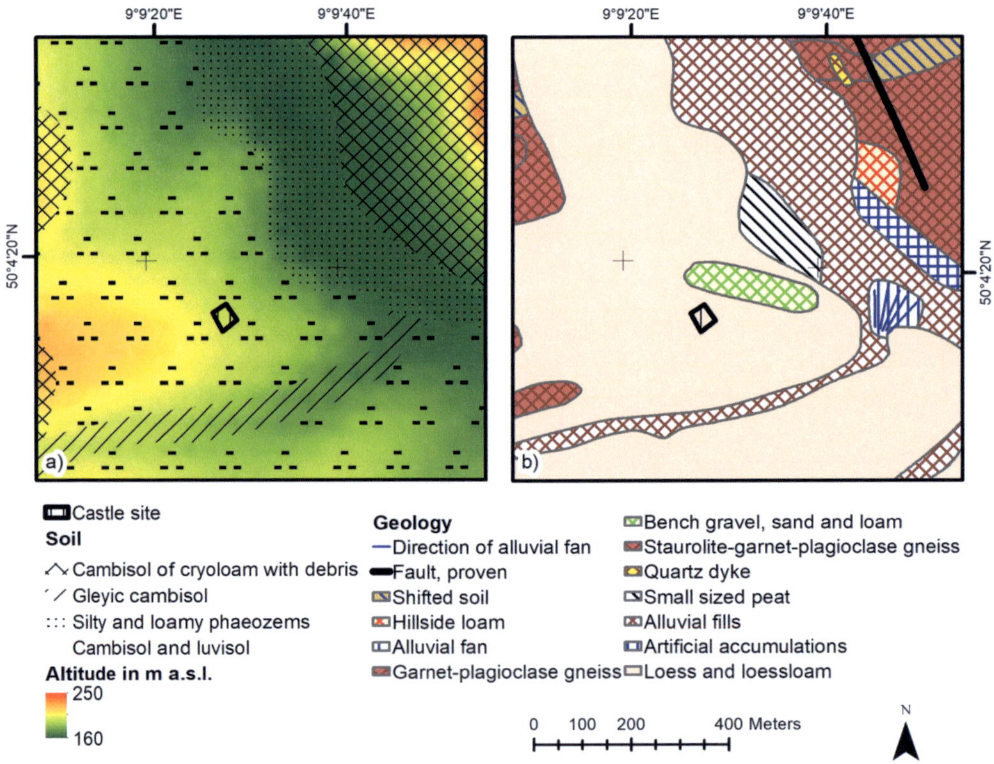

Sources: Okrusch et al., 1967; Bavarian Agency for Digitalisation, High-Speed Internet and Surveying, 2014; Bavarian Environment Agency, 2017a
Map creation: J. Trappe

(Rosmanitz *et al.*, 2020). The castle included a stone house and a proto-industrial company. The stone house underwent several reconstructions and even a deconstruction before the 15th century CE (Rosmanitz *et al.*, 2020).

The study site in Mömbris is located directly within the town of Mömbris, near and approximately underneath the graveyard. The current altitude of the hill associated with the former *Mömbris Castle* is about 180 m a.s.l. This height is on a northeastern exposed slope of the Kahl Valley (Figure 8a). In contrast to the V-shaped valley of the Aschaff (e.g. Chapter 2.2), the Kahl has created a U-shaped valley, including a distinct floodplain, appearing mainly as meadows.

Today, the castle hill is characterized as a bushy hill (Figures 9b, c) surrounded by steep slopes and fences or walls belonging to the graveyard or private houses (e.g. Figure 9a).

Geology

Surrounding the castle site of Mömbris, staurolite-garnet-plagioclase gneiss drops out as bedrock. A burden cap consisting of loess loam is located at the castle site itself and on parts of the northeast-exposed slope surrounding it. Additionally, bench

Figure 9: Field pictures – Mömbris: a) orthophoto of the study site; b & c) castle hill from different sites

Additional sources: own pictures, 2020
Map creation: J. Trappe

gravel, sand and loam are located north of the castle site. Only the valleys are filled with alluvial deposits and, in places peat, (Figure 8b).

Pedology

Due to its position in an area with thick loess loam deposits, cambisols and luvisols have developed. Located upslope of the loess loam deposits cambisols of mica schist's weathering products and gneiss's weathering products are located. On the plain of the Kahl, fluvial deposits have created phaenozems (e.g., Figure 8a).

2.5 Alzenau (A)

The early stages of *Rannenburg* or *Randenburg Castle* near Alzenau are associated with the authorities of "von Kälberau." When they ceased in 1227 CE, the castle most

Figure 10: Study site – Alzenau: a) altitude and hillshade combined with the soil map; b) geological map

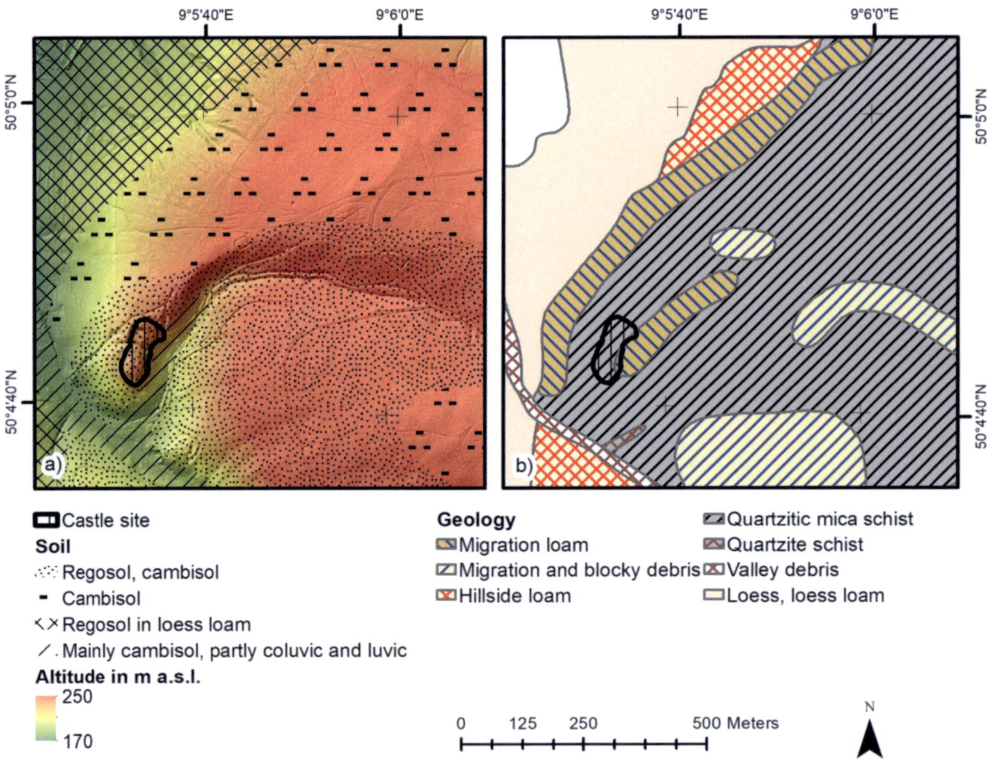

Sources: Okrusch et al., 1967; Bavarian Agency for Digitalisation, High-Speed Internet and Surveying, 2014; Bavarian Environment Agency, 2017a
Map creation: J. Trappe

likely went to the counts of Rieneck (Gröber & Karlinger, 1916). Scarcely anything is known about the role of *Rannenburg Castle* and its authorities in the feud. Therefore, it is described as the most enigmatic castle in the Spessart from an archaeological

Figure 11: Field pictures – Alzenau: a) overview; b) gully erosion north of the castle; c) the northern part of the castle; d) the main hill and a flattened area nearby; e) the main castle hill

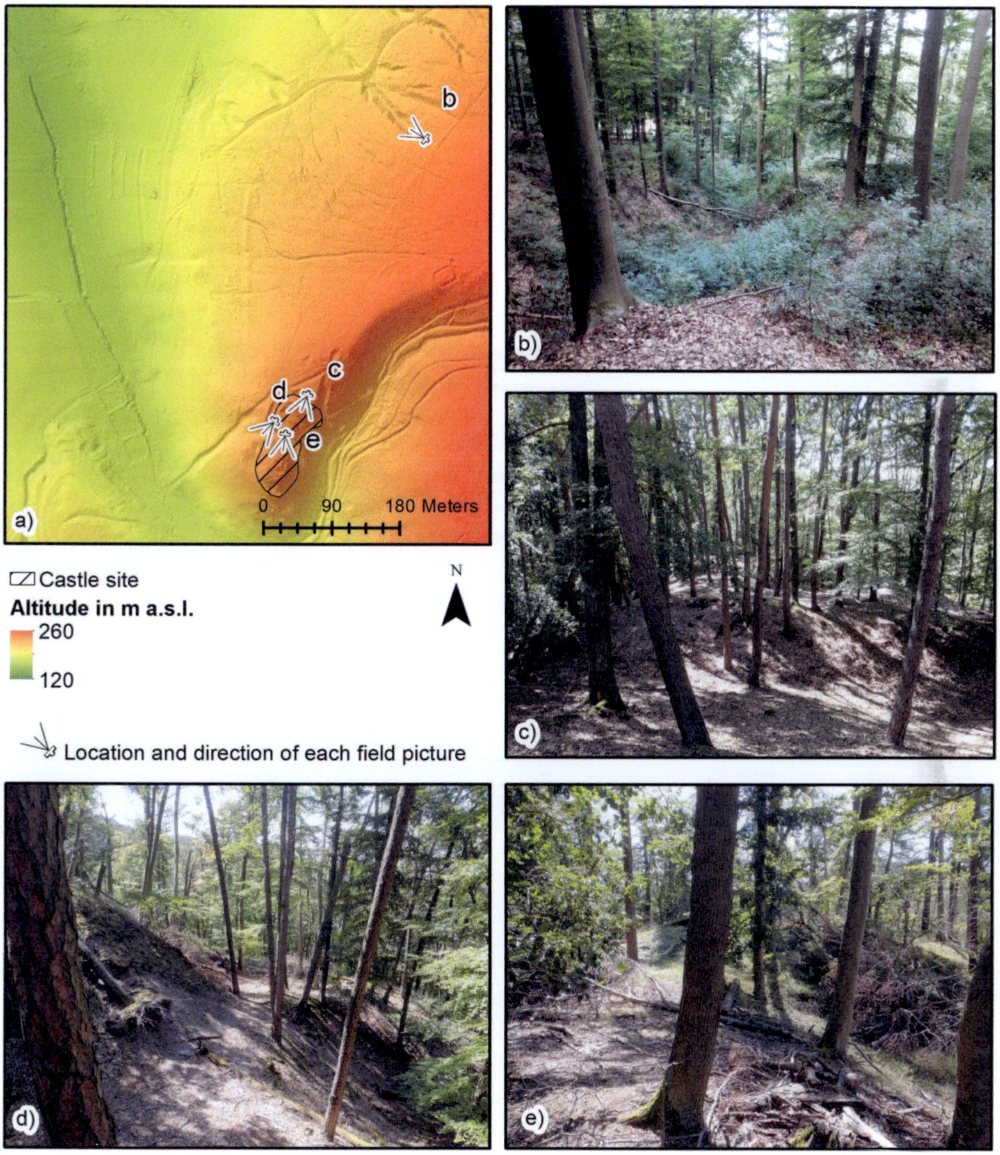

Sources: Bavarian Agency for Digitalisation, High-Speed Internet and Surveying, 2014; own pictures, 2020
Map creation: J. Trappe

perspective (Ruf, 2019). In 1386 CE, the castle was clearly associated with the authorities in Mainz. However, after 1398 CE, no further mention of the castle was found. This lack, combined with the founding of *Alzenau Castle* nearby in the city of Alzenau, leads to this period being estimated as the end of the usage of this castle (Gröber & Karlinger, 1916).

In contrast to the other study sites in Alzenau, no excavation has been conducted yet. However, after a prohibition in 2019, a new plan for excavating *Rannenburg Castle* in 2021 has been developed. Possibly, the results of this study site can be put in context and proven in the future by such excavation.

Rannenburg Castle is located on the last foothill of the Vorspessart (Figure 2a) on the southeastern slope of one of the central mountain ridges of the Vorspessart (Figure 10a), the so-called Hahnenkamm (Figure 2b). The position means it can be classified as a ridge castle.

Today, there are forest-covered areas with intense gully erosion (Figure 11b) in the north and at the castle site itself (Figure 11c, d, e).

Geology

In the center of the ridge of the Hahnenkamm, and additionally, at the castle site, mica schist mixed with quartzite is located as bedrock, surrounded by mica schist of the Geiselbach-Formation (Figure 10b). Only a few hundred meters away, the metamorphic rocks of the Vorspessart crop out, and the Lower Main Valley, primarily located in Hessen mainly comprising Quaternary sediments, is found. These two sections of the ridge's slopes near the castle are covered with up-to-several-meters thick loess or loess loam deposits (Okrusch *et al.*, 1967; Okrusch *et al.*, 2011). Additional moved and reprocessed material, such as hillside loam or migration loam, is located in the surrounding area. In the broader study site, the area was partly quarried for the metamorphic bedrock as building rocks (Figure 10).

Pedology

Surrounding *Rannenburg Castle* near Alzenau are cambisols or regosols described (Figure 10a). The cambisols become luvic and colluvial features in the valleys. Upslope, in the higher regions of the Hahnenkamm, the cambisols become podzolic. In the huge loess deposits, carbonate-rich regosols have developed.

2.6 Eisenberg (E)

The Eisenberg was an important mining area in the past. Findings from an excavated bloomery furnace allow for the dating of the iron mining activity to the 13th and early 14th centuries CE. However, the mining area is linked to Neudorf's deserted village, which was first archivally described in the 12th century CE and, thus, mining could have taken place over a more extended period. Nevertheless, many questions remain unclear regarding the economic history and even the lithological conditions that gave rise to this mining area (Hasenstein *et al.*, 2019).

Figure 12: Study site – Eisenberg: a) altitude and hillshade combined with the soil map; b) geological map

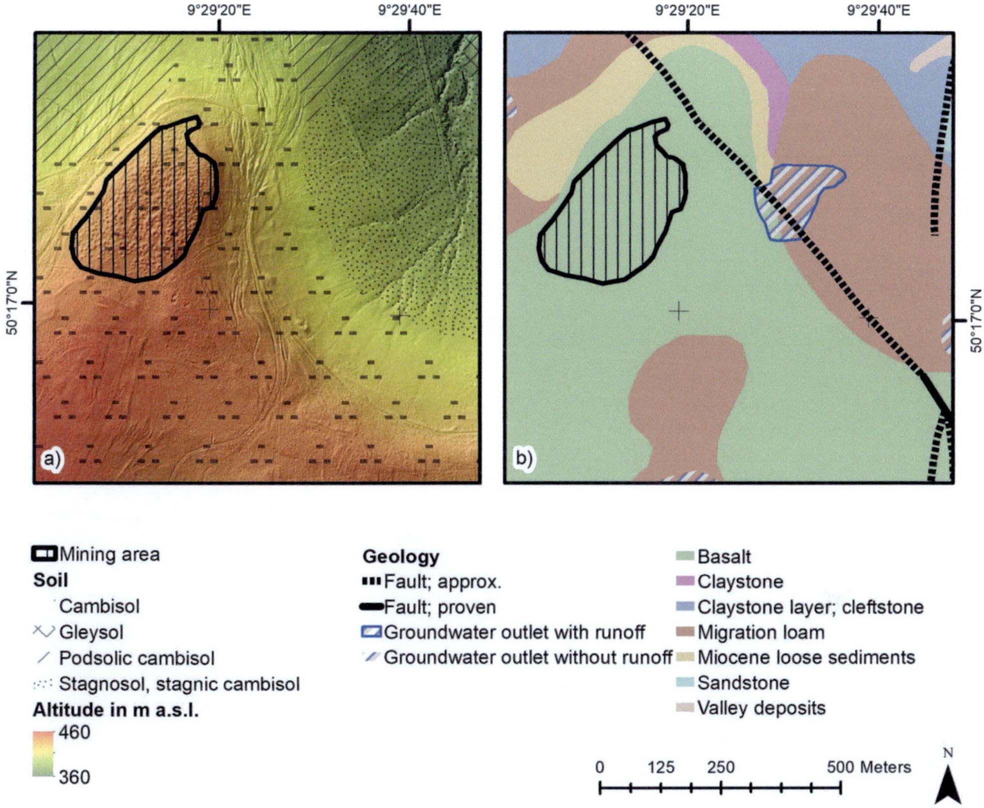

Sources: Diederich & Ehrenberg, 1998; Hessian Administration for Land Management and Geoinformation, 2007; Hessian Agency for Nature Conservation, Environment and Geology, 2014
Map creation: J. Trappe

The Eisenberg itself is a hill south of Steinau an der Straße, with an altitude of 469 m a.s.l. North of the peak is the Eisenkopf, with altitudes of about 440 to 450 m a.s.l. The Eisenkopf is characterized by over 100 depressions, identified as collapsed shafts (Figure 12a).

Today, the heights of the Eisenberg as part of the Eisenkopf are covered with forest. This is also true for the mining area (Figure 12b). Additionally, several small (Figure 12c) and large (Figure 12d, e) wet meadows are located in the area, such as the Neudorfwiesen.

Geology

Regarding the study site Eisenberg at the Eisenkopf, near Steinau an der Straße, the geological map (Figure 12b; Diederich & Ehrenberg, 1998) illustrates the Eisenkopf as being composed of Alkaliolivinbasalt. The Alkaliolivinbasalt is not well-segment-

Figure 13: Field pictures – Eisenberg: a) overview; b) collapsed shafts; c) small wet meadow north of the mining area; d & e) wet meadow ('Neudorfwiesen')

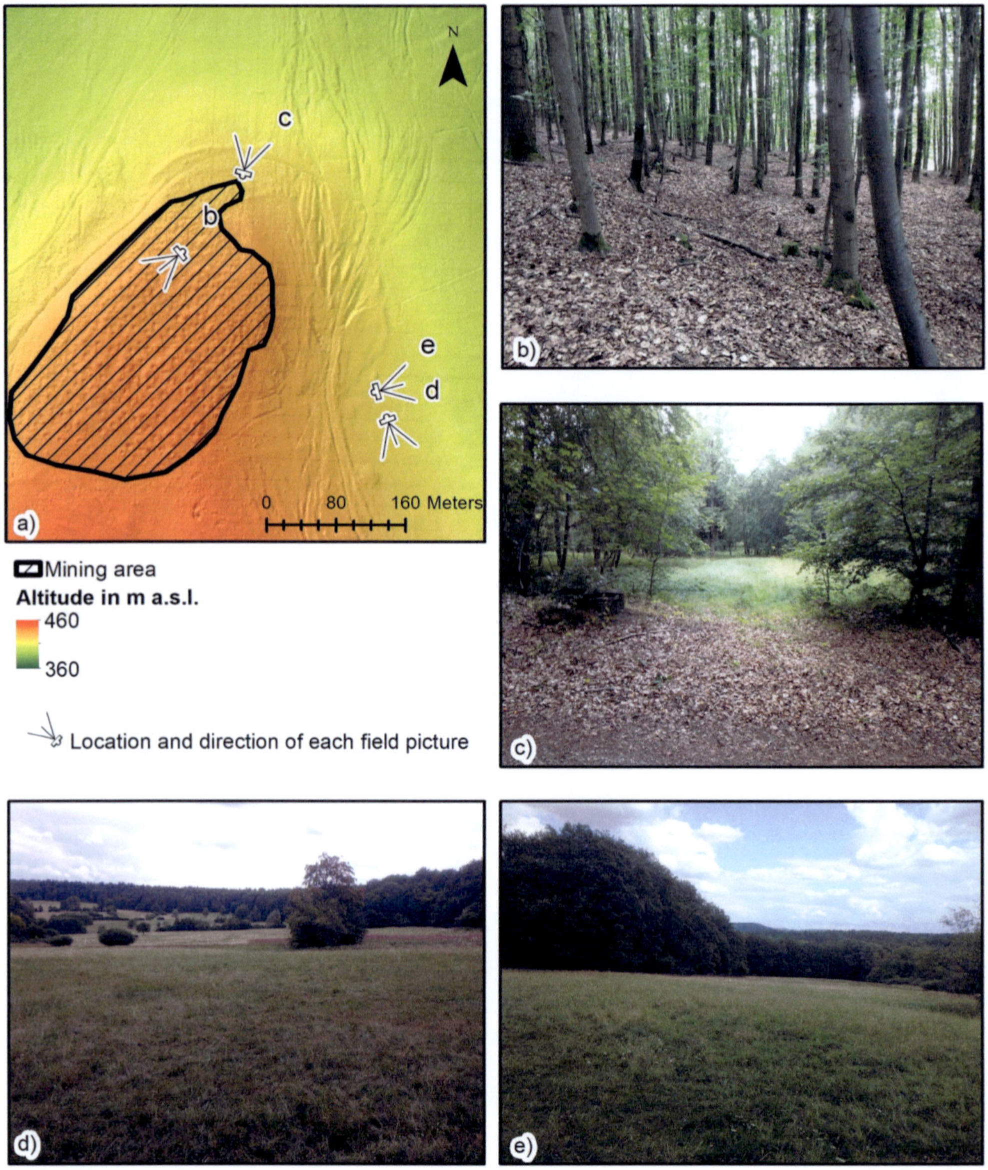

Sources: Hessian Administration for Land Management and Geoinformation, 2007; own pictures, 2020
Map creation: J. Trappe

ed from the tholeiitic basalt, and the edge is more or less diffuse. The magmatic activity within the area of investigation is associated with Vogelsberg volcanism, which had up to four phases of basalt and tuff production and is dated to different parts of the Miocene (Stegemann, 1964). The basalt was discharged in lava flows.

The flow itself comprises of three units: a compact unit, an intermediate unit and blistered, slag-like units on the top and in the middle. The basalt is surrounded by PCBs, including loess loam and fragments of the basalt. Additional, partly brackish or mainly freshwater sediments of the Miocene are also illustrated. These sediments are composed of sand, gravel and clay and are up to 40 m in size. The alternating of sandy and clayey layers often develops into sliding. The layering, especially within the dropping-out sections, is masked and difficult to differentiate (Ehrenberg & Hickethier, 1982). Hints at the existence of these clay-rich sediments in the subsurface are found in the field at wet-to-boggy positions. The Triassic sediments in particular, but also the basaltic layers are heavily faulted into blocks.

Pedology

At the Eisenberg, cambisols and podzolic cambisols are illustrated at positions with low loess loam content, while at positions with higher loess loam content, cambisols to luvic cambisols have developed. In contrast, the Neudorfwiesen and other meadows surrounding the hill more often have stagnosole and stagnic cambisols. Gleysols have developed only within the river valleys (Figure 12a).

3 Methods

First, a general workflow is presented in this chapter, followed by a description of the used methods.

The source load for the single study areas, pre-excavation, was often thin and not completely documented. To improve the scientific base of this work, GIS-based results were initially created. Subsequently, site-specific questions were developed from these results to follow the objectives of this work. Therefore, the structure of this work differs slightly from the usual scientific structure by presenting the GIS-based results first, followed by the creation of the site-specific questions. Figure 15 provides an overview of the performed steps.

However, the investigations at all individual study sites were conducted using the following general workflow steps: S1 to S6.

S1 – Archaeological importance

Archaeological research defined the wider study area and was already conducted by the ASP (Figure 15a) because they have the archaeological knowledge and background necessary for the preparation of the first workflow step.

S2 – GIS-based terrain analyses

GIS-based terrain analyses, such as morphometric analyses and wetness analyses, were used for a topographic interpretation of the study site in combination with thematic and topographic map analyses and archaeological scientific analyses, such as a detailed excavation (conducted by specialists other than, the author). This information raise different site-specific questions with a different focus, dividable into the following categories:

1. Surficial, natural features
2. Surficial, anthropogenic features
3. Buried, natural features
4. Buried, anthropogenic features
   4.1. On-site
   4.2. Off-site

The combination of site-specific questions allows insights into the interaction between paleo-relief and anthropogenic impacts during the period of interest. These categories are mentioned behind the capital letter of the study site (Figure 15b).

Figure 14 provides examples of the deduction from GIS-based terrain analyses for the site-specific questions, as well as a possible reconstruction of the landscape.

Only in Mömbris from this procedure was deviated because the wider study area is covered by the graveyard of Mömbris, which area extended in the early 21st century CE, and other building structures of the present town. Therefore, photographs were collected using a drone, and a precise digital elevation model (DEM) of the castle site itself was developed (prepared by Christian Büdel). This DEM was analyzed only descriptively to develop the site-specific questions.

Figure 14: Model of GIS-based relief analyses for the creation of site-specific questions: a) recent surface, partly of natural and partly of anthropogenic origin; b) site-specific questions are deducted from the recent relief and additional data; c) the former relief and the anthropogenic effect on it can be explained

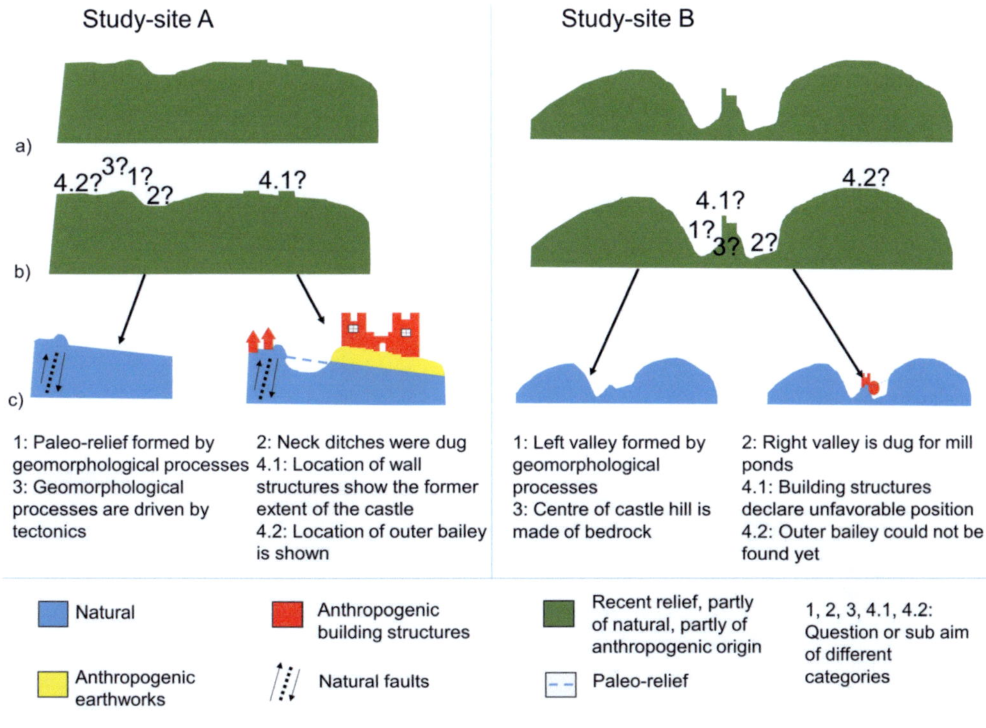

Sources: own illustration

S3 – Geophysical methods

Different geophysical methods allow one to create a spatial model of the distribution of physical parameters within the subsurface. These methods have different advantages and disadvantages. To minimize the disadvantages, different methods were used complementarily whenever necessary (Figure 15c). The results derived from geophysical methods remain ambiguous. Primary ERT data were described qualitatively. Guidelines and standardized analyses, combined with geometrical features, helped to reduce the ambiguity and allow for a more quantitative description of the data (Figure 15d).

S4 – Ground-truth

Selective invasive methods can be correlated with the geophysical data to reduce ambiguity (Figure 15e).

S5 – Interpretation of the data

By combining all the data from the above steps, it is possible to interpret them and derive the development of the landscape. This step is performed to utilize adequate

Figure 15: Model of the workflow

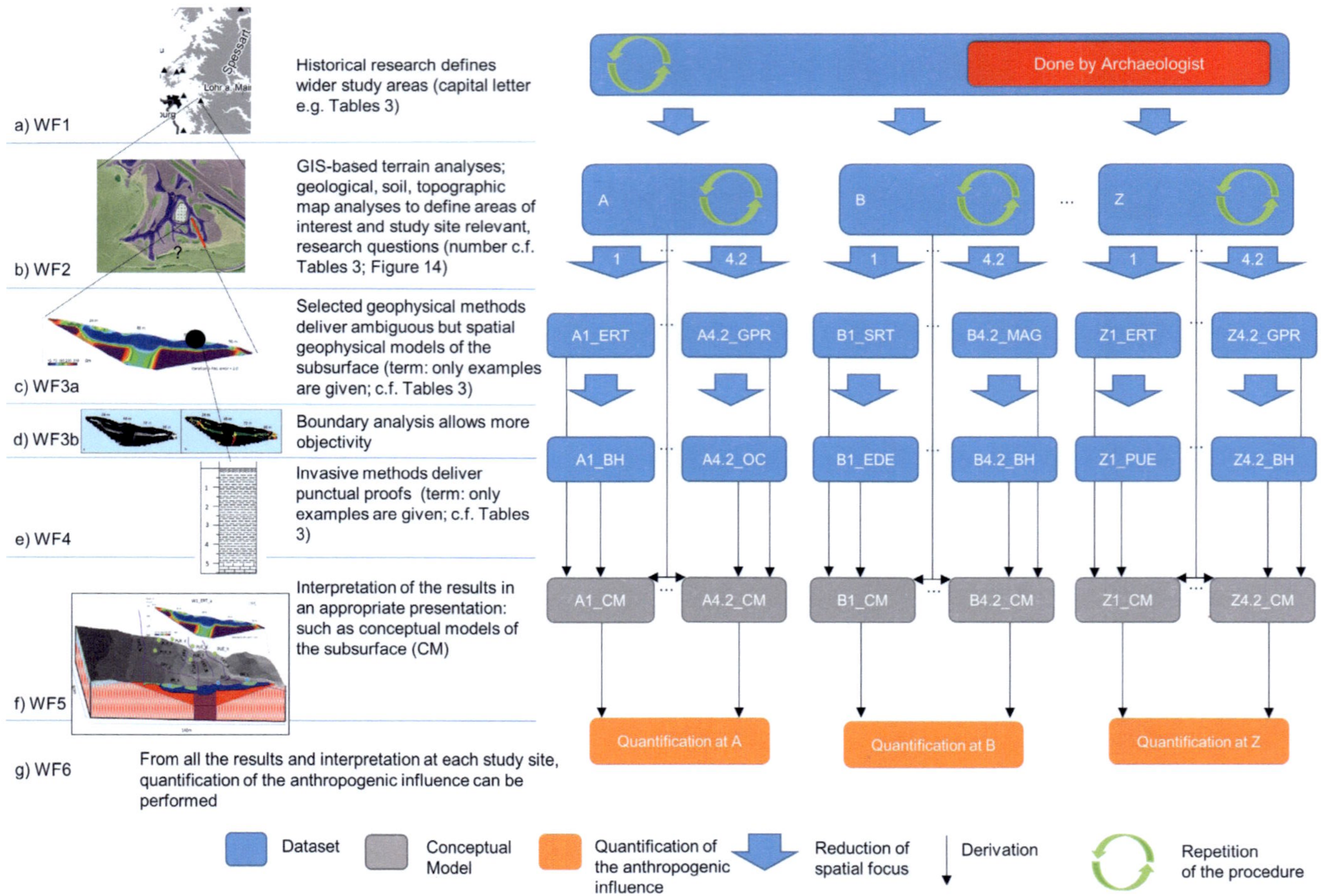

Sources: own illustration after: Hessian Administration for Land Management and Geoinformation, 2007; Bavarian Agency for Digitalisation, High-Speed Internet and Surveying, 2014; Bavarian Environment Agency, 2020a

methods of presentation of the interpretations, such as the creation of a conceptual model of the subsurface or an interpretation in combination with already excavated anthropogenic features, to answer the raised site-specific questions (Figure 15f).

S6 – Quantification of the anthropogenic influence

The interpretation of the data reveals various anthropogenic influences on relief development. These influences can be classified on an ordinal scale and mapped. Only in the Mömbris an intense anthropogenic influence at the entire study site was observed. Therefore, it was deviated from this procedure. (Figure 15g)

Subsequently to these steps, the results of the single study sites, the transferability of the results and this multi-method approach are discussed.

3.1 GIS-based relief analysis and morphometric description

This section describes the methods necessary for workflow Step 2 (S2) – *GIS-based terrain analyses.*

The GIS-based relief analyses helped to create site-specific questions by analyzing the complex, often dense vegetation-covered topography and to identify areas of interest that are often invisible in the field and not yet known. Additionally, this information helps to put the geophysical and ground-truth data in context with each other and helps to extrapolate the information obtained to a broader scale.

To calculate the morphometric data, a DEM with a 1 m range was used. With the basic tools of *ArcGIS* 10.6 (*ESRI;* Redlands, California, USA), the slope in degrees and the hillshade were calculated. To classify the landforms, the Relief Analysis Toolbox (Miller, 2014) was used. First, the topographic positioning index (TPI; Weiss, 2001) was calculated with a 25 m range (TPI [25]) and 500 m range (TPI [500]). The TPI puts the altitude of each point in the context of the altitude of the surrounding points in the 25 m or 500 m range. Positive values indicate that the point of interest is located, on average, higher than the surrounding points, whereas negative values reveal that the point is located lower, on average. Values near zero are either in flat terrain or in terrain with almost the same number of points higher and lower than the point of interest (for example, on slopes with a constant angle). Therefore, the TPI needs to be combined with the slope to create a simple and automated landform classification (Deumlich *et al.*, 2010). The classification checks how the described TPI values and the slope behave with each other. For example, a high TPI [25] value combined with a low TPI [500] value and no significant slope shows a local elevation in lowlands.

For the creation of hydrological knowledge about the surface condition, the topographic wetness index (TWI) was calculated using *SAGA-GIS* 6.4 (*SAGA User Group Association;* Hamburg, Germany). This index combines the upslope contributing area and the slope itself to quantify topographic control on hydrological processes (Sørensen *et al.*, 2006). In this study, the TWI was mainly used to reveal the potential of wetness from a topographic point of view. Additionally, this analysis helps to identify other important influences on wetness, such as a hydrogeological impact, if the observed wetness differs from the TWI. The base of all these Calculations is the DEM (digital elevation model) of Bavarian Agency for Digitisation, High-Speed Internet and Surveying (2014).

3.2 Geophysical methods

This section describes the methods necessary for workflow Step 3 (S3) – *Geophysical methods.*

In this study, ERT, ground-penetrating radar (GPR), seismic refraction tomography (SRT) and magnetic surveying were utilized as geophysical methods and are now described.

These geophysical methods create models of physical parameters through the collection, measurement and calculation of physical information and the properties of the subsurface material. The derived models are not unique and present only an approximation of the true physical condition within the subsurface (Reynolds, 2011). Therefore, whenever possible, different geophysical methods are used comparatively, or they are combined with ground-truthing methods to distinguish the best-fitting model for the subsurface. The theoretical basis of the different geophysical methods, as well as detailed methods of data processing and collection, are described in general in several textbooks, such as Everett (2013), Reynolds (2011) and Telford *et al.* (1990). Additionally, many other textbooks exist for one of these methods or their special applications, for example, Wagner (2007) and Conyers (2012) for archaeological use. Therefore only a short summary of the underlying theory and data processing of each realized method are provided. The focus lies on the explicit performed ways of processing, collecting and setting up the individual methods. Therefore this section has no claim of completeness.

The importance of geophysical methods in geomorphology, geoarchaeology and related studies has been increasing in the past couple of decades because they deliver a physical model of the subsurface in high spatial resolution or great detail without being invasive. Therefore, several studies have been published that are relevant to all the different disciplines this study touches on. At this point, the different applications of the utilized geophysical methods are briefly introduced.

All four methods have been utilized for geoarchaeological approaches. Numerous examples are provided in El-Qady and Matwaly (2019), Wagner (2007) or Posselt *et al.* (2007), to name a few. However, most of these examples are for geoarchaeology usage as defined by Rapp and Hill ("[As] any earth-science concept, technique, or knowledge base for the study of artifacts and the processes involved in the creation of the archaeological record"; 2006). It follows that several of these studies were applied to find yet-unknown or buried parts of buildings or other anthropogenic features as a geophysical prospection.

For geomorphological or other geoscientific questions, the application of the utilized methods changed slightly. Whereas SRT, ERT and GPR are often used for different applications, other types of magnetic analyses are more common. However, some reviews deliver an overview of different geophysical applications in geomorphology in general (e.g. Schrott & Sass, 2008) or for specific applications (Hauck & Kneisel, 2008).

Magnetic surveys are often used for large-scale archaeological prospection, especially for the prospection of negative building structures, such as ditches, fireplaces and post-holes. With these methods, Roman legionary campsites (Drahor, M., G., 2019), Neolithic rondel areas (Křivánek, 2020), the surroundings of Stonehenge (Gaffney *et al.*, 2012) or even entire Scythian cities (Becker & Fassbinder, 1999) can be prospected. On the one hand, magnetic surveys are the quickest of the applied methods, and it is possible to prospect hectares in good time. On the other hand, this method needs huge areas to be interpretable. Depth resolution, at least for most archaeological questions, is not provided, and magnetization changes need to be put into context.

Regarding necessary depth resolution, at smaller or rough study sites or at "magnetic-contaminated" (e.g. with iron fences) study sites ERT or GPR are usually chosen, for example, for finding buried building structures. Several studies have been published in which ERT and GPR were used for geophysical prospection, mostly creating 3D subsurface models. The scales at which GPR is used can be smaller than those of magnetic surveys, for example, to prospect single buildings underneath today's cities (Ristić *et al.*, 2020), to prospect where the internal structure is important (Linck & Fassbinder, 2014) or to prospect building complexes (Bianchini Ciampoli *et al.*, 2020). In addition, 2D prospection is especially performed in buildings (Atya *et al.*, 2012) or whenever the internal structure of the expected buried features is more important for the interpretation (Nuzzo *et al.*, 2009).

Although ERT is less commonly used than GPR, it is still a common method for geophysical prospection in a geoarchaeological context. Using ERT, measurements in 2D and 3D are usual, depending on the specific question and logistical background. Measuring huge areas in 3D with a high resolution and deep penetration depth is much more time-consuming than are the methods mentioned above. However, due to the ERT setup of electrodes connected to the ground, other advantages in steep or exposed positions occur. Chavez *et al.* (2017) measured, for example, on top of a pyramid in Mexico, to reveal its interior. Other authors have prospected vertical walls (Mol & Preston, 2010) or tombs (Deiana *et al.*, 2018).

In contrast, ERT has been successfully applied in several studies within the German low-mountain range, for example, in caves in karstified landscapes (Kaufmann, 2014; Kaufmann & Romanov, 2017), for extensive hydrological questions (Koch *et al.*, 2009) or at slopebogs (Trappe & Kneisel, 2019). Additionally, ERT is frequently used for prospecting and mapping faults, mainly on a large spatial scale (Giocoli *et al.*, 2011; Galli *et al.*, 2014). Other studies have combined ERT and GPR to map faulting and fracture zones at different scales (Jõeleht *et al.*, 2017; Bhat *et al.*, 2018). However, SRT could help as well in the detection of faults (Blecha *et al.*, 2018).

Seismic tomography methods are less common for archaeological usage. However, some applications can be found for void detection (Imposa *et al.*, 2018) or for striking substrate differences within archaeological structures, such as in graves of Native Americans (Nisengard *et al.*, 2008).

It is common to combine some of these methods for archaeological use to reduce their disadvantages and increase their reliability. Masini *et al.* (2017) combined ERT and GPR at an archaeological site in Henan, China, as did Correia (2019) in Portugal. In addition, a combination of magnetic surveying, GPR and ERT (Welc *et al.*, 2017; Li *et al.*, 2019), or even all these methods, has often been applied (Fassbinder *et al.*, 2017) for geophysical prospection.

For geomorphological questions, in addition to the complementary usage of GPR for the ERT measurements of the studies mentioned above (e.g. Kaufmann & Romanov, 2017; Trappe & Kneisel, 2019), as well as other complementary usages of GPR, for example, at landslides in the Swabian Alps (Sass *et al.*, 2008) in the German uplands, the size of different PCBs can be measured (Gerber *et al.*, 2010). However, most of these applications refer to other fields, such as analyzing permafrost regions (Kunz & Kneisel, 2020) or answering glaciological questions (Eisen *et al.*, 2011).

Additionally, these methods have also been used to reconstruct the paleo-relief in the context of archaeological land use. Thus, ERT is an important part of the multi-method approaches of Fischer *et al.* (2016) and Lange-Athinodorou *et al.* (2019) for reconstructing the paleo-relief of an anthropogenical influenced landscape. It has also proved useful for the reconstruction of archives in karst landscapes (Siart *et al.*, 2010). Furthermore, SRT has been used to reconstruct the subsurface underneath temples (Metwaly *et al.*, 2019).

Whereas GPR data can be interpreted in a standardized way (Neal, 2004), this is not common for ERT data yet.

3.2.1 Electrical resistivity tomography

The most important geophysical method in this work was ERT. From the wide range of electrical resistivity methods, direct current geoelectric was chosen. First, the basics, the implementation of the method, data processing and inversion procedures are described. Then, the new approaches of boundary analysis and the guidelines for ERT description are outlined.

Basics

Electrical resistivity methods are meant for determining the spatial distribution of the apparent electrical resistivity within the subsurface. It bases on the knowledge that different ground materials have different electrical conductivities as a diagnostic physical property (Reynolds, 2011). However, not only does the material itself play an important role, but also the porosity, the resistivity and the amount of pore water (Archie, 1942).

For electrical resistivity methods, a source injects a current into the subsurface with grounded point electrodes (Telford *et al.*, 1990). The injected current flows radially within a volume of earth with a homogeneous composition, the so-called homogeneous half-space (cf. Figure 16). Caused by a negative (to the current direction) potential gradient, the voltage drops between any two points on the surface. It is possible to calculate this voltage at a distance from a single current point source (Reynolds, 2011).

The true resistivity of material can only be measured within a uniform block (Reynolds, 2011). However, the measured resistivity values within the subsurface are the so-called apparent resistivity values because they are, in general, measured within an inhomogeneous medium and include a geometry factor (*k*) that is dependent on the used array of the measurement (cf. Figure 16). By definition, the apparent resistivity is not a subsurface's physical property anymore but a mixture of various true resistivity values of the inhomogeneous subsurface (Lowrie, 2011; Reynolds, 2011).

In the field, the direct electric current has to be injected (as mentioned above) into the ground with two so-called current electrodes (A and B) and, of course, a power source. This causes the creation of an electric field. In this manner, the potential can be measured at two different electrodes, the so-called potential electrodes (M and N; cf. Figure 17). These four electrodes are the so-called quadripoles for the single measurement. Using Ohm's Law with the abovementioned calculated voltage and a measured potential, the apparent resistivity can be calculated (Figure 17). Additionally, a geometric factor needs to be added due to the utilized array.

Four electrodes are needed for a single measurement (Figures 16 and 17). For a measurement in more than one dimension, there are over 100 (Reynolds, 2011) ways of arranging these electrodes geometrically, with different advantages and disadvantages (Dahlin & Zhou, 2004; Lange, 2005). However, only a few arrangements are common.

Figure 16: Homogeneous halfspace

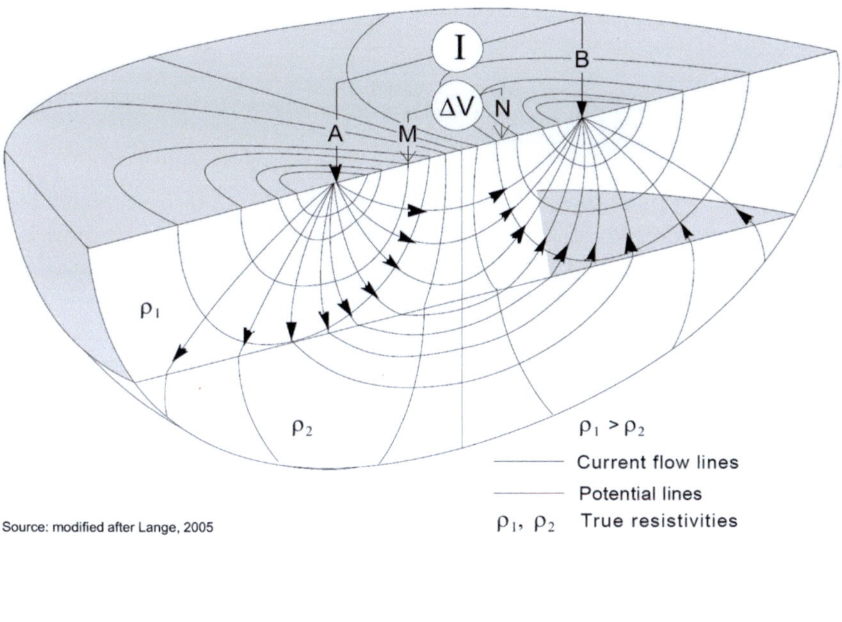

Source: modified after Lange, 2005

For this study, mainly the Wenner-Schlumberger array and partly the Dipole-Dipole array were used (cf. Figure 16). Both arrays have different advantages and disadvantages, especially regarding sensitivity (Barker, 1979; Furman *et al.*, 2003; Dahlin & Zhou, 2004; Kneisel & Hauck, 2008). The Wenner-Schlumberger array has a good resolution for horizontal and vertical geomorphological structures and is rather robust against noise. In contrast, the Dipole-Dipole array includes a better horizontal resolution and a higher number of data points near the surface. However, its vertical resolution and investigation depth are lower, and it is more vulnerable to noise due to the lower signal-to-noise ratio (Kneisel & Hauck, 2008).

The Wenner-Schlumberger array produces a more robust data set, vertical changes can be expected within the subsurface and deep data points seem to be necessary for the interpretation. Therefore, this array was more often used than the Dipole-Dipole array. Furthermore, test measurements with the Dipole-Dipole array resulted in unusable datasets at some study sites.

Figure 17: Electrode configuration, geometric value and calculation of apparent resistivity

Electrode configuration

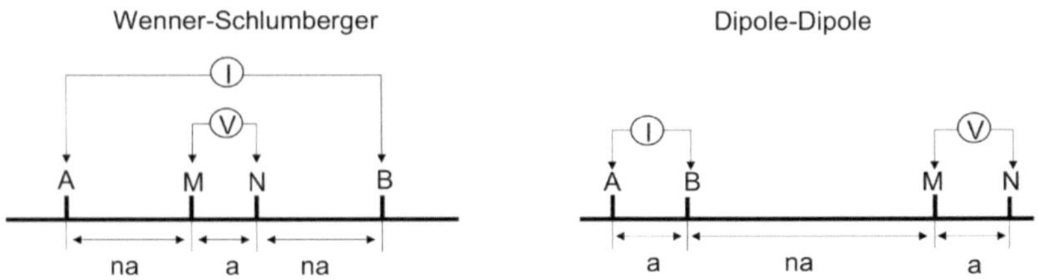

Calculation of the geometric value (k): Wenner-Schlumberger $k = \pi * n * a * (n+1)$; Dipole-Dipole $k = \pi * a * n * (n+1) * (n+2)$

Calculation of apparent resistivity (ρ_a) modified after Ohm's Law: $\rho_a = \frac{V}{I} * k$

A, B: Current electrodes
M, N: Potential electrodes
a: Spacing
I: Applied current
V: Measured Voltage

Sources: own illustration after: Telford et al., 1990; Lange, 2005; Reynolds, 2011

Implementation

All measurements in this study were measured in 2D or quasi 3D (described below) based on 2D measurements and which is essentially only an application for inverting these data. For measuring, *Syscal/Pro* or *Syscal Junior Switch 72* by *IRIS Instruments S.A.S.* (Orléans, France) were used. Both were utilized with one or two multi-core cables, each with 36 take-outs.

All electrodes, arranged linearly, were put into the ground with a defined spacing and coupled with the multi-core cable. Depending on the chosen electrode configuration, different electrodes were used as current and potential electrodes (cf. Figure 17). For measuring in 2D, these electrodes varied for every single measurement and were switched automatically according to the used array (Figures 16 and 17).

Data processing and inversion

Contrary to GPR and fluxgate magnetic data, the ERT data needed further processing steps to create the subsurface's geophysical models. The apparent resistivity values, measured at the different positions during the implementation, did not display an apparent resistivity value at a single position. Especially regarding measurements with an increased *n*-value (cf. Figure 17), the apparent resistivity values of the entire distance, following the homogeneous half-space (cf. Figure 16), were integrated.

However, first, quality assurance needed to be performed. The data quality was checked manually using *Prosys II (IRIS Instruments S.A.S.;* Orléans, France). Values with a deviation of 5% between the single, stacked measurements at each position were deleted. Such a large difference between measurements that approximate the same results were considered a result of technical constraints. In addition, a pre-inversion was performed. If a single calculated apparent resistivity value varied more than 100% from the measured one, it was deleted from further calculation (Loke, 2020). Additionally, negative apparent resistivity values were deleted. Although they can naturally exist (Reynolds, 2011), at these study sites, these values and those with more than a 100% difference were attributed to measurement errors. Those errors occurred, especially when measuring under high voltage and current flow, with decreasing battery voltage. (Incidentally, partly for this reason, the Dipole-Dipole array was used rarely; the high conductivity of the material at some study sites, in combination with the short distance between the current electrodes in this array, caused an intense current flow, producing a high number of unusual quadripoles due to low energy sources.)

To create a 2D model, including the spatial distribution of the true resistivity values of the subsurface, different modeling processes and algorithms can be used to calculate it from the apparent resistivity values. In this study, the forward modeling process of the software package of *RES2DINVx64* (*Geotomosoft;* Penang, Malaysia). Such forward modeling is a process of minimizing a misfit between the collected dataset and a calculated dataset via an iterative inversion of them. The inversion stops when either the error between the single measurements changes too slightly or a maximum number of seven iterations is reached (Loke, 2015). As a result, the calculated and the collected apparent resistivity distribution is displayed, and an inverse model resistivity section based on the calculated apparent resistivity distribution is displayed. Additionally, the error between the collected and the calculated model is shown. A major issue is the choice of iteration for the interpretation of these data. A small number of iterations causes a larger error. In contrast, too many iterations can cause overfitting of the data by overestimating so-called inversion artifacts. Thus, the error does not directly display the quality of the collected data, and a smaller error percentage does not necessarily mean a more realistic model (Hauck & Mühll, 2003). For the inversion, two different schemes can be used: the smooth "least-square inversion" (L2 norm) and the "blocky" or "robust inversion" (L1 norm). The least-square inversion scheme minimizes the square of differences between the measured and calculated apparent resistivity values. This method is very sensitive to outliers. In contrast, the robust inversion scheme uses the absolute error and is less sensitive to outliers. However, the robust inversion scheme delivers better results if there are sharp boundaries within the subsurface, such as the soil-bedrock boundary (Loke *et al.*, 2003; Loke, 2015). Therefore, the robust inversion scheme was chosen to calculate the displayed models for this study. Additional various damping factors and other configurations can be changed as well (e.g. for a precise smoothing of very noisy data). However, for this study, the results calculated with the default parameters (Loke, 2015) seem to fit best with the expected subsurface. As a result, 2D models of the subsurface were produced, including the x- and z-direction resolution.

For the inversion, the apparent resistivity values are assigned to cells or blocks. After the inversion process, these blocks still occur. These results are a smoothed version on the base of the block. The smoothed tomograms are easier to interpret but include an additional interpolation. Therefore, it is possible to display those model blocks as well. In addition, the datasets can be extended to the full size of the profile (Loke, 2015). This approach produces interpolated values at the edges of the profile, but also the possibility of more reasonable values at the edge of the collected data points.

Even though the measurement was tested with both inversions, different configurations, different models and different damping parameters, it was decided to use for all 2D results the fifth iteration of the robust inversion scheme, with default damping parameters and configuration as the best-fitting model. Nevertheless, massive changes had to be made in the described configuration to generate results, allowing an interpretation that differs from the displayed one. Therefore, it was decided not to change the values between the single measurements to maintain comparability regarding the guidelines for ERT description (see below in Subsection 3.2.1).

Not only 2D ERT models were produced, but also 3D models. To create these models, 2D-collected data were utilized. Therefore, this technique is called quasi-3D or 2.5D. However, in the following, quasi-3D is utilized. After quality assurance, the parallel collected 2D tomograms were combined using the "Collate data into RES3DINV format"-tool of *RES2DINVx64*. By using a script, *RES2DINVx64* creates a single 3D file out of the different *RES2DINVx64* files with the help of the distance between the single lines. Orthogonal lines can also be added (for further information on this procedure, see Loke, 2015, 2020).

Now the data can be inverted analog to 2D data, including an additional dimension, of course, using *RES3DINVx64* (*Geotomosoft;* Penang, Malaysia). The results include the error in apparent resistivity between the calculated and the collected data. In contrast to the 2D tomograms, which show the resistivity plotted with their x and z distribution, the resistivity values are displayed in their x and y distribution and as split into depth slices.

The topography influences the spread of the potential field (Lange, 2005). Especially ridges and valleys, are substantially affecting the current flow during measurement (Telford *et al.*, 1990). Therefore, it is necessary to include the topography in the measurement. However, the influence also depends on the direction of the current flow (Telford *et al.*, 1990). Therefore, an inclusion of the topography is necessary; it forms, on the one hand, a realistic surface for the measurements and, on the other hand, it reduces the potential error occurring throughout the topography.

In Mömbris, in addition to the layers of the 3D ERT, a real 3D model of the resistivity distribution was created. Initially, an average filter with a kernel size of 5 was chosen. This was necessary to reduce small anomalies at the top and to make the most significant parts visible. Second, a threshold value was estimated at 600 Ωm to display the most important anomalies within the data. Using the isosurface of the *Voxler* (*Goldensoftware*; Golden, Colorado, USA), all values above the threshold values were displayed in 3D.

Boundary analyses of ERT data

To derive additional quantitative data from the ERT results, boundary analysis was used. This analysis creates an additional dataset that can be used for data description and is derived from the inverted ERT data. It must be noted that these analyses represent a method of derivative function of the inverted ERT data that displays the spatial distribution of data points for a more detailed description. A generalized description of the performed working steps is provided in Figure 18. The data of the boundary analysis is necessary for a description according to the guidelines for ERT description (described in the following section).

Well-known for spatial analyses, ArcGIS 10.6 was chosen for this purpose. First, the already inverted ERT data were exported as "xy-data." For the following steps, it is important to use the ERT data, including topographic information. Later, normalization is performed, a process, which is sensitive to high maximum values. Therefore, some quality management is necessary. As such, all resistivity values with values higher than 3.5 times the 95% quantile had to be deleted (Figure 18a). However, this value-based only on a test with different threshold values and is a compromise between the data meaningfulness and deleting as few data points as possible to avoid distorting the dataset. Subsequently, the data must be transformed into a shapefile, including a transformation of the y and z values. This step is necessary for the following calculation steps.

With these data, the written *boundary analysis tool* can be run to process the sharpness of the r-layer boundary and its orientation. For generating comparable datasets, it is necessary to choose for each creation a raster dataset with the same cell size. For comparability in the following interpretation, a cell size of 0.5 m was chosen. For spacings smaller than or equal to 1 m, the cell size has to be transformed to 0.25 m. These values were chosen as a compromise between resolution, smoothness and interpolation of cell size.

First, the tool interpolates a raster dataset of the resistivity data (z value) using an inverse distance weight (IDW) algorithm. It was decided to interpolate with the IDW with its default values, including the power of 2 and a variable search radius of 8 (Figure 18b). Due to a homogeneous distribution of data points within the shapefile, other interpolation algorithms and other configuration values created raster datasets with weak differences in the significance area. The values were divided by the maximum resistivity value of the dataset for normalization in relation to the maximum of the dataset and then multiplied by 100 to obtain a percentage value (Figure 18b). With this method, the values were reclassified to values ranging from 0 to 100, which is necessary to obtain a similar range of the values in all datasets and a unique scale of boundaries, especially when using datasets with high values in the following steps. However, this method implies the creation of relative sharpness values for single profiles with a reference size of the maximum value of each profile, which is necessary to mention when comparing them with other datasets. Using the "slope" tool of the ArcGIS toolbox, the percentage difference for all cells was calculated (Figure 18c). The slope of the resistivity values was reclassified into five categories (Figure 18e). These values represent the intensity of the increase or decrease of the resistivity values.

Figure 18: Flowchart – boundary analysis: a) prepared input table; b) IDW-interpolation, normalization of the dataset; c) derivation of the slope in percent; d) derivation of aspect; e) reclassification of the slope, as described in Flowchart table 1; f) reclassification of the aspect, as described in Flowchart table 2; g) rectangle raster file is clipped to the extent of the model cells; h) for each classified slope pixel above 1, the aspect value is given and classified after the described classes in flowchart table 3; i) reclassified and clipped slope is exported as "Layer boundary"; j) the classes of the aspect are exported as an additional file on top of the "Layer boundary" as "Boundary orientation."

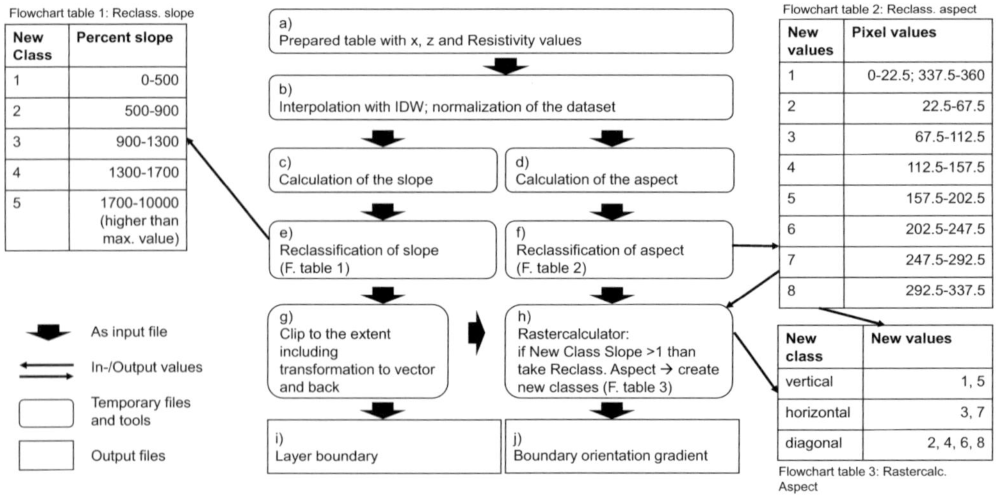

Source: own illustration

During the calculation of the slope, the change in values is divided by the distance between the cells. To reduce the spacing's sensitivity and to normalize the slope, the result has to be multiplied by the used spacing; otherwise, the slope value would be anti-proportional to the spacing and the same value changes measured with 0.5 m spacing would have been ten times the slope values of those measured with 5 m spacing. This point is further discussed in Section 5.3. To align the rectangle raster dataset's size, it was transformed into a vector dataset, clipped with the extent of the data points, and subsequently transformed again into a raster dataset (Figure 18g).

Regarding the orientation of the boundaries, first, the "aspect" was calculated (Figure 18d). The degree values of the aspect were reclassified into eight values (Figure 18f), one for each compass direction, including the intermediate ones (e.g.

northeast). (Since "north" is divided into the degrees 0–22.5° and 337.5–360°, nine classes were created.) Subsequently, the raster calculator was used to assign each pixel with a slope higher than Class 1, for example, the aspect to *horizontal* (1: north and south), *diagonal* (2: northeast, southeast, southwest, northwest) and *vertical* (3: east and west; Figure 18h). All these raster datasets were exported as images and can be interpreted, following the guidelines for ERT description (see Subsection below).

Guidelines for ERT description in addition to the resistivity distribution

According to workflow Step S3, guidelines for ERT description were developed.

The interpretation guidelines allow for a more standardized description of the ERT data. The different ERT datasets were collected from many different study areas, including different lithology, topography and various anthropogenic impact intensities. Resistivity values varied strongly, and their distribution seemed to be partly chaotic. Due to this high variability in resistivity distribution, standardized analysis and description of the collected ERT data seem feasible to reduce ambiguity. A visual presentation of the noted parameters is given in Figure 20.

The guidelines were developed for the most frequently used Wenner-Schlumberger configuration within this study and were only adapted to Dipole-Dipole to maintain comparability. A robust comparison between individual measurements is only guaranteed with the same measurement configurations. Nevertheless, this method can provide helpful information for any type of ERT 2D measurement, including when comparing different configurations.

First, the ERT data was inverted with *RES2DINVx64* by always using the robust inversion scheme, which was stopped after the fifth iteration (see the previous Subsection). Second, a boundary analysis was conducted (see the previous Subsection) for greater objectivity in the description of the data. The three collected images of ERT data were then described according to these guidelines (Figures 19 and 20).

Resistivity-layers (r-layers) are defined by areas that are mostly homogeneous, with comparable resistivity values. They are intersected by boundaries, calculated with the boundary analysis with values > 1, the ground surface or the edge of the tomogram (cf. Figure 19a). The boundary must be at least on more than half of the boundary between the r-layers. The minimum extent of an r-layer for this study is defined by a length of more than five times the electrode spacing in at least one direction.

Contrary to r-layers, resistivity-anomalies (r-anomalies) are smaller (< 5 times the electrode spacing) than r-layers and are surrounded by areas with different (higher or lower) resistivity values or the edges of the tomogram. To describe the collected data in greater detail, the following characteristics are stated in addition to the pure resistivity distribution (cf. Figure 19a).

The number of r-layers describes the number of layers visible in the tomogram. The number of layers is evident by the number of rows in the description tables.

For each layer, the following values are noted (cf. Figure 19b):
Within the *thickness of r-layers*, the size of the smaller axis for the r-layer is noted after Table 1. The homogeneity of the resistivity distribution within single r-layers is noted in *homogeneity*. *Horizontal changes* describe the opportunity and intensity of hori-

Figure 19: Flowchart – guidelines for ERT description: a) decision for layers and anomalies; b) parameters noted for each r-layer/pair of r-layers; c) parameters noted for each r-anomaly type

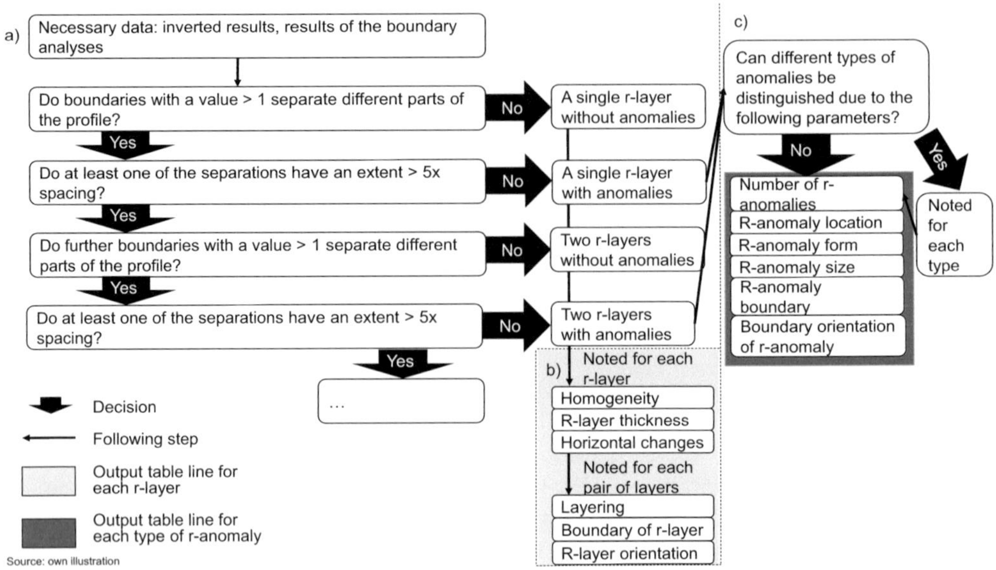

Source: own illustration

zontal changes. In the case of horizontal changes, first, the layer of the left-hand side of the tomogram is described. *Layering* describes the degree of differentiation between one r-layer and the r-layer below or on its right-hand side (in the tomogram) in the case of horizontal resistivity variations. In *boundary of r-layers,* the calculated sharpness of a boundary is noted. Additionally, the *r-layer orientation,* calculated using the boundary analysis, is added to the r-layer description.

Smaller resistivity differences within the tomograms are classified as r-anomalies. First, the position of the r-anomalies is noted in the r-anomaly location. Most layers of r-anomalies are defined as a single type of r-anomalies. In addition, different forms and sizes can cause different types of anomalies. For each type, the following parameters are noted (cf. Figure 19c):

The *number of anomalies* describes the number of anomalies relative to the length of the profile. Various ways of calculating the number of anomalies were tested, including different spacings, array types and the number of quadripoles per model block. However, the number of quadripoles per model block especially varies, even within a single measurement. To keep this formula as simple as possible, understandable and comprehensible, it was reduced and specified for a Wenner-Schlumberger array with 36 electrodes (Figure 20, upper right part). Therefore, this approximation needs to be treated with caution by comparing different data with different array and geometry parameters.

The *A* value can be transformed into the *number of r-anomalies* value, following Table 1.

The *r-anomaly form* describes the form of the r-anomaly basically in the following categories: *block:* x and y extent are almost the same; *quasi-layer:* small-sized layers

Figure 20: Guidelines for ERT description – overview

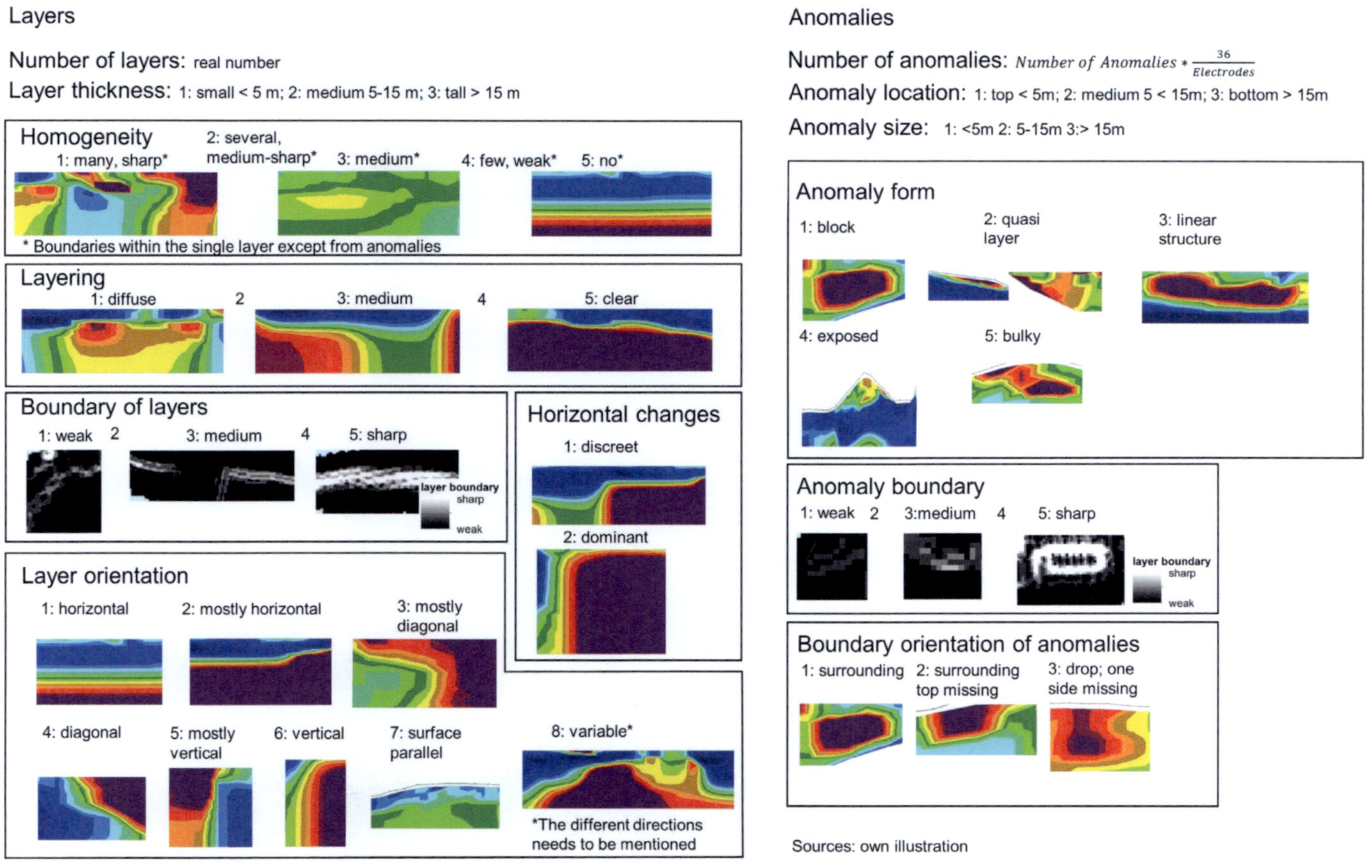

Sources: own illustration

Table 1: Guidelines for ERT description – values

Column name	Value description
Homogeneity	1: diffuse → 5: clear
R-layer thickness	1: small (< 5 m); 2: medium (5–15 m); 3: large (> 15 m)
Layering	1: diffuse → 5: clear
Horizontal changes	1: discreet; 2: dominant/clear; 0: no changes
Boundary of r-layers	1: weak → 5: sharp
R-layer orientation	1: horizontal; 2: mostly horizontal 3: mostly diagonal; 4: diagonal 5: mostly vertical; 6: vertical 7: surface-parallel; 8: variable
Number of r-anomalies	1: A < 1; 2: A > 1 < 2; 3: A > 2
R-anomaly location	1: top (< 5 m); 2: central (5 < 15 m); 3: bottom (> 15 m)
R-anomaly form	1: block, 2: quasi-layer; 3: linear structure; 4: exposed; 5: bulky
R-anomaly size	1: (A < 1); 2: (1 <= A < 2); 3: (A >= 2)
R-anomaly boundary	1: weak → 5: sharp
Boundary orientation of r-anomalies	1: surrounding; 2: surrounding, top missing; 3: drop, one side missing

Source: own research

are on top or possibly, poorly resolved layers are at the edge of the profile; *linear structure: x* and *y* extent vary; *exposed:* r-anomalies are at special positions; *bulky:* undefinable r-anomalies.

In *r-anomaly size,* the size is provided in three categories, following Table 1. Analog to the differentiation of the layers, the intensity of the *r-anomaly boundary* and the *boundary orientation of r-anomalies* is described.

All these parameters allow an accurate and less ambiguous description of the data. The exact positions of the described r-layers and r-anomalies relative to each other play a subordinate role. Nevertheless, a small allocation of the different r-layers and r-anomalies is provided in the results chapter.

3.2.2 Ground-penetrating radar

First, the geophysical basics of GPR are described. Then, the implementation of this study is mentioned.

Basics

GPR uses temporal electromagnetic impulses to create radio wavetrains. These wavetrains are injected by a transmitter (the so-called Tx) on a surface antenna and penetrate the ground at about one-third of the speed of light (Ullrich *et al.*, 2007). The waves are reflected and refracted whenever the dielectric constant of the material, which is penetrated, changes (cf. Figure 21). Therefore, the measurement is based on the differences in the dielectric constant of different materials or their condition, water content and compositions. The intensity of the reflection depends on the contrast in the dielectric constant of the materials (Reynolds, 2011). A wide range of dielectric constants and velocities of radar waves of/in different materials are provided in Reynolds (2011). However, the reflected part of the wavetrains is measured by a receiver (the so-called Rx) as amplitudes of the electric field strength and its phase (Blindow, 2005), and the two-way travel time in nanoseconds is also recorded (Conyers, 2012). For the measurement, two antennas, each with a transmitter or a receiver, are used. The wavetrains are not only refracted and reflected, but a part of the injected radio wavetrains penetrates farther. Therefore, further changes, in deeper parts in the dielectric constant, can cause another refraction.

In contrast to most geophysical methods, the physical parameter, in this case, the dielectric constant, is not measured, calculated or derived. The direction of the contrast is neither visible in the collected data nor the altitude of it nor even the sole values. The only received information from the dielectric contrast is provided by the electric field strength of the reflected wave, which is dependent on the contrast. The radiowaves' speed within the current material of interest is essential to define the depth of the reflections. Nevertheless, using the specific velocity of the radar waves, the material can be derived. Such a measurement can be performed using the common mid-point, so-called CMP, method (Reynolds, 2011). In addition, the polarity of the received waves for the assumption of whether the dielectric constant of a material is higher or lower than the material beneath it.

Figure 21: Setup of GPR: a) sedimentological layers; b) anthropogenic structures

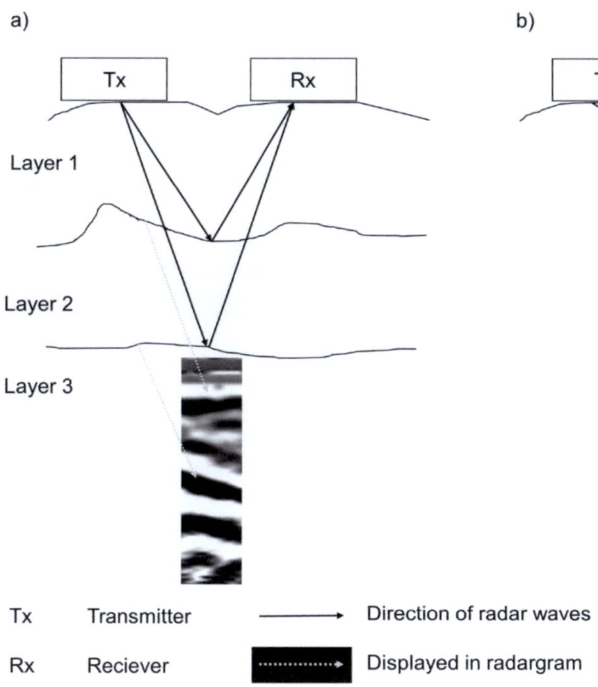

Sources: own illustration, based on Blindow, 2005; Reynolds, 2011

Implementation

GPR is measured in different frequencies. While the penetration depth decreases by increasing the frequency, the resolution increases (Reynolds, 2011). Therefore, the number of measurements per meter usually increases as frequency increases. Measuring with 30 MHz induced a radar wavetrain each meter, whereas with 200 MHz, a wavetrain every 0.125 m was induced, the so-called *step size* in the following work. The measurements of this work were performed using different frequencies and also different systems of the GPR. The antennas for 30, 50 and 100 MHz were utilized using the *PulsEKKO PRO System*, whereas the frequencies of 250 MHz were utilized using the *NOGGIN System. Sensors & Software Inc.* (*Mississauga*, Canada) produce both systems. The main difference (except for the frequencies) is the separation of the antennas. The *PulsEKKO PRO System* consists of single antennas, which can be used in different applications. In contrast, the *NOGGIN System* is a single block and must be used in a special wagon. Additionally, the *NOGGIN System* uses shielded antennas, the opposite of the unshielded antennas of the *PulsEKKO PRO System.* Shielded antennas focus the wavetrain in the measured direction and deliver a better signal and penetration depth, even at higher frequencies. Unshielded antennas are more flexible in use and allow a measurement even in rough relief. However, the antenna separation depends on the frequency; therefore, the system's choice is mainly for logistical reasons.

GPR's theoretical resolution is about one-fourth of the wavelength (Reynolds, 2011), which is a function of the velocity and the frequency. As such, a resolution with low frequencies and high velocities is higher than with high frequencies and low velocities. Bristow (2009) describes, for example, a theoretical resolution of 50-MHz antennas for 0.06 m/ns^{-1} (a relatively low velocity in the subsurface, e.g. Reynolds, 2011) of 0.3 m and for 0.15 m/ns^{-1} (a relatively high velocity in the subsurface) of 0.75 m. In contrast, for 200 MHz antennas, 0.075 and 0.1875 m as the theoretical resolutions, respectively, are described. Therefore, for different applications and aims, different frequencies are used.

The antenna separation is four times the step size, which is why two types of impulse signals were utilized in this work. On the one hand, the impulses are induced using a special button for low frequencies or rough terrain. On the other hand, the impulses are induced by an odometer useful for larger frequencies. Usually, measurements with 30 and 50 MHz antennas are conducted using the button, whereas higher-frequency antennas are measured using the odometer (Sensors & Software Inc., 2016).

These single measurement points are usually set up in a transect or grid to provide a 2D transect or a 3D grid (Conyers, 2012). However, the measurements are still 1D measurements strung together in opposition to the ERT 2D and quasi-3D results, even though this is not mentioned often.

As discussed above, radar wave' speed can be distinguished via a single reflection using the CMP method. However, the speed is massively dependent on the water saturation and condition (Blindow, 2005) and can change rapidly with depth (Reynolds, 2011). Additionally, such measurements provide a single value for the entire distance to the reflection, and a clear reflection is needed. The subsurface's expected changes and the lack of clear reflection mean such measurements were waived, and literature values were used for depth calculation.

The data were digitalized using *EKKO_Project 4* (*Sensors & Software Inc.;* Mississauga, Canada). Depending on the used frequency and system, different gains, intensities, filters and other applications were used. The signals of the unshielded antennas were especially slight and needed further processing for interpretation.

For all applications, background subtraction was utilized. This filter masks all surface-parallel signals. The so-called ground and air waves mainly cause these signals, which superimpose the other reflections. Additionally, a *Dewow* filter was used. This filter deletes a low-frequency signal that can occur when transmitting the wavetrain and superimposes the signal of the measurement (Sensors & Software Inc., 2016).

The GPR receiver receives all signals at different frequencies. Therefore, it is necessary to mask out all signals outside the range of interest. The *bandpass filter* deletes all information between 40% and 80% and between 120% and 160% of the frequency of the used value range.

The intensity of the radar wave decreases very sharply with depth. Nevertheless, there is still some signal left in deeper areas. To make all signals visible, the data is gained using a SEC2 (spreading and exponential calibrated compensation) gain, which is the preferred gain for GPR data (Sensors & Software Inc., 2016).

Additional to the data, the topography was noted. This is not only important for displaying a realistic surface and helping to interpret the data, but also, in the case of a non-surface-parallel dipping of the layers in the subsurface, the topography needs to be migrated (Bristow, 2009). Therefore, the reflection's dip is restored, and a static correction of the topography is applied automatically.

Due to the used gain and the dipole signal of the wave, no measured values in opposite to geophysical methods were documented. Therefore, and for better readability of the displayed data, the GPR line itself and another line with highlights are illustrated in the results chapter. The subsequently performed geometrical description of the GPR data is in line with Neal (2004).

Analog to the 3D ERT in Mömbris, in addition to the layers of the 3D GPR, a real 3D model of the resistivity distribution was created. First, an average filter with a kernel size of 9 was chosen. This was necessary to reduce small anomalies at the top of the grid and to make the most significant parts visible. Second, a threshold value was estimated at 8,600 mV to display the most important reflectors within the data. Using the isosurface of the Voxler, all values above the threshold values were displayed in 3D.

3.2.3 Seismic refraction tomography

First, the geophysical basics of SRT are described. Then, the implementation in this study is discussed.

Basics

The basis of all seismic measurements is the targeted excitation of elastic waves for subsurface penetration. Such excitation is performed at a known time, in our case, using a sledgehammer, but vibrators, drop weights and explosions are common. This excitation causes elastic waves excitation, including the primary (P) and secondary (S) waves. These waves are refracted, and the returning waves from the underground to the earth's surface are recorded using geophones pushed into the ground. This method allows conclusions to be drawn about the subsurface's lithological composition (Fertig, 2005).

Important for the interpretation of seismic waves is their velocity during propagation through the subsurface. The velocity depends on the elastic modulus and the density of the propagation material. Their calculation differs between a P-wave and an S-wave, which are caused by a different modulus that needed to be utilized because P-wave deformation parallels wave propagation, and S-waves are orthogonal to them (Schrott & Hoffmann, 2008).

In the case of two layers within the subsurface, including an increase of speed for a seismic wave ($V_2 > V_1$) and a specific incidence angle according to Shell's Law, the wave is refracted and travels along the substrate boundary (cf. Figure 20; Sandmeier & Krummel, 2005; Schrott & Hoffmann, 2008; Reynolds, 2011). Now, the refracted seismic wave travels along the boundary with the velocity of the lower material, producing head waves moving upward and being measured by the geophones (Sandmeier & Krummel, 2005; Schrott & Hoffmann, 2008). However, by adding more layers to the subsurface, refraction means that parts of the wave penetrate farther through the material. Thus, again refraction and traveling along the boundary happen at the next change of layers, producing more head waves.

Figure 22: Simplified setup of SRT design of a seismic survey and simplified raypath geometry in a layered subsurface with three layers

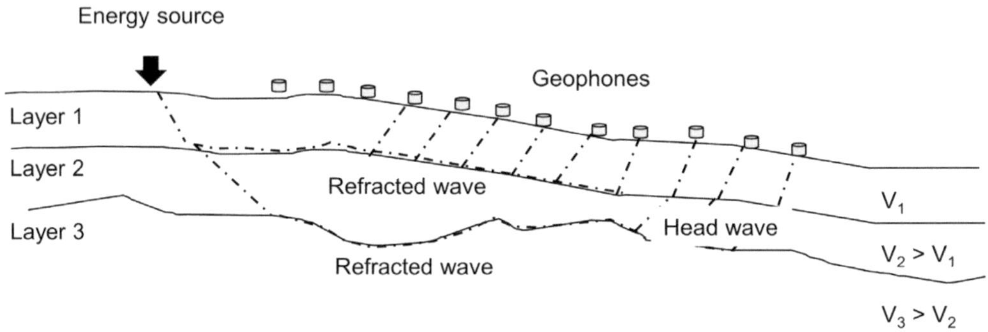

Source: modified after Schrott & Hoffmann, 2008

Implementation

SRT measurements were made using a *Geode seismograph* (*GEOMETRICS Inc.*; San Jose, California, USA) and 24 geophones in 3 and 5 m spacings. The number of shot points depends on the subsurface's expected complexity (Schrott & Hoffmann, 2008). Due to more or fewer planar layers, but also vertical changes, at least between every second geophone, a shot was performed. At each shot point, the seismic signal was generated using a 5-kg-sledgehammer, stringing a metal plate and stacking it ten times.

Analog to ERT, the data of SRT needed to be inverted. However, two more steps needed to be performed utilizing the inversion with the software package of *SeisImager/2D* (*GEOMETRICS Inc.*; San Jose, California, USA).

The first arrivals were picked using *PICKWIN*. This step is necessary to create time-distance plots by combining the first arrivals of the signal at all geophones. Only clear first arrivals were used for quality management. Especially in the broader distances from the shot point, the first arrivals are more challenging to identify.

Subsequently, the time-distance plot was inverted using *PLOTREFA*. First, a starting model is defined using a layered model with a positive gradient in velocity with depth: a model with 20 m, a minimum value of 0.3 km/sec and a maximum value of 3.0 km/sec, including ten layers was chosen. These parameters are proposed in Geometrics (2009). Additionally, the topographic information needs to be added before the inversion is started. Rather similar to the inversion of the ERT, the tomographic inversion tries to fit the minimum travel time between the single shot points and geophones iteratively. This process is repeated, targeting a reduction of the root-mean-square-error between the modeled and the calculated travel times by modifying the model with each iteration step by comparing the calculated travel times with the measured ones (Geometrics, 2009). This process stops after a predefined value of iterations.

For quality management, only values with a reciprocal time error of less than 5% were utilized to minimize the effect of faulty measuring points or errors in the arrival picks. If values are above 5%, then they can be corrected automatically.

Subsequently, the tomographic inversion was performed using 20 iterations, five nodes, no smoothing passes and a smoothing weight of 0.5 for horizontal and vertical smoothing, the latter with five layers. For minimum and maximum velocity, 0.3 km/sec and 3.0km/sec were used. These parameters are proposed in Geometrics (2009). In all radargrams, the 20th iteration was used, and the ray traces were displayed to differentiate between the measured and the pure interpolated data.

3.2.4 Magnetic surveying

First, the basics of magnetic surveying are described and then, its implementation in this study is described.

Basics

Magnetic surveying is used to measure the magnetic field and its spatial distribution at a site. Principally, three magnetic fields influence the measurement of magnetic distribution everywhere on earth. The first field is the earth's magnetic field, with its poles close to the geographic poles and an origin in the interior of the earth, only changing over the course of years or decades. This field overlaps with the external magnetic field sourced beyond earth, which changes over the course of split seconds or days. Additionally, there is a temporally constant magnetic field, the so-called anomaly field, with high spatial variance, produced by the magnetization of the different materials in the upper part of the earth's crust. This last field allows for conclusions to be drawn about subsurface conditions (Knödel, Krummel & Lange, 2005).

Within the first two types of magnetic fields, every mineral within the anomaly field is located and becomes magnetized by each field's different specific intensities. All materials are also susceptible to this magnetization in different ways. Often, a permanent magnetization of the single material has to be added to this induced magnetization. Both are possibly forcing in different intensities and directions. Consequently, the resultant magnetization, measured at one position, is a possible mixture of different magnetization types, which has to be considered in interpretations (Reynolds, 2011).

Different materials have different susceptibilities and forms of magnetization how they react in an external field. Some minerals act diamagnetically, such as marble, quartz and graphite. These minerals reduce the magnetic field when placed in them. Other minerals are paramagnetic. The ions of those minerals orientate in the direction of the magnetic field when placed in it. Paramagnetic ions, such as iron, cobalt and nickel, have such a strong magnetic interaction that all ions in a local region, the so-called domain, orientate parallel to the earth's magnetic field, so-called ferromagnetism. The amplitude of ferromagnetism is up to 10^6 times stronger than diamagnetism or paramagnetism (Telford *et al.*, 1990). Ferrimagnetic minerals, such as magnetite, create anti-parallel moments, in which one of the moments is stronger, causing a net remnant magnetization even outside of magnetic fields (e.g.; Reynolds, 2011; Fassbinder, 2015). There are a couple of more ways for minerals to act when magnetized. However, these are less important for magnetic surveying in the present area of investigation.

In summary, each measured point of magnetization in nanotesla (nT) is a function of the intensity of all three field types. Therefore, the susceptibility of material surrounding the measurement, the way the material acts in a magnetic field and the content of permanently magnetized material within the area of interest influencing the measured point.

Caused by the reduction of magnetization with distance to a source (Knödel, 2005), all different ground-based magnetic surveying methods try to measure the magnetization of the anomaly field. As the content of minerals influencing the anomaly field is important, their distribution near the surface plays a major role. Important for the identification of geomorphological or archaeological sites is the remnant or permanent magnetization. After the magnetic field's absence, magnetization remains and is still measurable, sustained by permanently magnetic particles' internal field strength (Reynolds, 2011; Everett, 2013). Therefore, moved material, for example, by gravitational mass movements or earthworks, such as refilled or fresh ditches, can be identified (Fassbinder, 2015).

However, human activity affects the contribution of the magnetic field strength either actively or passively. For example, in the topsoil, bacteria with ferromagnetic minerals in their cells can be observed (Fassbinder *et al.*, 1990). Such bacteria could explain the detectability of post-holes by magnetic surveying. Although the wood of the post-holes usually has no influence on the magnetic field, the enrichment of the described bacteria, due to the wood's rotting, could (Becker, 1987). Additionally, on the surface, fire lets maghemite occur, a ferrimagnetic mineral (Le Borge, 1960; Fassbinder, 2015). This mineral accumulates due to relocation processes laterally and in deeper areas. The fire initially causes a reduction of hematite to magnetite, followed by a reduction to maghemite. This effect is called the Le Borgne effect (Graham & Scollar, 1976). Artifacts, especially ones made of iron and other ferrimagnetic minerals, strongly impact the measured magnetic field, too, and can be observed with this method (Fassbinder, 2015).

Implementation

The magnetic surveying was processed using a *FEREX DATALOGGER 4.032* (*Dr. Foerster GmbH and Co. KG;* Reutlingen, Germany*)*. The used magnetometer belongs to the so-called fluxgate magnetometers.

The Fluxgate magnetometer was developed during World War II to detect submarines. They are essentially made of at least one or (in the present case) two cores. These cores consist of magnetic material and are wound by two coils. The coils are wounded anti-parallel. Therefore, whenever a magnetic signal is received, the two cores detect it on opposite sides and produce the measuring point (Telford *et al.*, 1990).

For the measurement, the area of interest is usually divided into quadratic or rectangular quadrants. The data were visualized and interpreted using *DATA2LINE* (*Dr. Foerster GmbH and Co. KG;* Reutlingen, Germany*)*.

For the measurement of the anomaly field, the magnetization of the main field and the fluctuating fields have to be eliminated within the measurement (Knödel, 2005). In practice, the zero point of the measurement has to be calibrated and is shifted to the local (in space and time) zero point. This calibration is not always absolutely precise. Therefore, a zero-mean filter was used to delete possible mistakes within the calibration.

3.3 Ground-truth

This section describes the methods necessary for workflow Step 4 (S4) – *Ground-truth.*

3.3.1 Invasive methods

Pürckhauer and Edelman soundings, as well as percussion core probing, are considered invasive methods. The description of the soil samples was mainly made in line with "Bodenkundliche Kartieranleitung 5" (BKA, Ad-hoc-Arbeitsgruppe Boden, 2005) in the field and subsequently translated in line with "World reference base" (WRB, Fao, 2015). Later, the description was simplified, focusing on sedimentological features, whereas pedological features are more generalized. For the grain-size analysis in the laboratory, the samples were taken after DIN ISO 10381. The distribution was measured in the soil laboratory of the University of Würzburg using a pipette, according to Köhn (DIN ISO 11277). The three samples collected in Hauenstein for AMS-^{14}C-analysis were sent to *Poznan Radiocarbon Laboratory* for dating.

Pürckhauer

As the most common method for checking the first two meters of the subsurface, numerous Pürckhauer soundings were performed. For this method, a soil sampler is penetrated into the subsurface using a considerable plastic hammer. The first meter had a diameter of about 18 mm, and the second was smaller. Therefore, soil samples could only be taken of small layers, and massive layers or those consisting of coarse sediments could not be penetrated.

Boreholes

Whenever it was necessary to drill farther than 2 m, percussion core probing with a demolition hammer *BH 65* (*Wacker Neuson SE*; München, Germany) was utilized whenever possible and allowed. Due to the number of samples, the substrate was easy to describe, and lamination, clay cutaneous and other sedimentological and pedological features were visible within it. The boreholes were rammed with open samples being 80 to 30 mm in diameter and 1 m in length. The boreholes drilled with a demolition hammer delivered samples meter by meter.

Edelman

The engine-driven boreholes made by the demolition hammer, including several hundred kilograms of equipment, delivered well-describable results. However, the Edelman (*Eijkelkamp*; Giesbeek, Niederlande) drill set is sometimes a more practical way of drilling in rough relief, for a single person or in areas where motor drilling is forbidden (e.g. in protected natural areas). In contrast to the meter samples of the engine-driven demolition hammer, Edelman boreholes are drilled and cleaned up subsequently. In general, the received substrate is put in a row while noting the borehole depth. The boreholes are not rammed but drilled. Therefore, layering and other sedimentological features are mainly disturbed and can usually not be described nor distinguished. However, the Edelman drilling equipment is a well-practiced method for describing the substrate beneath 2 m of depth or in coarse sediment. Different samples were used according to the behavior of the substrate.

3.3.2 Outcrop

Sometimes opportunities arose to generate ground-truth data in addition to the abovementioned invasive methods. Whenever the lithology was visible at the surface, such as in creek beds or at breaking-off edges, dug holes with excavators or prospection pits, these results are summarized within this section. These varied for each study site and, therefore, were divided after each study site.

At *Wahlmich Castle* in Waldaschaff, a prospection pit was dug to validate the position of a magnetic anomaly at one of the ridges. Additionally, the bedrock crops at the breaking-off edge are described.

The bedrock visible within the creek bed near *Hauenstein Castle* is described and displayed in this study.

In August 2019, a heavy storm hit the northern Vorspessart. During the storm, gale-force gusts uprooted several trees. Therefore, in *Rannenburg Castle's* surroundings, at the roots of these trees, fresh sediment outcrops allowed for a description of the underlying sediments without acting invasively. Additionally, the bedrock could be observed in one section and is described to validate the geophysical data.

At the study site of Eisenberg, an excavator dug outcrops. One within a shaft for prospecting its depth and behavior, and another one a few meters outside the mining area. Within the first outcrop, a Prückhauer sounding was conducted to reach a greater depth. However, the outcrop and sounding are described as one.

3.4 Positioning

To display all the measured geophysical data and the positions of the ground-truth, different ways of measuring were undertaken due to the global positioning system (GPS) signal and the existence of fixpoints belonging to the excavations. These measurements are necessary for the workflow Steps 3 and 4 (S3 and S4) – *Geophysical methods and Ground-truth.*

3.4.1 Global positioning system

Essentially, GPS measures the difference in time between each satellite and the antenna. Including the time offset, four satellites are necessary to determine an unambiguous point on the earth's surface. GPS has an accuracy of about up to 1 m without using Real-time kinematics (RTK) on free plains. Several factors influence accuracy, especially the speed of the waves emitted by the satellites in the ionosphere and the atmosphere, but also relativistic effects, time offsets and geometrical reasons, to name a few, can force GPS offset. Most of these effects and mistakes can be reduced by using RTK, which can be created using a correction signal via a mobile SIM card.

Simply put, the SIM card receives signals from fixpoints nearby. These signals include the offset of each satellite to the previously measured fixpoint. With this information, the position can be theoretically distinguished with millimeter precision. In practice, all points, are used up to 5 cm error, are utilized without further correction. A second method for receiving a correction signal is utilizing a basis station, a second GPS point with a fixed position. Using a basis station, the dataset of the measured points still has the previous offset, but within the dataset, meaning the offset within a single dataset is minimized to a few centimeters. The measurements were performed using a *GS14* or *GS16* (*Leica Camera AG*; Wetzlar, Germany).

However, with both ways of measuring, some geometrical mistakes forced by the surroundings cannot be reduced. Feedback effects especially occur in an area surrounded by trees or houses, reducing accuracy by up to several meters (for more detail see e.g. Torge, 2002; Groves, 2013).

3.4.2 Tachymeter

Working with a *Multistation* (*Leica Camera AG;* Wetzlar, Germany), tachymeter fixpoints of the excavation were used. These points are usually set with GPS for excavation and are necessary for the position of the archaeological description of a site. They were used whenever it was possible and necessary to combine archaeological and geographical results. The exact position of the tachymeter can be calculated by the angles and distance of at least two known points. Between the tachymeter and a reflector at the position of interest, there needs to be visual contact. Therefore, this method could not be used everywhere (e.g. Freeden & Rummel, 2017).

3.4.3 Manual positioning

Start-, and endpoint of geophysical transects, the position of ground-truth methods and the edges of 3D grids were measured with a higher offset in dense vegetation and out of the range of fixpoints. For a precise topographic migration of the geophysical data, the changes in altitude along the profiles or grids were estimated, noted manually and proved later using a DEM.

3.5 Quantification of the medieval anthropogenic influence

This section described the necessary works according to workflow Step 6 (S6) – *Quantification of the anthropogenic influence.*

To summarize the findings and interpretations of the single study sites, an ordinal quantification of the different anthropogenic influences on the relief and relief development during the medieval period was made. The relief is considered in most cases at the microscale.

From the results and interpretation of the used methods, including their possibility of quantification, an ordinal scale was derived. Contrary to interval-scaled approaches (Evrard *et al.*, 2008; Rózsa *et al.*, 2020), the applied methods, types of impact, duration since the time of interest and rare archives complicated a more objective scale than the created one. It must be noted that the chosen approach only indicates impacts that can be actually seen at the site and not those already eroded or removed. Additionally, the anthropogenic effects need to be linked to the time of interest. In areas that are not mapped, no measurements were taken. However, that does not mean that there is no anthropogenic influence. In addition, the strength of anthropogenic influence is divided into active and passive on the relief development – active influence includes a direct change of the landscape, and passive influences change the type and intensity of natural processes (Herget, 2006; Schütt, 2006b). Four categories of medieval anthropogenic influence were derived for this study (cf. Table 2):

Category 1 – Predominantly natural origin: No evidence for anthropogenic impact within the data. Human beings obviously used the areas of investigation but the indications for the influence have already been eroded and weathered, are undetectable with the used methods or happened in a different period.

Category 2 – Slightly passive and active influence of the relief: Slight evidence for anthropogenic usage is visible within the geophysical or invasive data (e.g. occupation layers or relocated fragments of walls) that can be associated or assumed within the period of interest. However, no obvious, massive impact is visible within the relief.

Category 3 – Massively passive influence on the relief: Geophysical and invasive methods reveal a massive effect on natural processes, including a visible effect on the recent topography (e.g. change in sedimentation/accumulation system or massive erosion processes that can be identified).

Category 4 – Massively active influence on the relief: Geophysical and invasive data reveal a visible, massively active effect on the recent topography (e.g. earthworks or wall structures).

Table 2: Classified anthropogenic influences on the relief during the medieval period

Influence Category	*Description*	*Characteristics*
1	Predominantly natural origin	No evidence for anthropogenic impact within the data.
2	Slightly passive and active influence on the relief	Slight evidence for anthropogenic usage within the geophysical or invasive data.
3	Massively passive influence on the relief	Geophysical and invasive methods reveal a massive effect on natural processes
4	Massively active influence on the relief	Geophysical and invasive data reveal a visible, massively active effect on the recent topography (e.g. earthworks or wall structures)

Source: own research

4 Results and interpretation

In this chapter, the results are described and interpreted. The individual results are designated with a letter for the study site, a number for the site-specific question and a shorthand for the method (cf. Tables 3), followed by sequentially ordered letters for each result.

Table 3: Abbreviation code: study site, question type and method

Abbreviation study site	*Study site*	*Abbreviation question type*	*Question type*	*Abbreviation method*	*Method*
W	Waldaschaff	1	Surficial, natural feature	ERT	Electrical resistivity tomography
H	Hauenstein	2	Surficial, anthropogenic feature	GPR	Ground-penetrating radar
A	Alzenau (Rannenburg)	3	Buried, natural feature	SRT	Seismic refraction tomography
M	Mömbris	4.1	Buried, anthropogenic feature on-site	MAG	Magnetic surveying
E	Eisenberg	4.2	Buried, anthropogenic feature off-site	PUE	Pürckhauer sounding
				BH	Borehole (percussion core probing)
				OC	Outcrop (variable)
				EDE	Edelman

Source: own research

4.1 Waldaschaff

First, site-specific questions were derived with the help of the GIS-based terrain analyses at Waldaschaff. The results for each question are provided and interpreted. Then, an estimation of quantification of the anthropogenic influence is made. A technical overview of the used application of most of the geophysical methods is given in Appendix 1 to Appendix 3.

4.1.1 Morphometric analyses for the creation of site-specific questions

The morphometric analyses for creating site-specific questions are associated with workflow Step 2 (S2) – *GIS-based terrain analyses.*

Figure 23: Field picture – Breaking-off edge Waldaschaff

Photographer:
J. Trappe, 2020

Figure 24: Morphometric analyses – Waldaschaff: a) landform classification and slope; b) topographic wetness index

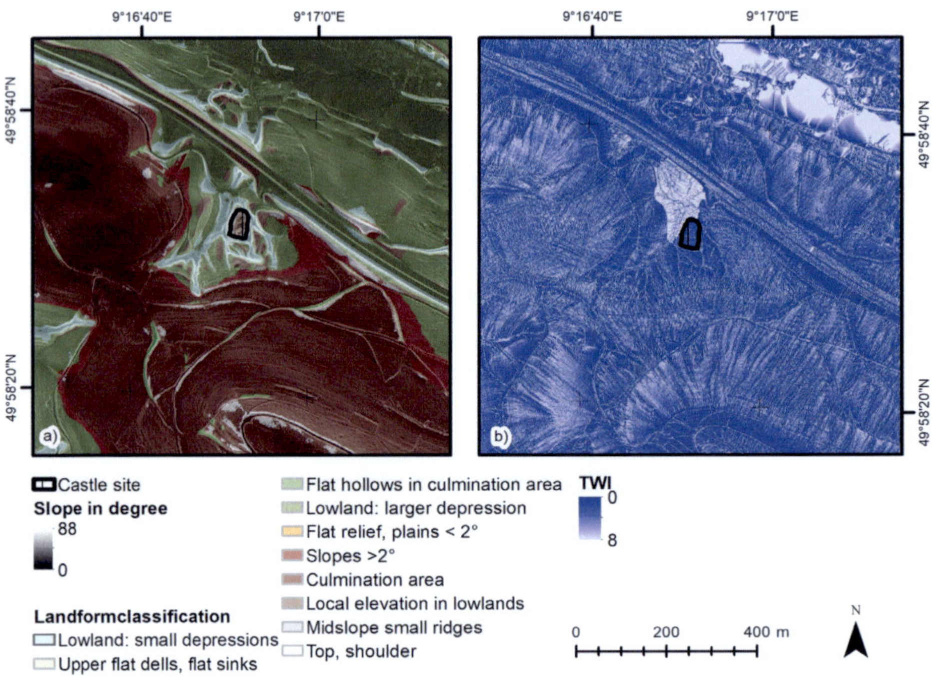

Source: own calculations, based on Bavarian Agency for Digitalisation, High-Speed Internet and Surveying, 2014
Map creation: J. Trappe

Being on a local elevation in the lowlands within the TPI landform classification (Figure 24a), the castle was constructed on top of a prong. This specific location affords an overview of the Aschaff Valley. South and southeast of the castle hill, there are several small valleys and ridges. These valleys are classified as lowlands, small depressions. They are surrounded by areas described as lowlands, larger depressions, with the small ridges being classified as local elevation in lowlands. This area ends sharply at a huge breaking-off edge (cf. Figure 23).

Runoff water (Figure 24b) accumulates in the northwestern part of the castle hill and within multiple depressions in the area of investigation; south and southeast of the castle hill, an episodic water flow toward these depressions developed, with a higher frequency at the western and eastern depressions.

Regarding its specific position within a rough landscape and by assessing its local importance, the following questions were derived (an overview is in Table 4):

Question W1 – surficial, natural features

What forces caused the development of the rough relief south of the castle hill?

The rough and exposed terrain south of the castle hill in Waldaschaff comprises small depressions and local elevations within these lowlands. Caused by its location directly at the castle hill, with dug neck ditches distributed in the area, this area's origin is still unclear and needs further investigations.

Question W2 – surficial, anthropogenic features

In what materials were the neck ditches dug?

As a manmade effect on the exposed terrain south of the castle hill, the dug neck ditches need to be further investigated. Especially south of the castle hill, they seem to be of anthropogenic origin. An excavation at the southern neck ditch in 2016 found a planar border at the (grano-)diorite, indicating that all bedrock types located at the study site were removed, not only the softer Bröckelschiefer (Rosmanitz & Bachmann, 2017).

Question W3 – buried, natural features

How do the mapped faults affect the development of relief?

The geological map (Weinelt, 1962) indicates a faulting system affecting the area of interest. However, for example, the prominent breaking-off edge indicates a more complex system than the one on the map does. If the area south of the hill is of natural origin, this fault system needs further prospecting and investigation as one of the important driving factors for relief development. Subsequently, these data can be combined with those of W1 to comprehend relief development at the study site.

Question W4.1 – buried, anthropogenic features – on-site

Do unknown building structures exist at the castle site?

On the castle hill, several parts remain untouched by excavations and are buried. Due to the complexity of different building phases and forms, the positions of buried manmade structures, earthworks, and occupation layers at the castle site are partially unclear. The site needs to be prospected for further archaeological interpretation and to estimate the anthropogenic impact. This question was prepared during the excavation to generate targeted excavation cuts.

Question W4.2 – buried, anthropogenic features – off-site

Where is the location of the outer bailey?

Commonly medieval castles included an outer bailey with buildings for farming, industry and trade. In Waldaschaff, such an outer bailey has not yet been found (Rosmanitz & Bachmann, 2017; Rosmanitz *et al.*, 2019b). Therefore, spatial prospecting was performed in possible positions for the bailey.

Table 4: Site-specific questions in Waldaschaff

Category	*Questions*
W1	What forces caused the development of the rough relief south of the castle hill?
W2	In what materials were the neck ditches dug?
W3	How do the mapped faults affect the development of relief?
W4.1	Do unknown building structures exist at the castle site?
W4.2	Where is the outer bailey located?

Source: own research

4.1.2 W1 – Origin of the rough relief: results

What forces caused the development of the rough relief south of the castle hill?

It is necessary to distinguish between the different lithological materials and their spatial distribution to answer this question. Therefore, various methods were used. All measurements of W1 were made more or less south of the castle hill in the described rough relief (Figure 25).

4.1.2.1 Geophysical data

The data in this section for W1 are associated with workflow Step 3 (S3) – *Geophysical methods.*

ERT

First, the inverted resistivity distribution of the profiles is described. Then, their relative distribution is discussed according to the guidelines for ERT description.

W1_ERT_a resistivity distribution

W1_ERT_a is located at the most dominant of the elevated areas within the lowlands southeast of the castle. This elevated area is split into three parts by small valleys. The profile ends up being almost at the elevation of the street leading to the castle.

W1_ERT_a (Figure 26) illustrates a low-resistive area with resistivity values up to 70 Ωm between meters 20 and 94. Its thickness is between 10 m at meter 24 and 2 m at meter 94. Below this area, higher resistivity values over 300 Ωm occur. Between

Figure 25: Location of measurements – W1

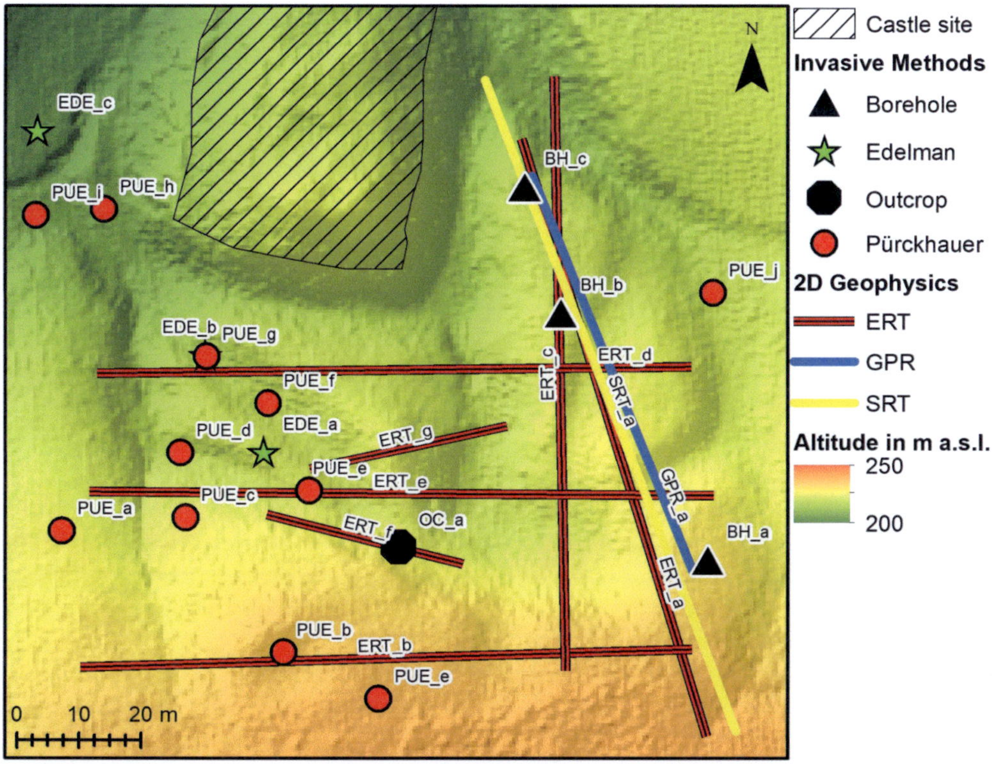

Source: Bavarian Agency for Digitalisation, High-Speed Internet and Surveying, 2014
Map creation: J. Trappe

Figure 26: Electrical resistivity tomography – W1_ERT_a

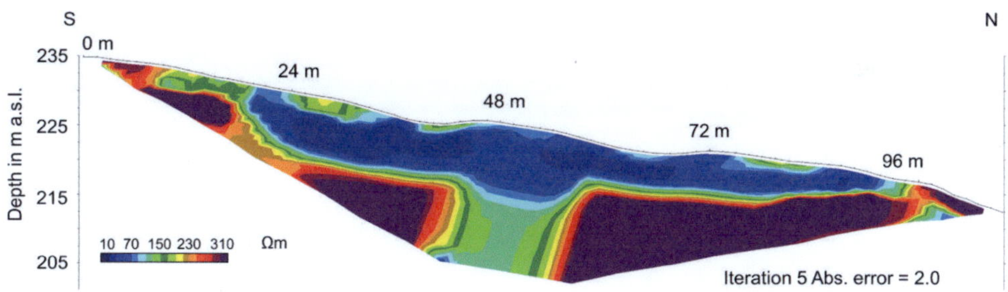

Source: own illustration

meters 45 and 60, the low-resistive area is underlain by resistivity values of about 100 Ωm. At about meter 60, the surface area makes a step of about 5 m and ends in an area of higher resistivity values. At both ends of the profile, resistivity values near the surface are between 150 and over 300 Ωm.

W1_ERT_a description after the guidelines

Combining the resistivity distribution with the boundary analysis (Figure 27), two r-layers become visible (Table 5).

Figure 27: Boundary analysis – W1_ERT_a: a) sharpness of r-layer boundary including the position of r-layers and r-anomalies; b) orientation of the boundary gradient

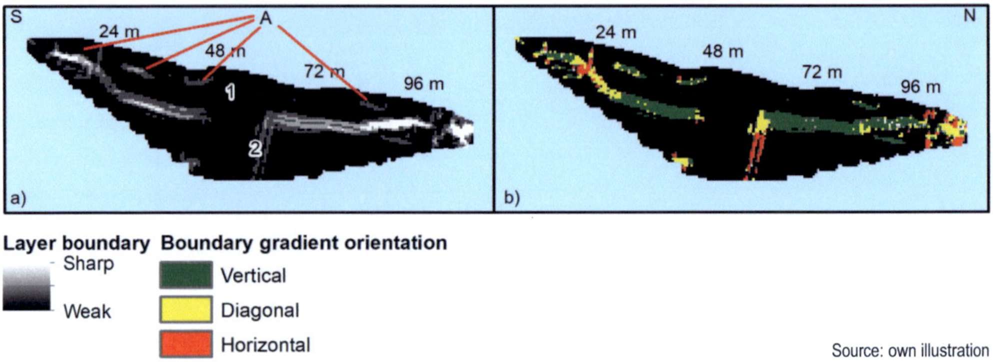

Source: own illustration

The layering is medium clear, revealing some variation in size, length and intensity. Horizontal changes partly surround the top r-layer, but also the bottom r-layer in the central part. The boundary is mostly sharp; nevertheless, it is weak in the middle part. The orientation of the boundary is mostly horizontal, with some differences at the bottom, but also vertical at the edges. Although the top r-layer (1) has a mostly high level of homogeneity, only displaying small boundaries, the bottom r-layer (2) has less homogeneity, especially within the central part. Four r-anomalies (A) were detected within the top r-layer, with a value of 2 for the number of r-anomalies. The anomalies act as layer fragments at the top. They were surrounded by a weak-to-slightly-weak boundary, which surrounds them, except at the top. The size of them is mostly up to 10 m.

Table 5: ERT guidelines – W1_ERT_a

R-layer	*Homogeneity*	*R-layer thickness*	*Layering*	*Horizontal changes*	*Boundary of r-layers*	*R-layer orientation*
R-layer 1	4	2	3	0	3	2
R-layer 2	3	3	-	2	-	-
R-anomaly	*Number of r-anomalies*	*R-anomaly location*	*R-anomaly form*	*R-anomaly size*	*R-anomaly boundary*	*Boundary orientation of r-anomalies*
R-anomaly Type A	2	1	2	1–2	1–2	2

Source: own research

W1_ERT_b resistivity distribution

W1_ERT_b is located on top of the breaking-off edge, south of the castle hill, and finishes western of profile W1_ERT_a.

Figure 28: Electrical resistivity tomography – W1_ERT_b

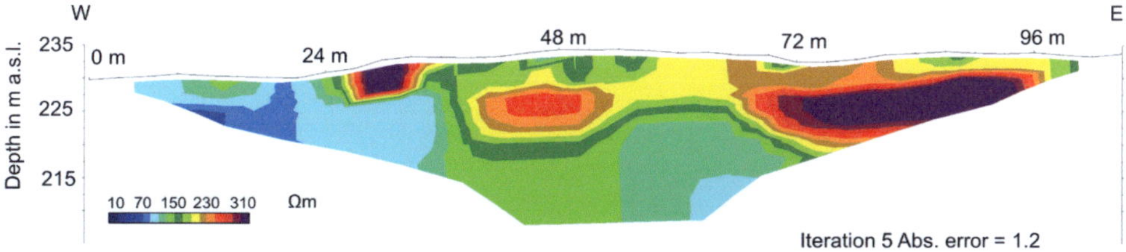

Source: own illustration

W1_ERT_b (Figure 28) illustrates two areas with different resistivity values. In contrast to W1_ERT_a, the relatively high-resistive area, with resistivity values around 300 Ωm, exists only within the eastern part of the profile. The low-resistive area, with resistivity values below 100 Ωm, occurs within the western parts of the profile. Within the central part, between meter 24 and meter 72 of the profile, a few lenses with resistivity values around 300 Ωm exist. However, the surrounding matrix has resistivity values in a lower range of about 150 Ωm.

W1_ERT_b description after the guidelines

Adding the boundary analysis' information (Figure 29) to the resistivity distribution, there are two r-layers visible in the eastern part, separated with a horizontal, diagonal and vertical border within W1_ERT_b (Table 6).

Figure 29: Boundary analysis – W1_ERT_b: a) sharpness of r-layer boundary including the position of r-layers and r-anomalies; b) orientation of the boundary gradient

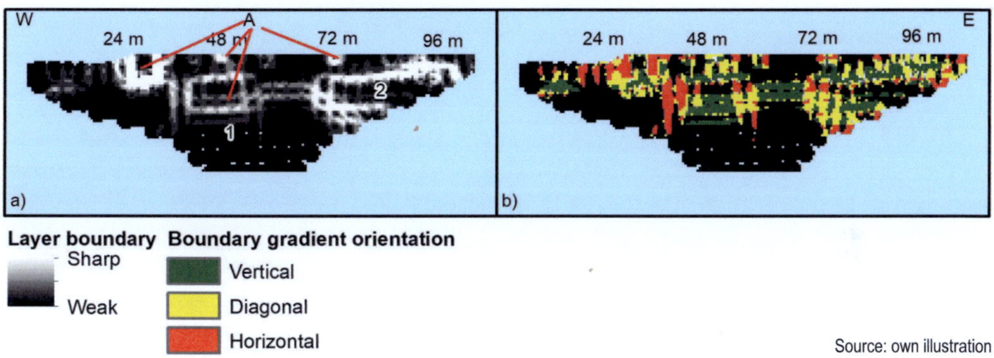

Source: own illustration

This part is well-separated from the other r-layer and surrounded horizontally and vertically by a sharp boundary. Layering is well-developed in W1_ERT_b. The homogeneity within the top or western r-layer (1) is low. In contrast, the eastern bottom r-layer's (2) homogeneity is more pronounced, and only a few boundaries are visible. One r-layer of r-anomalies (A) is visible at the top. A sharp boundary surrounds the western, medium-sized r-anomaly. Other, smaller r-anomalies are visible within this r-layer as well. In deeper areas, additional r-anomalies occur with less sharp boundaries surrounding them.

Table 6: ERT guidelines – W1_ERT_b

R-layer	*Homogeneity*	*R-layer thickness*	*Layering*	*Horizontal changes*	*Boundary of r-layers*	*R-layer orientation*
R-layer 1	1	1	2	2	5	3
R-layer 2	4	3	-	2	-	-
R-anomaly	*Number of r-anomalies*	*R-anomaly location*	*R-anomaly form*	*R-anomaly size*	*R-anomaly boundary*	*Boundary orientation of r-anomalies*
R-anomaly Type A	3	1–2	1	2–3	4	4

Source: own research

W1_ERT_c resistivity distribution

W1_ERT_c is at the edge of the most dominant of the elevated areas, finishing on top of the breaking-off edge about 3 m south of meter 85 of W1_ERT_b.

Figure 30: Electrical resistivity tomography – W1_ERT_c

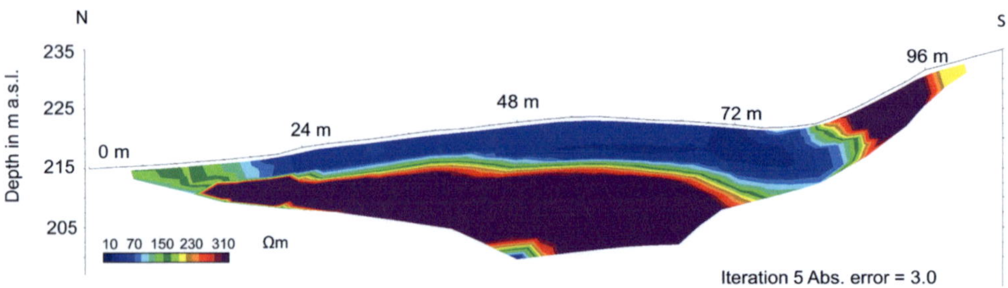

Source: own illustration

In W1_ERT_c (Figure 30), a low-resistive area, with resistivity values of up to 70 Ωm exists underneath the surface. This area has a thickness between approximately 4 m in the southern part and 10 m at meter 78. Underneath this area, resistivity values reach over 300 Ωm. In the first 20 meters of the profile, resistivity values at the uppermost part are between 100 and 200 Ωm. Uphill, at meter 78, higher resistivity values occur at the surface and at the bottom r-layer, with resistivity values o f over 300 Ωm. The two high-resistive parts are separated or shifted around meter 78.

W1_ERT_c description after the guidelines

Combining the resistivity distribution with the boundary analysis (Figure 31), three r-layers are evident, with a clear layering between them (Table 7).

The boundary of the first r-layer (1) to the uphill high-resistive r-layer (2) is sharp and has a diffuse orientation, mainly with a diagonal or horizontal orientation. In comparison, the boundary of r-layer 1 in the bottom part is sharper, orientated vertically and only a little diagonal or horizontal. In the northern part, a low-resistive

Figure 31: Boundary analysis – W1_ERT_c: a) sharpness of r-layer boundary including the position of r-layers and r-anomalies; b) orientation of the boundary gradient

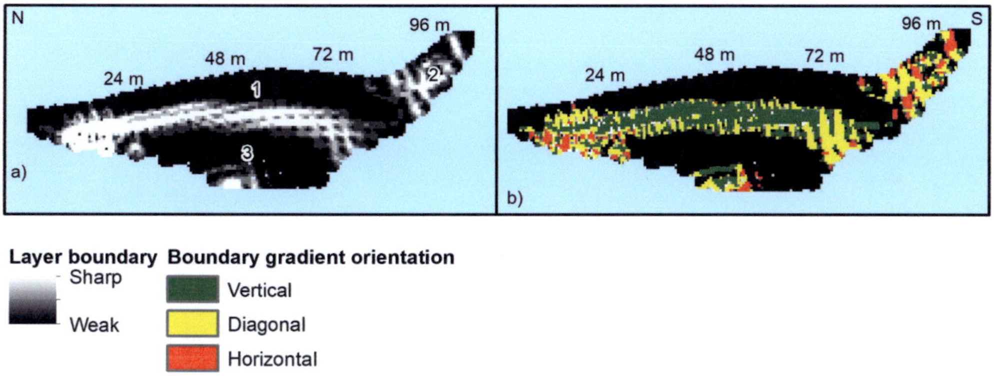

Source: own illustration

r-layer exists, with 3 to 12 m in size on top of an r-layer (3) with higher resistivity values. Additionally, in the southern part, a horizontal change to an r-layer with higher resistivity values within the uppermost r-layer occurs. This r-layer is of medium size. There is no clear boundary separating the different resistivity values in the northern part, at about meter 18 of the profile. Except for the northern part, the top r-layer has a high homogeneity. The southern r-layer has various boundaries within it, indicating low homogeneity. In contrast, the bottom r-layer is more or less homogeneous but contains an intense boundary at the bottommost part at meter 48.

Table 7: ERT guidelines – W1_ERT_c

R-layer	*Homogeneity*	*R-layer thickness*	*Layering*	*Horizontal changes*	*Boundary of r-layers*	*R-layer orientation*
R-layer 1	4	1–2	5	2	5	5
R-layer 2 (horizontal split)	2	1	-	-	-	-
R-layer 3	4	2	5	1	5	2
R-anomaly	*Number of r-anomalies*	*R-anomaly location*	*R-anomaly form*	*R-anomaly size*	*R-anomaly boundary*	*Boundary orientation of r-anomalies*
R-anomaly Type A	0	-	-	-	-	-

Source: own research

W1_ERT_d resistivity distribution

Parallel to the breaking-off edge and downhill of it, W1_ERT_d is located. This profile passes some smaller elevated areas in the western part and finishes at the dominant elevated area in the east.

Figure 32: Electrical resistivity tomography – W1_ERT_d

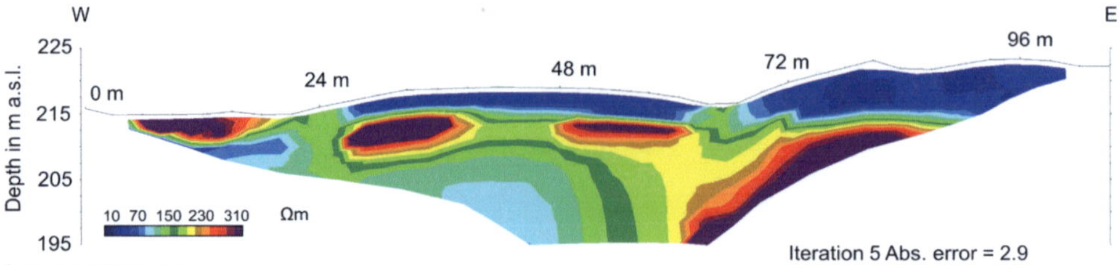

Source: own illustration

Except for a small area with higher resistivity values of up to 330 Ωm until meter 24, the resistivity values of the top part of W1_ERT_d (Figure 32) are up to 70 Ωm. This low-resistive area varies in size and ends at about 210 m a.s.l. In the eastern part, underneath, beginning at meter 60, an area with higher resistivity values of over 300 Ωm occurs. Below 214 or 210 m a.s.l., until meter 60, the resistivity values vary between 100 and 250 Ωm. The highest of these resistivity values occur in lenses between meters 36 and 48, in a depth of 214 to 205 m a.s.l.

W1_ERT_d description after the guidelines

Following guidelines for ERT description (Table 8), three r-layers can be differentiated by combining the resistivity distribution with the boundary analysis (Figure 33).

Figure 33: Boundary analysis – W1_ERT_d: a) sharpness of r-layer boundary including the position of r-layers and r-anomalies; b) orientation of the boundary gradient

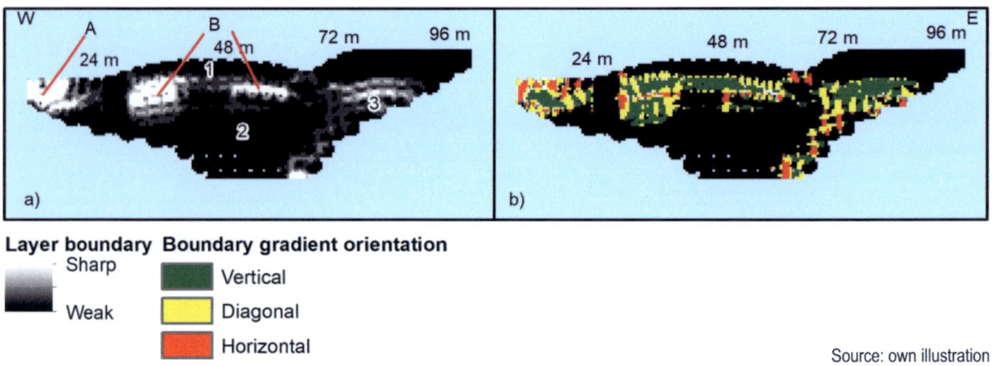

Source: own illustration

A single uppermost r-layer (1) occurs separated with a medium-weak but varyingly intense horizontal boundary, meaning the bottom r-layer is separated diagonally into eastern (2) and western sections (3). The top and eastern r-layers are mostly homogenous, with the western r-layer displaying a lower homogeneity. The layering of the boundaries between all three r-layers is distinct medium-well. Within the top r-layer of the profile (A) and the central one (B), r-anomalies are located. Although sharp boundaries surround the uppermost and central r-anomalies, the top crops out at the uppermost ones. Thus, this is classified as a quasi r-layer, and the central ones have a blocky structure. The size of both r-anomalies is between 5 and 15 m.

Table 8: ERT guidelines – W1_ERT_d

R-layer	*Homogeneity*	*R-layer thickness*	*Layering*	*Horizontal changes*	*Boundary of r-layers*	*R-layer orientation*
R-layer 1	1	1	2	2	5	3
R-layer 2	4	3	-	2	-	-
R-anomaly	*Number of r-anomalies*	*R-anomaly location*	*R-anomaly form*	*R-anomaly size*	*R-anomaly boundary*	*Boundary orientation of r-anomalies*
R-anomaly Type A	3	1–2	1	2–3	4	4

Source: own research

W1_ERT_d resistivity distribution

Parallel to the breaking-off edge and W1_ERT_d, W1_ERT_e is located and passes some smaller elevated areas.

Figure 34: Electrical resistivity tomography – W1_ERT_e

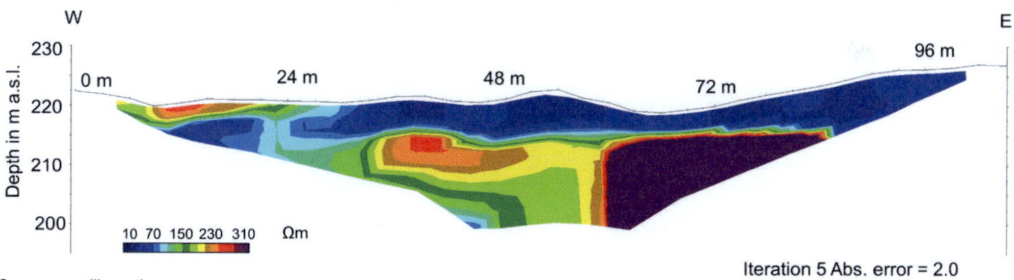

Source: own illustration

The resistivity distribution within W1_ERT_e (Figure 34) looks quite similar to the one at W1_ERT_d. Except for a small area with higher resistivity values up to 270 Ωm until meter 24, within the top part, the resistivity values are up to 70 Ωm. This low-resistive area has a varying size and ends horizontally at about 215 m a.s.l. In the eastern part, underneath the upper area, beginning at meter 60, an area with higher resistivity values of over 300 Ωm occurs. Below 214 m a.s.l., until meter 60, the resistivity values vary between 100 and 250 Ωm. The highest of these resistivity values occurs in a lens between meters 36 and 48 at a depth of 214 to 205 m a.s.l.

W1_ERT_e description after the guidelines

With the addition of the information of the boundary analysis (Figure 35) to the resistivity distribution, three r-layers exist in W1_ERT_e (Table 9).

There is an uppermost r-layer (1), with sizes between 3 and about 13 m and a weak boundary up to meter 60 and a sharp boundary in the eastern part. The

Figure 35: Boundary analysis – W1_ERT_e: a) Sharpness of r-layer boundary including the position of r-layers and r-anomalies; b) Orientation of the boundary gradient

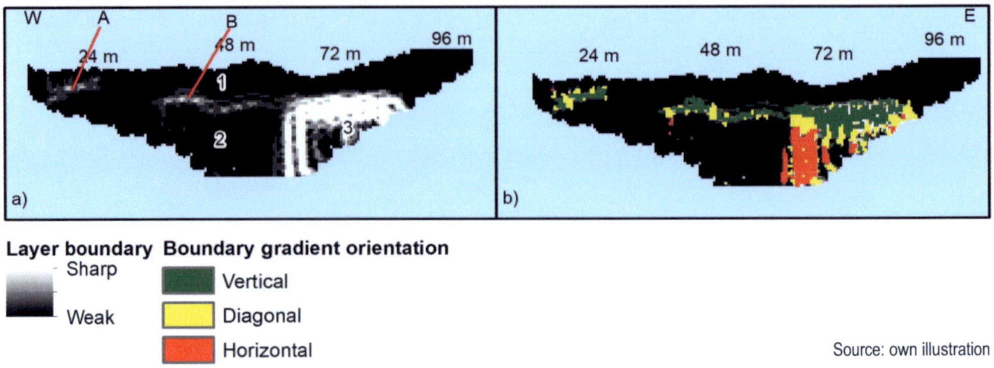

Source: own illustration

layering between this r-layer and the one below is developed to a medium extent, and the boundary is horizontally orientated. Underneath this layer, the other layers occur: a western r-layer (2) with medium resistivity values and an eastern r-layer (3) with higher values, split by a horizontal boundary. All three r-layers are homogeneous. Only within the lower western r-layer do small boundaries

Table 9: ERT guidelines – W1_ERT_e

R-layer	*Homogeneity*	*R-layer thickness*	*Layering*	*Horizontal changes*	*Boundary of r-layers*	*R-layer orientation*
R-layer 1	5	1–2	3	0	W: 2; E: 5	1
R-layer 2	4	3	5	2	5	6
R-layer 3 (horizontal split)	5	3	5	-	-	-
R-anomaly	*Number of r-anomalies*	*R-anomaly location*	*R-anomaly form*	*R-anomaly size*	*R-anomaly boundary*	*Boundary orientation of r-anomalies*
R-anomaly Type A	1	1	2	2	3	2
R-anomaly Type B	1	2–3	1	2	3	1

Source: own research

Figure 36: Electrical resistivity tomography – W1_ERT_f

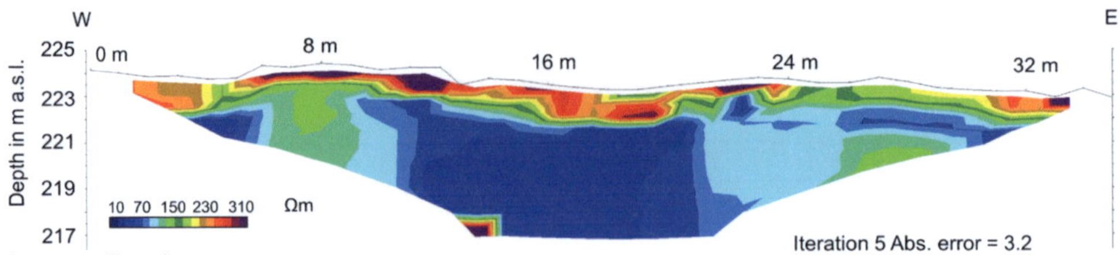

Source: own illustration

occur. Within the top (A) and the western (B) r-layer, r-anomalies occur. Both are surrounded by a medium-weak boundary, whereas the bottom boundary of the bottom r-anomaly is weak.

W1_ERT_f resistivity distribution

W1_ERT_f is located directly at the breaking-off edge.

In W1_ERT_f (Figure 36), a thin area with resistivity values between 150 and over 300 Ωm occurs. This area has a thickness of about 1 m and covers an area with lower resistivity values, mainly below 100 Ωm.

W1_ERT_f description after the guidelines

With the combination of the resistivity distribution and the boundary analysis (Figure 37), two r-layers were classified, a top (1) and a bottom (2) one (Table 10).

Figure 37: Boundary analysis – W1_ERT_f: a) sharpness of r-layer boundary including the position of r-layers and r-anomalies; b) orientation of the boundary gradient

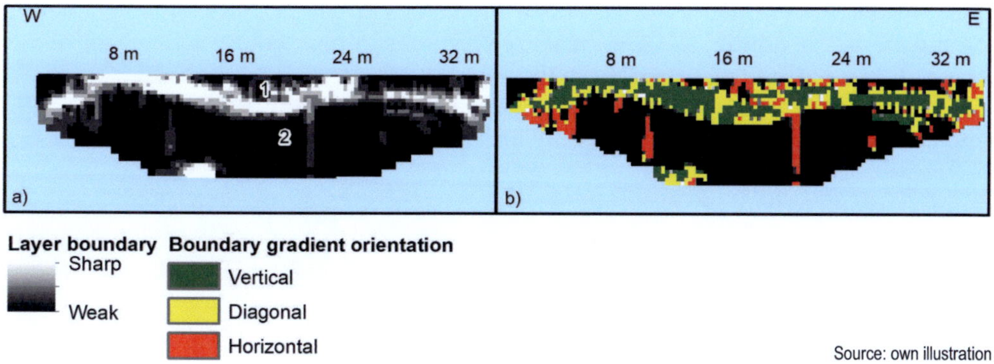

Source: own illustration

The boundary between these two r-layers is sharp and mostly orientated vertically, whereas the other boundaries are weaker and diffusely orientated. The homogeneity of both r-layers is medium-to-low. Especially in the top r-layer, different resistivity values with various boundaries occur. Nevertheless, no clear r-anomalies were classified, although the resistivity distribution varies a great deal.

Table 10: ERT guidelines – W1_ERT_f

R-layer	*Homogeneity*	*R-layer thickness*	*Layering*	*Horizontal changes*	*Boundary of r-layers*	*R-layer orientation*
R-layer 1	2	1	4	0	3–5	7
R-layer 2	3	2	-	0	-	-
R-anomaly	*Number of r-anomalies*	*R-anomaly location*	*R-anomaly form*	*R-anomaly size*	*R-anomaly boundary*	*Boundary orientation of r-anomalies*
R-anomaly Type A	0	-	-	-	-	-

Source: own research

W1_ERT_g resistivity distribution

W1_ERT_g crosses the largest of the smaller elevated areas.

Figure 38: Electrical resistivity tomography – W1_ERT_g

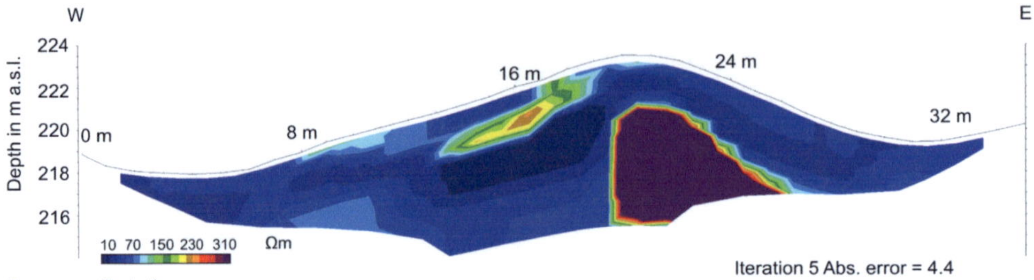

Source: own illustration

In W1_ERT_g (Figure 38), one lens with higher resistivity values within a matrix of lower resistivity values occurs. The matrix has resistivity values of a few tens of Ωm. The more prominent area has resistivity values up to 1,000 Ωm. Between meters 20 and 26 of the profile, the high resistivity values appear below a depth of about 2 m, but this depth level increases in the eastern direction. The lens has a breadth of about 1 m and occurs within a depth of 1 m between meters 12 and 18 of the profile. The resistivity values of this r-anomaly are between 120 and 200 Ωm.

W1_ERT_g description after the guidelines

Adding the boundary analysis' information (Figure 39) to the resistivity distribution, it appears that two r-layers are located within W1_ERT_g (Table 11).

Figure 39: Boundary analysis – W1_ERT_g: a) sharpness of r-layer boundary including the position of r-layers and r-anomalies; b) orientation of the boundary gradient

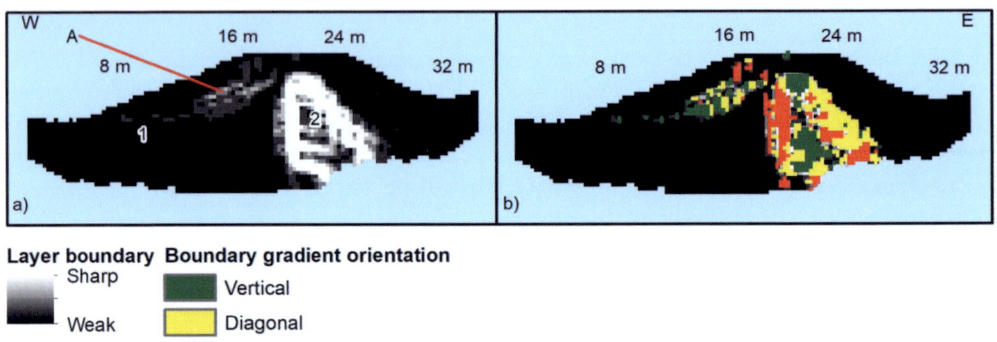

Source: own illustration

The low-resistive top r-layer (1) is split by sharp boundaries from a high-resistive bulky r-layer. Due to its size, this r-layer (2) cannot be considered an r-anomaly. The orientation is horizontal in the western part and diagonal in the eastern part, almost surrounding the bulky r-layer. There is a small r-anomaly (A) within W1_ERT_g with a weak boundary surrounding it.

Table 11: ERT guidelines – W1_ERT_g

R-layer	*Homogeneity*	*R-layer thickness*	*Layering*	*Horizontal changes*	*Boundary of r-layers*	*R-layer orientation*
R-layer 1	4	2	5	2	5	8; 20 m: 6; 21–27 m: 4
R-layer 2 (horizontal split)	3	1	5	-	-	-
R-anomaly	*Number of r-anomalies*	*R-anomaly location*	*R-anomaly form*	*R-anomaly size*	*R-anomaly boundary*	*Boundary orientation of r-anomalies*
R-anomaly Type A	1	1	1	1	1*	1

Source: own research

GPR

W1_GPR_a is located parallel to the profile W1_ERT_a.

Figure 40: Ground-penetrating radar measurement – W1_GPR_a: a) data; b) data with marked changes in signal intensity and structure

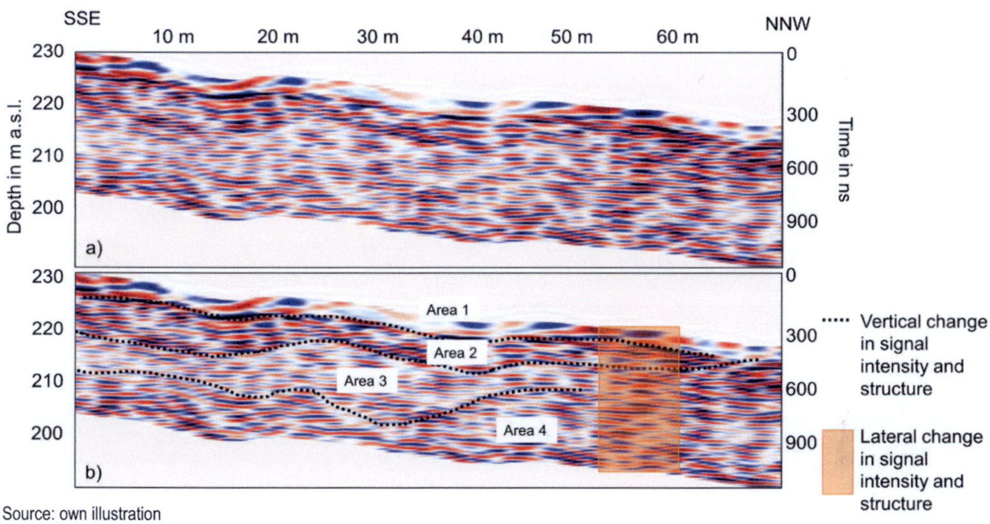

Source: own illustration

W1_GPR_a (Figure 40) is located almost at the same location as W1_ERT_a. Within this profile, three to four mostly laterally located areas are visible in the southern part. In the uppermost part (1), several small reflectors are narrow at the surface. In the depressions of the profile, these reflectors reach their maximum extent. Underneath this area, a second area (2) with strongly visible but interrupted small fragments of reflectors is visible. This area reaches a maximum extent of about 10 m and follows the topography slightly by narrowing northwards. Between meters 50 and 60, this area can not be clearly distinguished. In the central part, an area (3) with lower energetic reflectors exists. Here, the difference in permittivity between the individual parts of it is low. Analog to the second area, the third area is interrupted between 50 and 60 meters. Again, the bottom area (4) has better-segmented reflectors and reflects a stronger signal, but there is also no undisturbed layering visible.

As described for the central two areas, a zone with a change in the signal intensity and its distribution is visible between meters 50 and 60.

With the radargram described in Neal (2004), the upper and lower boundaries are more or less concordant, to a not well-developed extent. Part erosional truncation is indicated at the upper boundary. A well-developed shape of the reflections is not visible. However, they are most likely planar/even with a horizontal dipping. The relationship between the reflections is subparallel, and the reflection continuity is discontinuous.

SRT

The location of W1_SRT_a is almost the same as that of W1_ERT_a and W1_GPR_a. However, due to a different setup of the methods, their lengths and locations are not precisely the same.

Figure 41: Seismic refraction tomography: W1_SRT_a

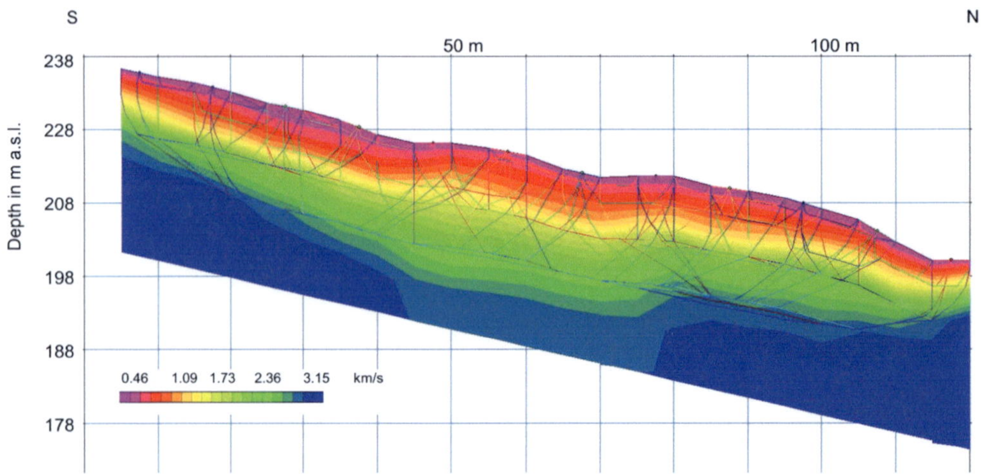

Source: own illustration

W1_SRT_a (Figure 41) illustrates the increase of velocity of seismic waves with depth, typical for SRT. At a depth of about 10 m, the SRT data form a tub-like structure with velocities around 2 km/s, comparable to the low-resistive layer in W1_ERT_a. The start and end of this structure cannot be differentiated in the SRT, in contrast to the ERT and GPR data. This structure, especially within the central part, looks similar to the boundary between the first and second layers in the W1_ERT_a.

4.1.2.2 Invasive methods and outcrops

The results within this subsection invasive methods and outcrops are associated with workflow Step 4 (S4) – *Ground-truth.*

Borehole

The borehole results are described, generalized and simplified to focus on the scientific question.

While W1_BH_a is located on the largest of the elevated area, east of the breaking-off edge, W1_BH_b and W1_BH_c are located at the second and third intersections northward, respectively.

Figure 42: Boreholes – W1: a) W1_BH_a; b) W1_BH_b; c) W1_BH_c

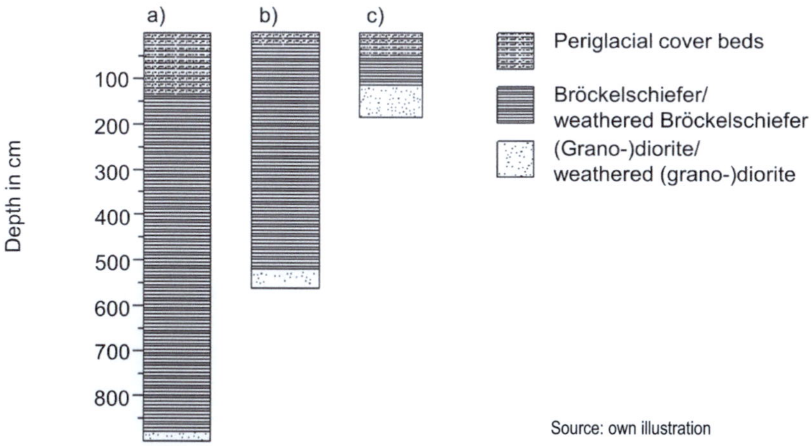

Source: own illustration

W1_BH_a (Figure 42a) illustrates PCBs with a size of more than 1 m on top of weathered Bröckelschiefer. The PCBs are not well-developed, and single layers are difficult to distinguish. In some parts of the core, the Bröckelschiefer is totally weathered to compact clay layers, but is less weathered and eponymously crumbling in other parts. In addition, siltstone and sandstone are in different weathering conditions, which are visible depending on the pores' volume and hydrologic coupling. However, those can act either aquiferous or dense. A dark red color dominates the entire core profile. An aquiferous layer partly causes a reduction of the iron ions and gray color. More often, the gray color is caused by the weathering of gray claystone within the usually red claystone of the Bröckelschiefer. Within the bottom part of the core profile, dioritic rocks are mixed into the clayey matrix of the weathered Bröckelschiefer and are associated with the abovementioned (Section 2.2) basal-breccia. Even though this rock type is described as the border between these two rock types, it is classified as (grano)diorite here because it offers the first evidence of the existence of (grano-)diorite.

Opposite in characteristics of W1_BH_a, the size of the PCBs is smaller, with about 20 cm in W1_BH_b (Figure 42b). This smaller size could be caused by the exposed position of the profile close to a ditch. Underneath this layer, Bröckelschiefer, or its weathering products, crops out in this profile almost solely existing of red claystone and its weathering products is observed. Especially at the bottom, slightly below 5 m of depth, siltstone and sandstone occur. Dioritic rocks are mixed into this matrix, too. This area fits quite well with the described basalbreccia. At 5 m depth, the content of (grano-)diorite increases and finishes in sole (grano-)diorite.

W1_BH_c (Figure 42c) was drilled to a depth of only 1 m. Within the uppermost 20 cm, PCBs have been deposited but are not well-developed anymore. As in the other boreholes of W1, the single layers cannot be distinguished. Underneath the PCBs, heavily weathered Bröckelschiefer occurs. Due to the weathering level of this layer, it was likely eroded upslope and accumulated at this position as clay. The bottom layer of the profile consists of weathered (grano-)diorite.

No drilling progress was possible in all three boreholes due to the increasing density of the less-weathered dioritic rocks.

Pürckhauer

W1_PUE_a is located at the westernmost part of the elevated area of the relief south of the castle. The positions of W2_PUE_b, -c and -d were set up as a western catena. While -b and -c are located upslope of the breaking-off edge, W2_PUE_d is located at a downslope position in one of the small depressions. W1_PUE_e, -f, -g, -h and -i represent a second downhill catena. Whereas W1_PUE_e is located uphill of the breaking-off edge, east of the position of W1_PUE_b, the position of W1_PUE_f is to the east but still within the same small depression that W1_PUE_d is located in. W1_PUE_g is located within the second small valley, and -h and -i are within the floodplain outside of the relief. In contrast, W1_PUE_j is located in the western part of the area of investigation outside the elevated areas and small depressions within the lowlands (cf. Figure 25).

The penetration of the Pürckhauer causes a compression of the profile. Therefore, in some profiles, the uppermost core is lost.

In W1_PUE_a (Figure 43a), a layer consisting of loamy sand occurs on top of pure sand and gravel-sized grus consisting of dioritic bedrock.

Within the first layer of W1_PUE_b (Figure 43b), silty clay loam occurs until a depth of 45 cm. Between 45 and 60 cm, the sediment can be described as clay loam, while the bottom layer is classified as sandy loam. Within the first layer, small un-weathered clay fragments of Bröckelschiefer are located within the matrix. There is grus of (grano-)diorite in the bottom layer.

The texture of W1_PUE_c (Figure 43c) is classified as sandy loam to loamy sand. Only between 35 and 50 cm does loam occur. The complete profile includes grus of diorite, especially flakes of mica.

In W1_PUE_d (Figure 43d), the texture mainly consists of loamy sand with clay and silt content. The location in a local valley ensures the accumulation of all the material within the catchment belonging to aeolian sediments, Bröckelschiefer and mainly (grano-)diorite. Due to its location close to a creek, hydromorphic features are visible.

Underneath the core loss within W1_PUE_e (Figure 43e), there are some fragments of clayey Bröckelschiefer within the layer of clay loam. At a depth of 40 cm, the sizes of the grus fragments increase to the size of gravel. Underneath this layer, the size of these fragments decreases, and there is a higher sand content. This layer is classified as sandy clay loam. In contrast, within the lowermost 5 cm, the clay content increases again and causes stagnic conditions with hydromorphic features on top. However, there are (grano-)diorite fragments mixed in with Bröckelschiefer in this layer, too.

The first layer of W1_PUE_f (Figure 43f) consists of loamy sand, including a high content of clay in the loamy part and grus fragments of both bedrocks. The bottom layer consists of slightly weathered Bröckelschiefer.

Figure 43: Pürckhauer soundings – W1: a) W1_PUE_a; b) W1_PUE_b; c) W1_PUE_c; d) W1_PUE_d; e) W1_PUE_e; f) W1_PUE_f; g) W1_PUE_g; h) W1_PUE_h; i) W1_PUE_i; j) W1_PUE_j

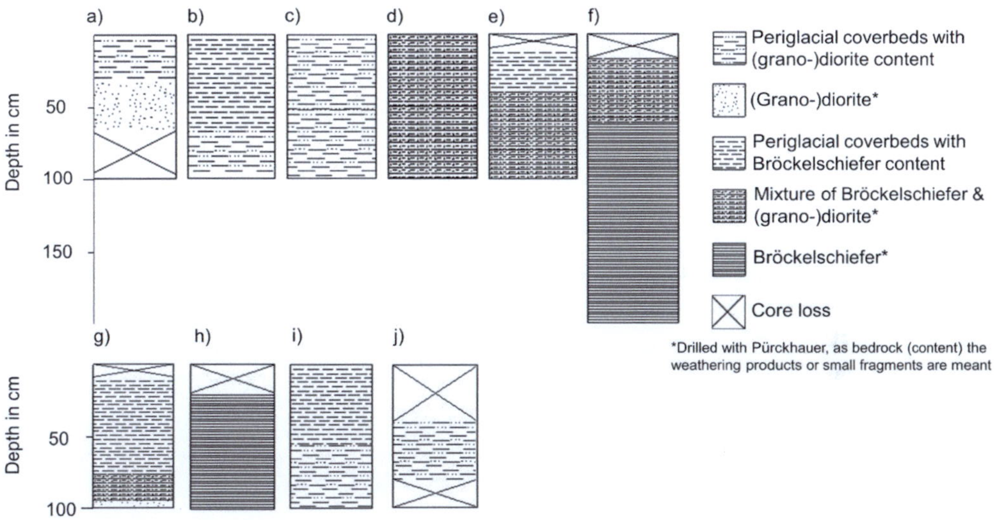

Source: own illustration

Underneath the core loss in W1_PUE_g (Figure 43g), until a depth of 40 cm, the texture consists of clay with some silt. This layer is stagnic and reduced. On top of this layer, water was standing. Underneath this layer, the sand and silt content increases. Additionally, some grus of the (grano-)diorite, especially flakes of mica, were found. Starting at a depth of 85 cm, the sediments are more sorted. While there were sand layers alternating with smaller-sized clay layers, including grus of (grano-)diorite, at the bottom of the profile saprolite of the (grano)diorite was visible.

The grain-size of W1_PUE_h (Figure 43h) consists of clay, partly including other textures. However, silt and sand occur in lenses. Therefore, the full profile is assigned to weathered Bröckelschiefer.

The top of W1_PUE_i (Figure 43i) consists of clay loam with fragments of Bröckelschiefer. Underneath this layer, the sand and gravel content increases, and the source of this sediment could be clearly assigned to (grano-)diorite.

Due to its location at a creek, with high water saturation, parts of the core of W1_PUE_j (Figure 43j) were washed from the Pürckhauer. The remaining part ranges from sand to loamy sand, including fragments of diorite that are up to the size of gravel and loess loam.

Edelman

The Edelman soundings are located in a downslope catena. Whereas W1_EDE_a is located at the first small, elevated area, downslope of the breaking-off edge, the location of W1_EDE_b is within the second of the small depressions, and W1_EDE_c is at the floodplain outside the focus area.

Figure 44: Edelman soundings – W1: a) W1_EDE_a; b) W1_EDE_b; c) W1_EDE_c

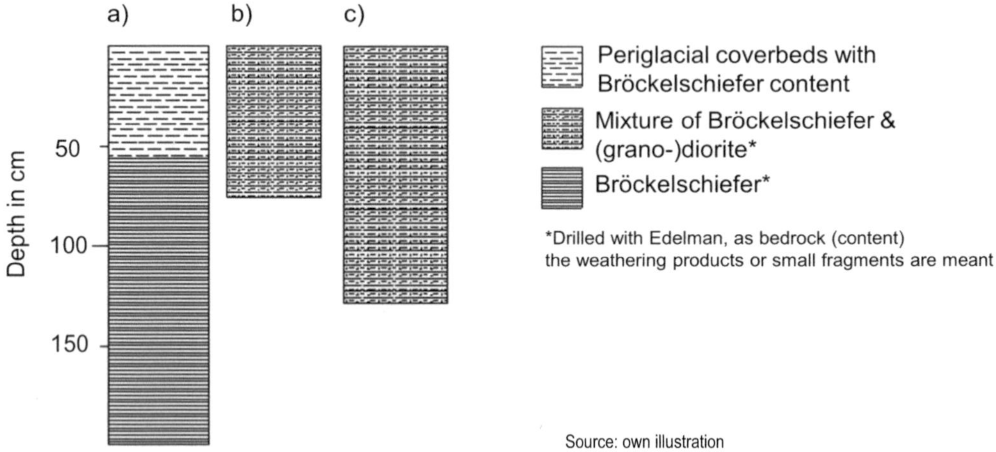

Source: own illustration

The uppermost layers of W1_EDE_a (Figure 44a) have a silty clay loam to silt loam texture, caused by a difference in the silt-clay ratio. However, the sand content is low. Within these layers, claystone of Bröckelschiefer and even small, burned sandstones are located. Underneath this layer, only clay or claystone occur.

In W1_EDE_b (Figure 44b), the texture consists purely of the weathering products of Bröckelschiefer and (grano-)diorite. The content of clay and sand or gravel varies heavily. However, a dominant silt component was not identified in any of these layers.

The texture of W1_EDE_c (Figure 44c) mainly consists of loam with different content of different grain-sizes. Compared with W1_EDE_b, there is a relatively high content of silt. The contents of clay, silt and sand vary a great deal.

Soil development

Due to the position of the profiles on slopes or within local valleys, gravitational processes dominate the local development of soils. Depending on the initial substrate, only regosols on loose sediments such as the PCBs or leptosols in bedrocks have developed. These soil types are classified by a low pedological development and have only small-sized topsoil on top of solely weathered loose sediments and bedrock. Notably, within the local valleys, the large amount of clay causes stagnic conditions, including hydromorphic features, such as redoximorphic-influenced layers.

Outcrop

The breaking-off edge is located south of the southernmost depression (e.g. Figure 45b). On this steep step, OC_a is located. Here, (grano-)diorite crops out (e.g. Figure 45a).

Figure 45: Field pictures – Outcrop W1_OC_a: a) diorite outcrop; b) breaking-off edge

Source: own illustration

4.1.3 W1 – *Origin of the rough relief:* interpretation

The interpretation of the data was performed according to workflow Step 5 (S5) – *Interpretation of the data.*

Sythesis of the results

Combining all the geophysical data, three to four different lithological units can be distinguished. Often at the uppermost part of the subsurface, a layer of resistivity values between about 100 and 200 Ωm is located. This layer is classified by weak boundaries surrounding it, medium homogeneity and small sizes. The layer often only occurs as fragments and is classified as a "quasi-layer"-formed r-anomalies. The GPR shows in the uppermost reflectors cropping out to the surface, which can be associated with it. However, due to the resolution, these structures cannot be resolved with the SRT. The invasive methods often show PCBs as uppermost layers, include those with varying content of silt mixed into the weathering products of the bedrock. Therefore, the resistivity values are higher than those with the high clay content of the second lithological unit (Reynolds, 2011).

This second unit is classified by resistivity values below 100 Ωm and a sharp boundary with the layer underneath. This second layer occurs mostly homogeneous. Mostly, the boundary has a vertical gradient, but sometimes horizontal gradients are visible as well. Some small reflectors are visible within the GPR in this area, and the SRT found values between 1 and 2.5 km/s. The invasive methods show the Bröckelschiefer at this position. The weathering intensity of this rock varies, and partly aquiferous layers, sandstone and siltstone, are embedded into the claystone. Such layers can explain the small but recognizable reflectors in the GPR data. However, both the resistivity and the seismic velocities fit quite well with those of clay and claystone (Reynolds, 2011).

Underneath this lithological unit, higher resistivity (> 300 Ωm) values occur. In addition to their existence as a bottom layer within many profiles, especially in the western part, this lithological unit occurs as blocky r-anomalies. In the southern part, it occurs at the top as well. The boundary surrounding this layer, or the r-anomalies, is mostly sharp. In the SRT, the layer is classified with higher velocities than those of the Bröckelschiefer, and the GPR displays an area without any strong reflections. By comparing the geophysical information with the invasive methods, these parts can be identified as the basal-breccia located at the bottom of the Bröckelschiefer and as the weathering products of the (grano-)diorite. Even though the resistivity values are still low for this type of bedrock, they decrease in some parts massively during weathering (Reynolds, 2011).

Below the third layer, the GPR indicates another lithological unit, classified by more visible reflectors in area 4. Additionally, in some exposed positions of the ERTs, their resistivity values increased to over 1,000 Ωm. At these positions, or in deeper areas not resolved in most ERTs, the un-weathered bedrock crops out. The resistivity values fit more to those described for such a bedrock (Reynolds, 2011). Within the SRT, a tub-like structure of Bröckelschiefer is evident. This structure, with velocities of about 2.0 km/s is comparable to the low-resistive layers of the ERT and is within the range of typical velocity values of clay (Reynolds, 2011). The higher velocity values at the bottom are associated with those of magmatic rocks. This is possibly not exactly the border between the Bröckelschiefer and the (grano-) diorite due to a depth of over 20 m. Perhaps, the saprolitized (grano-)diorite cause this difference, and it follows the boundary between the Bröckelschiefer and the (grano-)diorite. Combined with the GPR, the form is changing, but there are still two massive lateral changes visible in signal structure and intensity. The uppermost part of it can approximately be associated with the boundary between the Bröckelschiefer and the (grano-)diorite or the basal-breccia. The border between area 3 and area 4 is associated with the transition of saprolite to un-weathered (grano-)diorite. All these changes within the reflections are weak. However, depending on its degree of weathering and water content, the dielectric constant of clay or claystone varies massively (2–40 ε_0) and can be similar to that of diorite (6.5–7 ε_0; Zheng *et al.*, 2005; Reynolds, 2011). It is expected that the dielectric constant of the bedrock in the field, especially of the (grano)diorite, varies even more due to different weathering status, mineral and water content.

Interpretation

What forces caused the development of the rough relief south of the castle hill?

In all the ERT profiles of W1, except W1_ERT_f, the (grano-)diorite exists in layers or anomlies with values of about 300 Ωm, mostly well-layered from the low-resistive Bröckelschiefer. The low-resistive Bröckelschiefer generally lies on top of the high-resistive areas. The border between the granodiorite and diorite cannot be distinguished using the ERT data and is difficult to distinguish visually with the field description. However, the border is described as diffuse and continuous (Okrusch *et al.*, 2011). Therefore, no further focus is placed on differentiating these rock types, especially since such a differentiation would have no influence on the interpretation.

Especially in the profiles with smaller spacings, it is evident that the Bröckelschiefer layer is often not located directly underneath the surface. The slightly higher resistivity values close to the surface in some of these transects are associated with PCBs or the culminated materials of them, including a high content of weathered bedrock fragments and partly anthropogenic debris.

The texture within the shallow subsurface changes fundamentally within the western part of W1. Additionally, in the local valleys, the accumulated deposits reveal that the catchment sometimes changes. The layers of the weathering products of Bröckelschiefer and those of (grano-)diorite sometimes alternate and fragments of these two rock types are mixed with materials of the PCBs in other profiles.

Figure 46: Conceptual models – W1: a) CM_a south-north orientated; b) CM_b east-west orientated

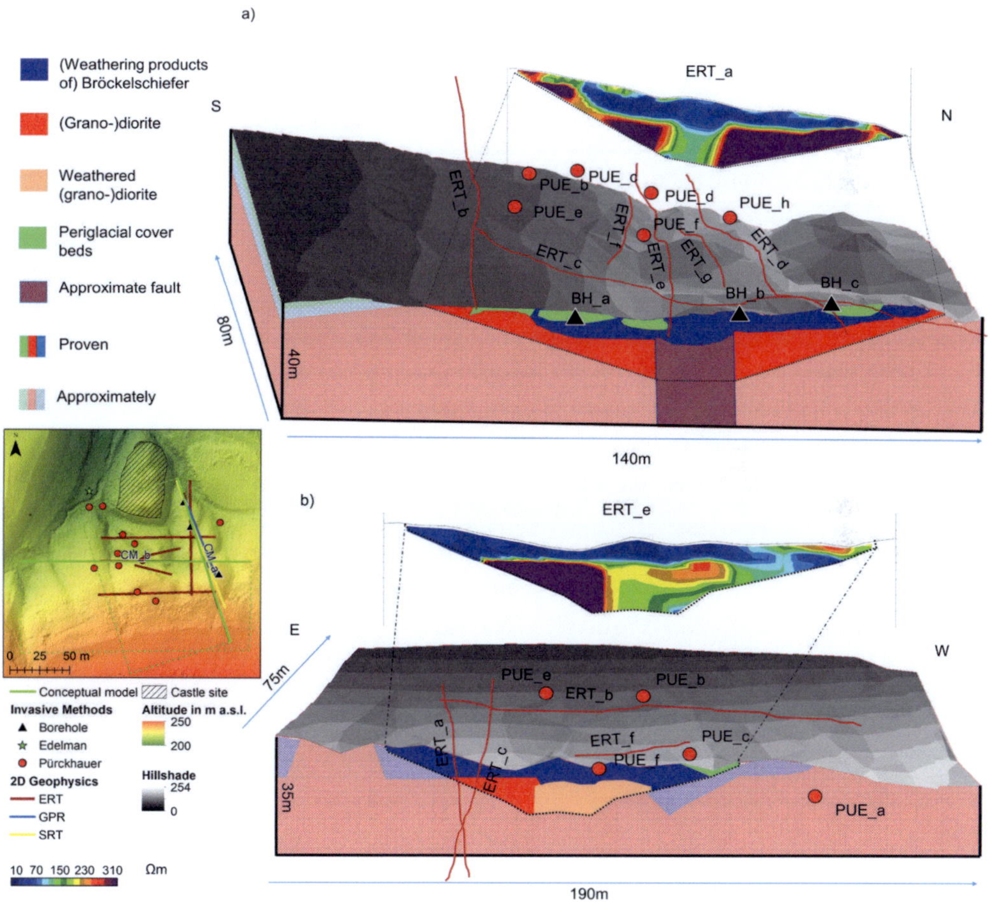

Source: own illustration, based on Bavarian Agency for Digitalisation, High-Speed Internet and Surveying, 2014

With the results of all the measurements combined with the conceptual models (Figure 46), it appears that all the small ridges consist of Bröckelschiefer. Only within deeper parts and at the most western ridge does (grano-)diorite occur on top. Upslope, the breaking-off edge W1_ERT_b has the (grano-)diorite at the top; but further west, Bröckelschiefer is located at the top instead. W1_ERT_f, with its position at the breaking-off edge, has only a small dioritic layer on top, followed by Bröckelschiefer.

PCBs only occur in a few exposed positions or as small-sized layers. However, most of this material seems to be eroded in the area of investigation. The ERT results particularly indicate that the surface of the (grano-)diorite body at the bottom of the profiles is independent of the recent topography, at least in the area of interest for W1.

All these data point toward the origin of the rough relief south of *Wahlmich Castle* being of natural origin. The Bröckelschiefer within this area is broken into different segments. The typical layering of this sedimentary claystone is partly visible within the borehole, and partly the bedrock is fully weathered. Combined with the lack of Bröckelschiefer at this position on the geological map (Weinelt, 1962) and its position further upslope, together with the mapped faulting system, geomorphological and tectonic reasons were most likely the driving force for the development of this relief. The (grano-)diorite, with its relative position to the Bröckelschiefer, underlines this theory. When the photographs of the (grano)diorite (Figure 45) are combined with W1_ERT_c and -f (Figure 30 and 36), it is evident that (grano-)diorite crops out on top and underneath the Bröckelschiefer at the breaking-off edge. This indicates a dipping of the structures in the slope direction, followed by accumulation of Bröckelschiefer below the edge. It seems that the breaking-off edge is caused by a down-lifting of the entire investigation area. Thus, the area of investigation is of natural origin, driven by a faulting system, which is further investigated in W3. Indeed, the castle site, made of (grano-)diorite, seems to be the northern end of this lifted area.

4.1.4 W2 – *Neck ditch:* results

In what materials were the neck ditches dug?

To address this research question, three ERT profiles were conducted in the ditch or at the small ridge south of it (Figure 47). The results are interpreted together with the results of the excavation from 2016 and the data addressing question W1.

4.1.4.1 Geophysical data

The geophysical data of W2 are associated with workflow Step 3 (S3) – *Geophysical methods*.

ERT

First, the inverted resistivity distribution of the profiles is described. Then, their relative distribution is discussed according to the guidelines for ERT description.

Figure 47: Location of measurements – W2

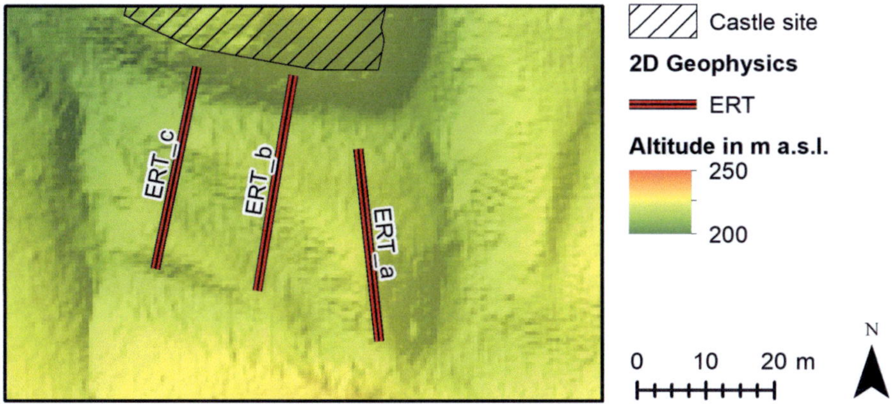

Source: Bavarian Agency for Digitalisation, High-Speed Internet and Surveying, 2014
Map creation: J. Trappe

W2_ERT_a resistivity distribution

W2_ERT_a is located within the neck ditch, starting from a small, elevated area in its western part.

Figure 48: Electrical resistivity tomography – W2_ERT_a

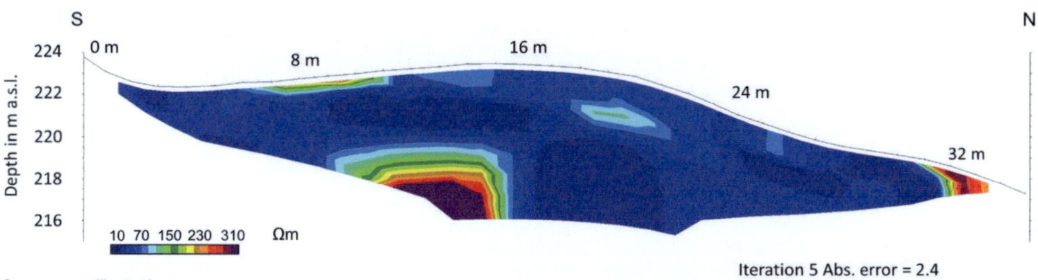

Source: own illustration

Throughout the entire profile of W2_ERT_a (Figure 48), a low-resistive matrix with resistivity values up to 70 Ωm is visible. Two lenses with resistivity values up to 300 Ωm occur. One is located between meters 9 and 15 of the profile at a depth of about 3 m, and the other is close to the surface at meter 32 of the profile. Two more lenses with resistivity values between 100 and 200 Ωm appear around meter 18 of the profile, at a depth of 1 m, with a size of 1 m, and between meters 6 and 11, close to the surface.

W2_ERT_a description after the guidelines

Using the ERT description guidelines (Table 12) and a combination of the resistivity distribution with the boundary analysis (Figure 49), W2_ERT_a displays a single, homogeneous r-layer (1).

Figure 49: Boundary analysis – W2_ERT_a: a) sharpness of r-layer boundary including the position of r-layers and r-anomalies; b) orientation of the boundary gradient

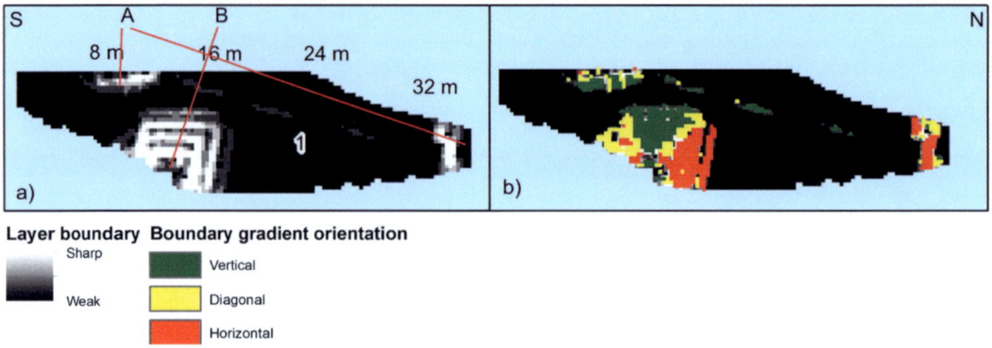

Source: own illustration

Within this r-layer, at the top and bottom, r-anomalies are embedded. At the top (A), the southern r-anomaly boundary is sharp and lacks a clear orientation, whereas the northern one is weak-to-medium and horizontally orientated. However, at the bottom (B) of the profile, another r-anomaly occurs with a sharp boundary. Nevertheless, the extent of the bottom r-anomaly is almost at the edge of being an r-layer.

Table 12: ERT guidelines – W2_ERT_a

R-layer	*Homogeneity*	*R-layer thickness*	*Layering*	*Horizontal changes*	*Boundary of r-layers*	*R-layer orientation*
R-layer 1	4	-	-	-	-	-
R-anomaly	*Number of r-anomalies*	*R-anomaly location*	*R-anomaly form*	*R-anomaly size*	*R-anomaly boundary*	*Boundary orientation of r-anomalies*
R-anomaly Type A	2	1	2	1	2	2
R-anomaly Type B	1	1	1	2	4	1

Source: own research

W2_ERT_b resistivity distribution

W2_ERT_b starts at a small, elevated area, crosses another elevated area and the neck ditch, finishing at the castle hill.

Starting in the northern part, until meter 8 of the profile W2_ERT_b (Figure 50), a low-resistive area with resistivity values below 70 Ωm and a size of up to 4 m occurs. At the bottom of this area, the resistivity values increase to 100 Ωm. At meter 8 of the profile, the low-resistive area diminishes, and the resistivity values at the bottom increase to 150 Ωm. At meter 6, a lens with resistivity values over 300 Ωm exists; southwards, the resistivity values are around 150 Ωm.

Figure 50: Electrical resistivity tomography – W2_ERT_b

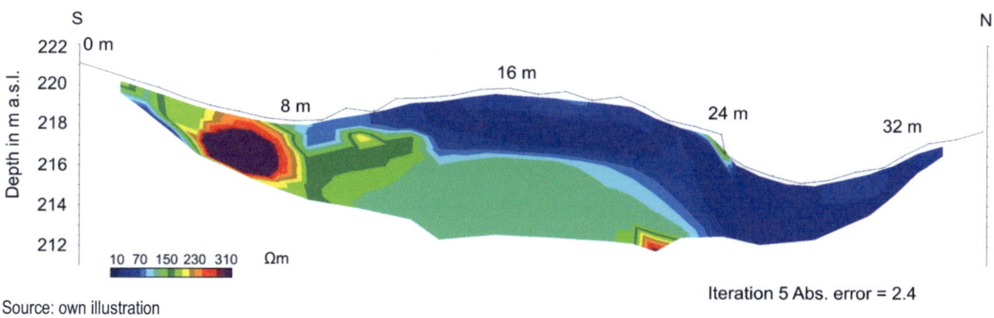

Source: own illustration

W2_ERT_b description after the guidelines

Using the information from the boundary analysis (Figure 51) added to the distribution of the resistivity values (Table 13), two shallow r-layers are visible, an upper (1) and a lower (2) r-layer.

Figure 51: Boundary analysis – W2_ERT_b: a) sharpness of r-layer boundary including the position of r-layers and r-anomalies; b) orientation of the boundary gradient

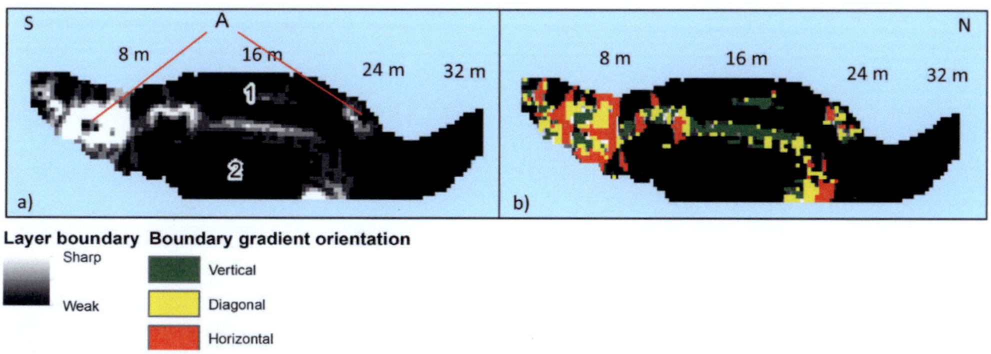

Source: own illustration

Table 13: ERT guidelines – W2_ERT_b

R-layer	*Homogeneity*	*R-layer thickness*	*Layering*	*Horizontal changes*	*Boundary of r-layers*	*R-layer orientation*
R-layer 1	4	1	3	0	2	2–3, 7
R-layer 2	3	1	-	-	-	-
R-anomaly	*Number of r-anomalies*	*R-anomaly location*	*R-anomaly form*	*R-anomaly size*	*R-anomaly boundary*	*Boundary orientation of r-anomalies*
R-anomaly Type A	2	1	1	1	5	1

Source: own research

The r-layers are separated by a weak-to-medium-weak boundary of various orientations. The layering between these two r-layers is developed to a medium extent, and both r-layers are mostly homogeneous but have a few boundaries in between their matrix. The r-anomaly (A) in the southern part has a fully surrounding sharp boundary with various orientations in all directions. At meter 24, another r-anomaly was identified.

W2_ERT_c resistivity distribution

W2_ERT_c crosses two small, elevated areas and finishes in the eastern part of the neck ditch.

Figure 52: Electrical resistivity tomography – W2_ERT_c

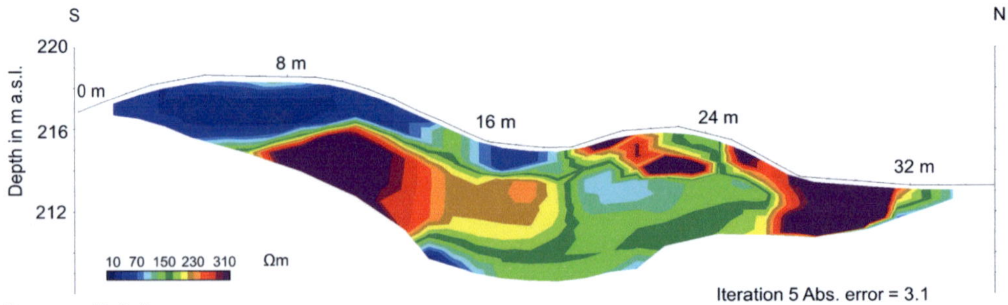

Source: own illustration

Close to the surface, visible between meters 6 and 18 of W2_ERT_c (Figure 52), is a low-resistive area with a size of up to 4 m. At meter 14, the resistivity values increase from a few tens of Ωm up to over 100 Ωm. As far as the ERT's geometry and penetration depth could tell, underneath this low-resistive area is a second area with resistivity values between 200 and over 300 Ωm. At meter 14 of the profile, starting at a depth of 4 m, there is the third area with lower resistivity values than the second one. At meter 19 of the profile, two lenses with higher resistivity values over 300 Ωm occur. The smaller lens at meter 19 of the profile starts near the surface with about 2 m of size. The second lens between meters 25 and 32 has a resolved depth of at least 4 m. The matrix around these two lenses has resistivity values between 70 and 150 Ωm.

W2_ERT_c description after the guidelines

Using the combination of this somewhat chaotic resistivity distribution and the boundary analysis (Figure 53), two r-layers were distinguished (Table 14): a top r-layer (1) with a heterogeneous resistivity distribution and a lower r-layer with a slightly higher homogeneity within the bottom r-layer (2).

Although the differentiation between these two r-layers is easy to identify in the southern part, it becomes more complicated northward. The boundary between these two r-layers is weak-to-medium and vertically-to-diagonally orientated. While there are smaller r-anomalies (A) with a weak-to-medium and diffusely orien-

Figure 53: Boundary analysis – W2_ERT_c: a) sharpness of r-layer boundary including the position of r-layers and r-anomalies; b) orientation of r-layer boundary gradient

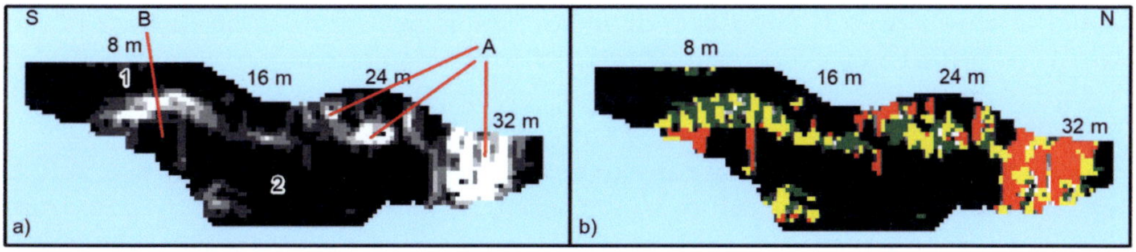

Source: own illustration

tated boundary, there is a more prominent r-anomaly (B), at meter 25, with a sharp boundary and a horizontal-to-diagonal orientation. The southern r-anomaly is not clearly separated from the bottom r-layer. Thus, it is not classified as a single r-layer.

Table 14: ERT guidelines – W2_ERT_c

R-layer	*Homogeneity*	*R-layer thickness*	*Layering*	*Horizontal changes*	*Boundary of r-layers*	*R-layer orientation*
R-layer 1	4	1	3	0	2	2–3, 7
R-layer 2	3	1	-	-	-	-
R-anomaly	*Number of r-anomalies*	*R-anomaly location*	*R-anomaly form*	*R-anomaly size*	*R-anomaly boundary*	*Boundary orientation of r-anomalies*
R-anomaly Type A	2	1	1	1	5	1

Source: own research

4.1.4.1 Invasive methods and outcrops

For W2, no invasive investigation was conducted by the author. Nevertheless, at this position, an archaeological excavation was undertaken in 2016 (Rosmanitz & Bachmann, 2017). According to the excavation, the entire, destroyed wall surrounding the castle had formerly been moved into the ditch. Therefore, any kind of drilling was interrupted by massive sandstone blocks. However, the results of the excavation were used as ground-truth to interpret the following workflow Step 4 (S4).

4.1.5 W2 – Neck ditch: interpretation

The interpretation of the data was performed according to workflow Step 5 (S5) – *Interpretation of the data.*

Synthesis

Combining the ERT data of W2 with the knowledge of the excavation and the invasive methods of W1, three to four major lithological units were distinguished. Analog to W1 in some exposed positions, resistivity values between 100 and 200 Ωm occur as r-anomalies at the top. These are surrounded by medium-sharp (in some parts sharp) boundaries and correlate with the PCBs.

Underneath the PCBs or in less-exposed positions at the top, resistivity values below 100 Ωm occur. These values are separated mostly with sharp boundaries from r-layers or r-anomalies with higher resistivity values. Comparing the lithological information of W1, the weathered claystone of the Bröckelschiefer is associated with these r-layers.

In the southern parts, high-resistive r-layers and r-anomalies occur. Due to their position and resistivity values, they can be associated with the (grano-)diorite. The resistivity values and especially the boundaries of them appear quite similar to the northern r-anomalies.

However, the excavation (Rosmanitz & Bachmann, 2017) states that, in the position of this r-anomaly, are destroyed wall structures made of Buntsandstein that formerly surrounded the castle. Most likely, these r-anomalies can be associated with those.

The stratigraphical situation of W1 and the literature values (Reynolds, 2011) are in line with these arguments. Even if the values themselves somewhat vary from the described ones, their relative placement fits well.

Interpretation

In what materials were the neck ditches dug?

The resistivity values of W2_ERT_a and W2_ERT_b are low. The material probably consists of relocated clayey material in the northern parts: the location of the neck ditch. Additionally, the uppermost meters of the first ridge southward consist of weathered, disturbed claystone Bröckelschiefer, according to the invasive methods of W1. Underneath the southern, elevated area, the (grano-)diorite crops out. Westward, only the burden cap of the elevated area consists of low-resistive Bröckelschiefer, according to W2_ERT_c. Within the neck ditch, the described wall structures (Rosmanitz & Bachmann, 2017) are located at this position. However, the Edelman cores of W1 indicate that the high-resistive area could also be the dioritic bedrock at this position. In the eastern part, the described wall structure could not be detected anymore with this method. With the digging or deposition of the material, the anthropogenic impact slightly changed the erosion regime within the floodplain. In the uppermost 55 cm, W1_PUE_i has a Bröckelschiefer content and a (grano-)diorite content in the bottom 45 cm. Both contents are mixed with the silt-rich PCBs, indicating a relocation of the material. Most likely, before the neck ditch building, the diorite-rich material was accumulated. After the building of the ditch, other material, including the described content of Bröckelschiefer, eroded further east.

To the question, it is assumed that the Bröckelschiefer was removed along the entire neck ditch. A very sharp boundary between the southern high-resistive (grano-)diorite underlines this theory, first established in the excavation in 2016 (Rosmanitz & Bachmann, 2017). However, some higher-resistive r-anomalies in the northern area can be associated with the described fragments of the collapsed walls.

4.1.6 W3 – Fault system: results

How do the mapped faults affect the development of relief?

The geological map indicates several faults southeast of the castle site and the rough relief. According to the results of W1, south of the rough relief, further faults are expected. The location on a slope (cf. Figure 4e) includes good conditions with smooth topography for measuring by using larger electrode spacings. The higher penetration depths of these applications resolve the faults as large features. Additionally, various geophysical profiles are placed south and southeast of the castle site (Figure 54).

Figure 54: Location of measurements – W3

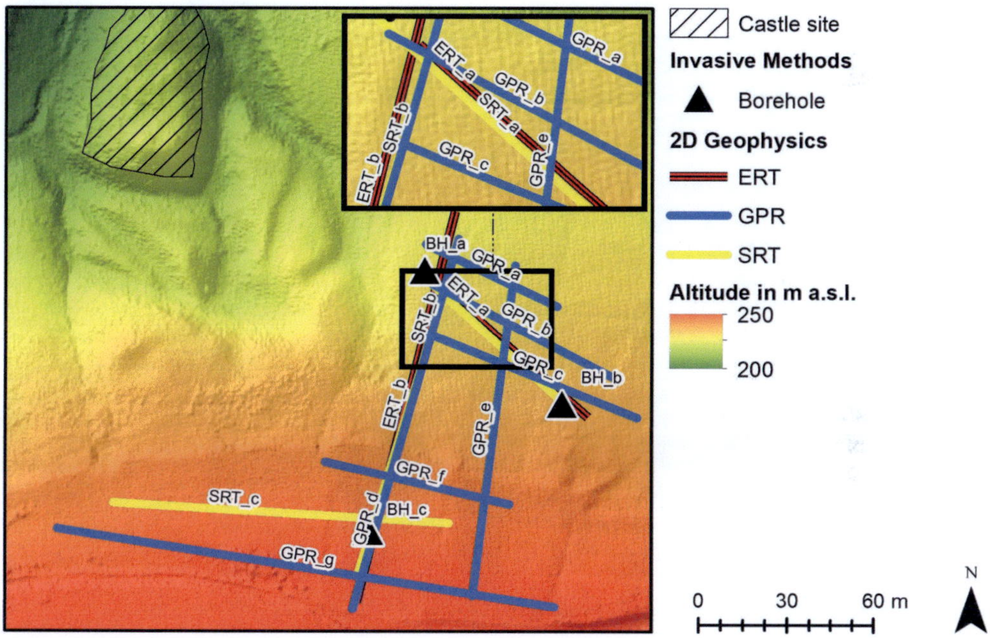

Source: Bavarian Agency for Digitalisation, High-Speed Internet and Surveying, 2014
Map creation: J. Trappe

4.1.6.1 Geophysical data

The geophysical data of W3 are associated with workflow Step 3 (S3) – *Geophysical methods*.

ERT

First, the inverted resistivity distribution of the profiles is described. Then, their relative distribution is discussed according to the guidelines for ERT description.

W3_ERT_a resistivity distribution

W3_ERT_a is located southeast of the most dominant of the small, elevated areas, at the northern meadow. The beginning is located within a small apple orchard.

Figure 55: Electrical resistivity tomography – W3_ERT_a

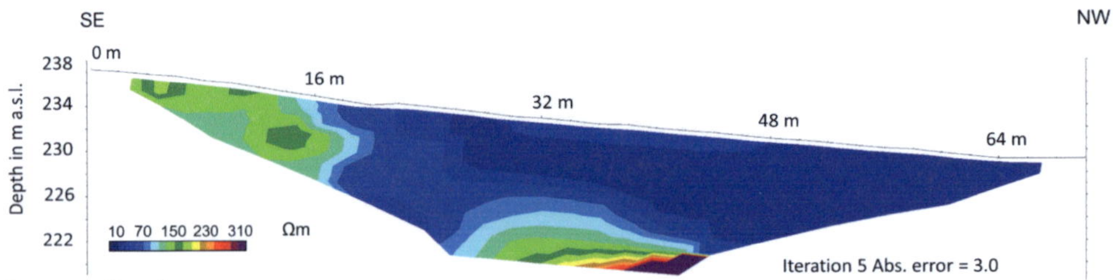

Source: own illustration

At meter 16 in W3_ERT_a (Figure 55), the profile is split into two parts. The southeastern part has resistivity values between 100 and 170 Ωm, but in the northwestern part, resistivity values under 70 Ωm to a depth of 10 m below the ground surface occur. Underneath this, the resistivity values increase again.

W3_ERT_a description after the guidelines

Adding the information from the boundary analysis (Figure 56) to the resistivity distribution, three r-layers are distinguished (Table 15).

Figure 56: Boundary analysis – W3_ERT_a: a) sharpness of r-layer boundary including the position of r-layers and r-anomalies; b) orientation of the boundary gradient

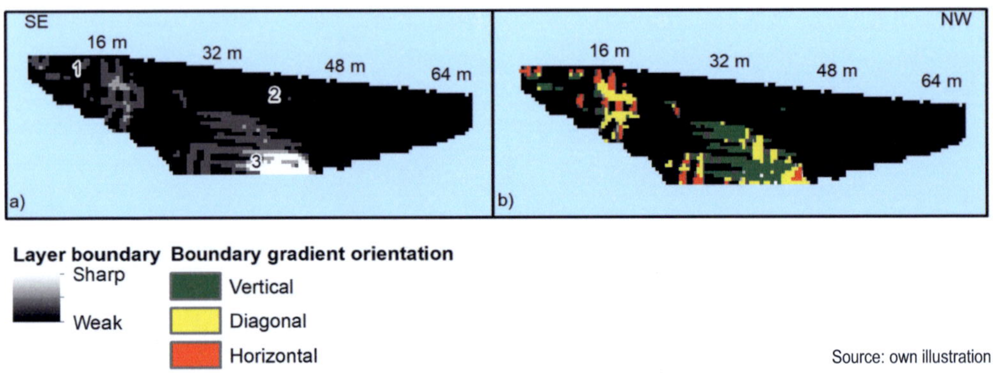

Source: own illustration

The top part is split into two r-layers, a southeastern (1) one and a northwestern (2) one, including a weak-to-medium boundary with a horizontal-to-diagonal orientation. In addition to the profile, the increase in resistivity at the bottom of the profile starts weakly and ends in a sharp boundary, mostly vertically or diagonally orientated; therefore, this is classified as an additional r-layer (3). The layering between all three r-layers is mostly well-developed, and the r-layers seem more or less homogenous; only a few weak boundaries occur within them. No r-anomalies are visible within W3_ERT_a.

Table 15: ERT guidelines – W3_ERT_a

R-layer	*Homogeneity*	*R-layer thickness*	*Layering*	*Horizontal changes*	*Boundary of r-layers*	*R-layer orientation*
R-layer 1	4	2	4	2	2	6
R-layer 2 (horizontal split)	4	2	3	0	3	2
R-layer 3	2	1	-	-	-	-
R-anomaly	*Number of r-anomalies*	*R-anomaly location*	*R-anomaly form*	*R-anomaly size*	*R-anomaly boundary*	*Boundary orientation of r-anomalies*
R-anomaly Type A	0	-	-	-	-	-

Source: own research

W3_ERT_b resistivity distribution

W3_ERT_b is located orthogonally to W3_ERT_a. They meet at about meter 116 of W3_ERT_b. The profile is located in a meadow. The meadow is split into two parts by a small step in topography at meter 56 of the profile.

Figure 57: Electrical resistivity tomography – W3_ERT_b

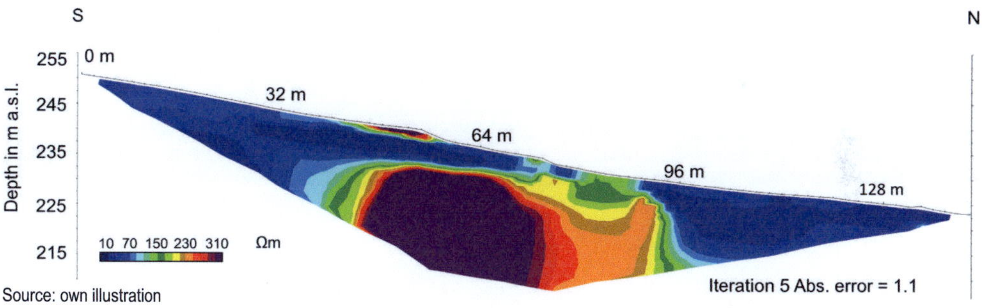

Source: own illustration

W3_ERT_b (Figure 57) has a strong vertical resistivity contrast. The northern part, starting from meter 96, has resistivity values under 70 Ωm throughout the complete model depth. Within the southern part, similar resistivity values occur. Underneath parts of the southern area and in the central part of the profile, resistivity values increase greatly to over 300 Ωm. The resistivity distribution seems to be uncoupled from the surface topography. Between meters 40 and 56, higher resistivity values occur at the top of the low-resistive area, too.

W3_ERT_b description after the guidelines

Within W3_ERT_b, two r-layers become visible by combining the resistivity distribution with the boundary analysis (Figure 58; Table 16).

Figure 58: Boundary analysis – W3_ERT_b: a) sharpness of r-layer boundary including the position of r-layers and r-anomalies; b) orientation of the boundary gradient

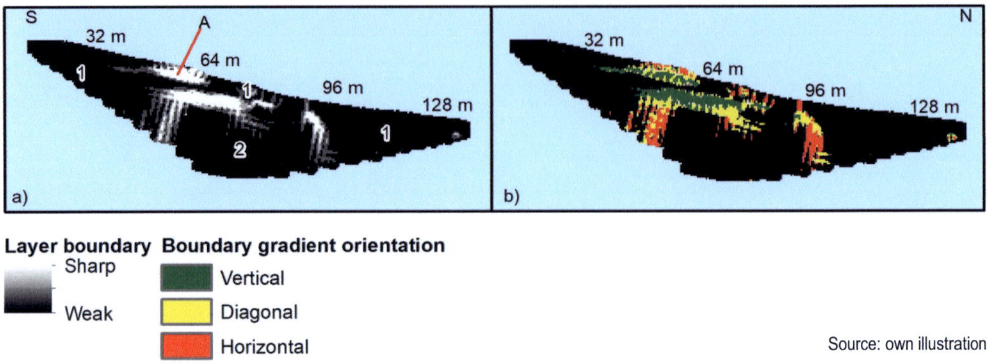

Source: own illustration

A sharp boundary delimits the high-resistive (2) r-layer from the low-resistive one (1). On the top part, the boundary is mostly orientated vertically, with the sides having horizontally orientated boundaries. The high-resistive area between meters 40 and 56 has a sharp, diffusely orientated boundary. Although the bottom r-layer is homogeneous, a few heterogeneities are visible in the top r-layer, especially between 64 and 96 meters. The higher resistivity values on top of the low-resistive r-layer are considered an r-anomaly even though this "quasi-layer" r-anomaly (A) is almost an additional r-layer by definition and size.

Table 16: ERT guidelines – W3_ERT_b

R-layer	*Homogeneity*	*R-layer thickness*	*Layering*	*Horizontal changes*	*Boundary of r-layers*	*R-layer orientation*
R-layer 1	4	2–3	2	2	5	8: < 50 & > 96 m: 6 51–95 m: 7
R-layer 2	5	3	-	-	-	-
R-anomaly	*Number of r-anomalies*	*R-anomaly location*	*R-anomaly form*	*R-anomaly size*	*R-anomaly boundary*	*Boundary orientation of r-anomalies*
R-anomaly Type A	1	1	2	3	5	2

Source: own research

GPR

W3_GPR_a, -b and -c are located at the northern meadow and within the small apple orchard. They are parallel to each other and are arranged almost slope-parallel.

W3_GPR_a, -b and -c

W3_GPR_a, -b and -c (Figures 59, 60 and 61) are located in an area where the geological map depicts a complex faulting system. In all three profiles, the same features are visible.

Figure 59: Ground-penetrating radar measurement – W3_GPR_a: a) data; b) data with marked changes in signal intensity and structure

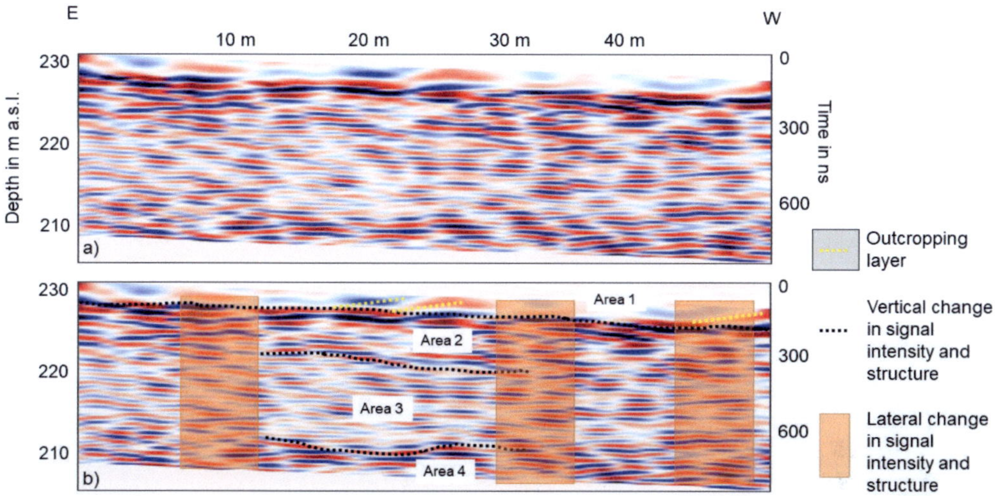

Source: own illustration

In all three radar measurements an uppermost area (1), once masked by the background noise and poorly developed in the data, but with some continuous layers cropping out.

Figure 60: Ground-penetrating radar measurement – W3_GPR_a: a) data; b) data with marked changes in signal intensity and structure

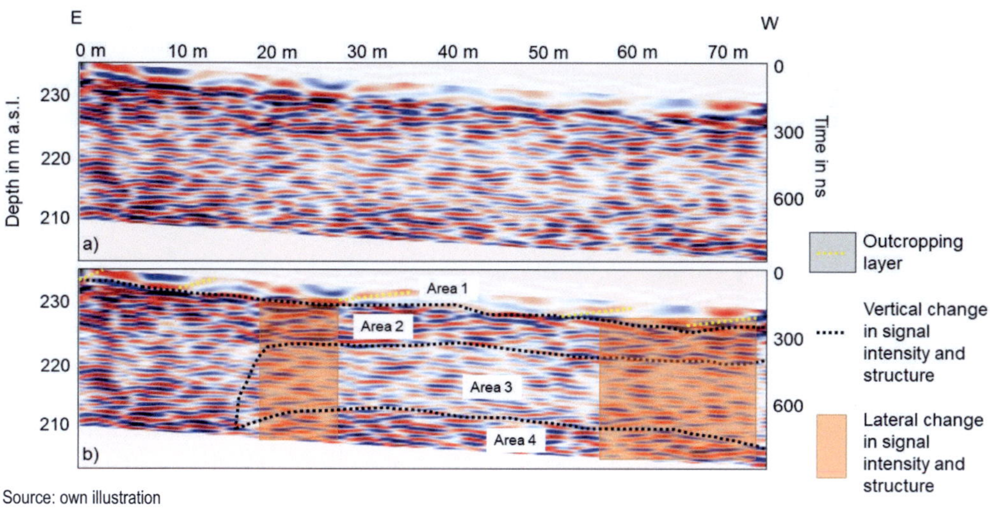

Source: own illustration

An area (2) of stronger reflections is illustrated below. This area is partly well-developed and easy to define compared with those above and below. In other parts, this layer is disturbed, poorly developed, and mixed with the surroundings.

Figure 61: Ground-penetrating radar measurement – W3_GPR_c: a) data; b) data with marked changes in signal intensity and structure

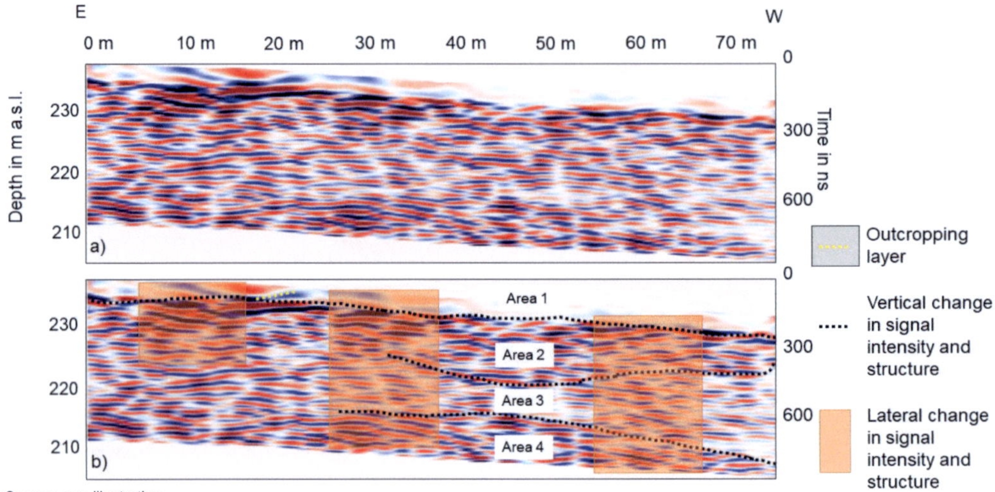

Source: own illustration

The third area (3) includes reflectors of lower intensity, but the fourth area (4) had higher intensity of reflections. All these layers are comparable to those described in W1_GPR_a in detail. However, most layers are better developed within W1_GPR_a. Nevertheless, in contrast to the profile of W1, clear, single reflectors are visible, several decameters in size. In all three GPR lines where areas with a high frequency of reflectors are located, these reflectors have an intense signal. These lateral changes in signal intensity and structure are not necessarily orthogonally located to the surface.

W3_GPR_e and -d

W3_GPR_d and -e are arranged downslope, covering the southern and northern meadows. W3_GPR_d and -e reach the faulting area from uphill and intersect the profile of W3_GPR_a, -b and -c.

The four described areas are visible within parts of these two profiles (Figures 62 and 63), too. Especially within the uphill 30 m of W3_GPR_e, the typical features of areas are well-developed and distinguished.

At the steep part of this slope, the uppermost area (1) is marked by intense reflectors. A few meters down the steep slope, an area of destructed layers is located. Farther downslope, especially in W3_GPR_e, the last 40 m have intense layering with strong reflectors, including several vertical structures, analog to W3_GPR_a, -b and -c.

Figure 62: Ground-penetrating radar measurement – W3_GPR_d: a) data; b) data with marked changes in signal intensity and structure

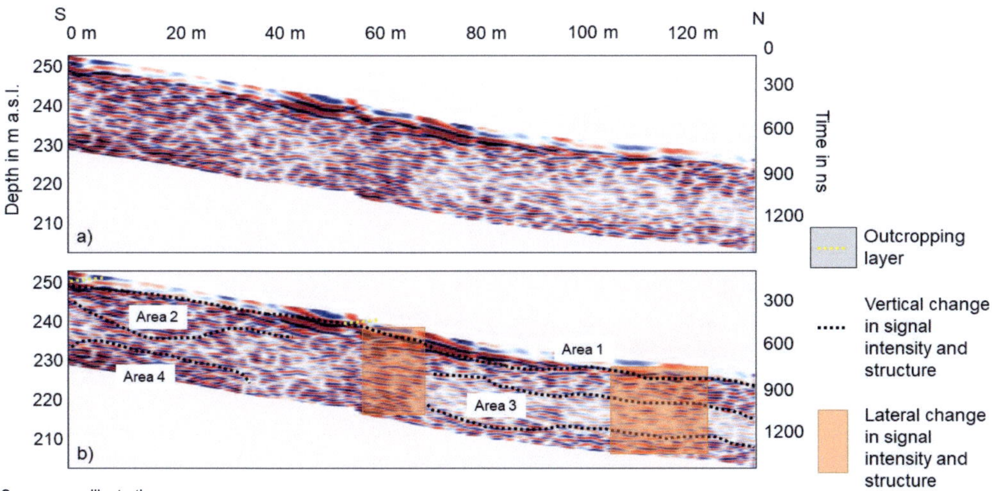

Source: own illustration

Figure 63: Ground-penetrating radar measurement – W3_GPR_e: a) Data; b) Data with marked changes in signal intensity and structure

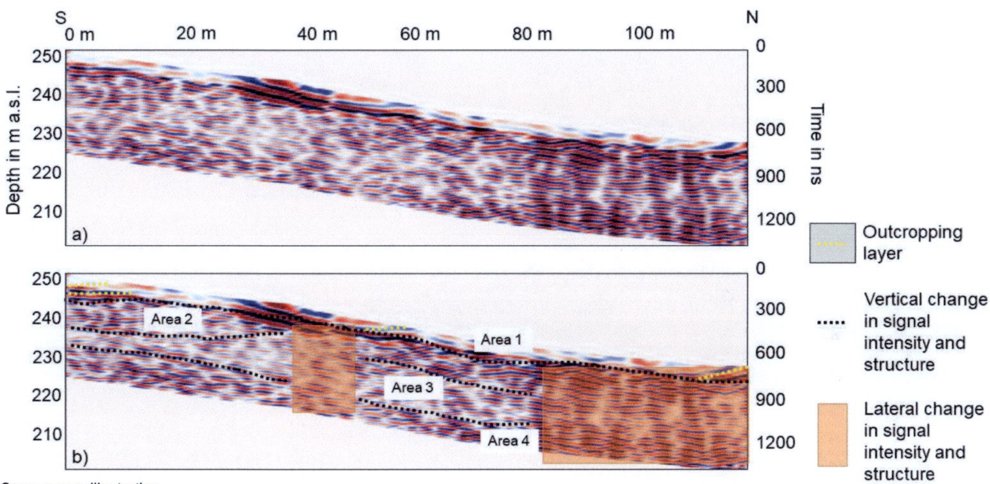

Source: own illustration

W3_GPR_f and -g

At the uphill, southern meadow, W3_GPR_f and -g are located. They are parallel to each other and almost parallel to the slope, intersecting with W3_GPR_d and -e.

Upslope from the area of expected faults, W3_GPR_f (Figure 64) and -g (Figure 65) are located. The four described areas are well-developed, and the already described vertical structures are visible as well. The sizes of the single layers vary greatly in comparison with the other GPR profiles.

Figure 64: Ground-penetrating radar measurement – W3_GPR_f: a) data; b) data with marked changes in signal intensity and structure

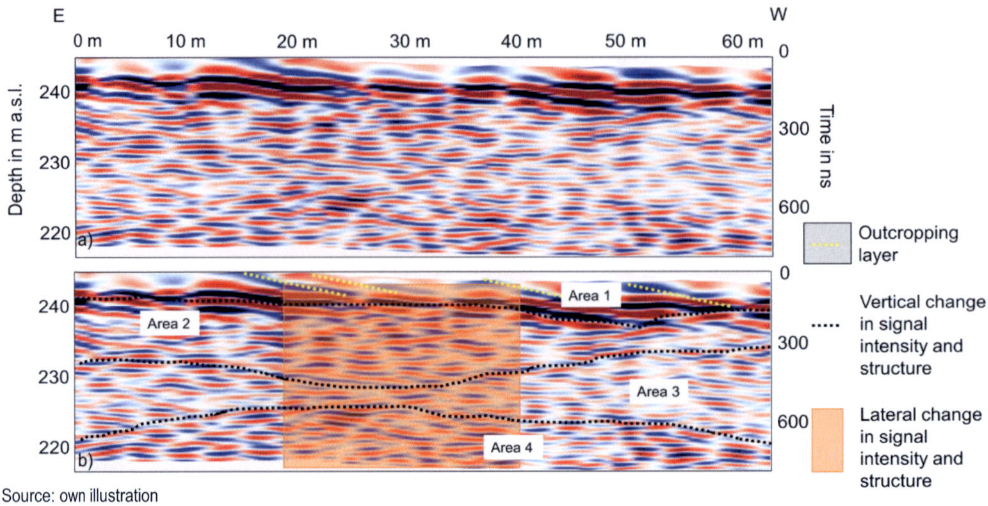

Source: own illustration

Figure 65: Ground-penetrating radar measurement – W3_GPR_g: a) data; b) data with marked changes in signal intensity and structure

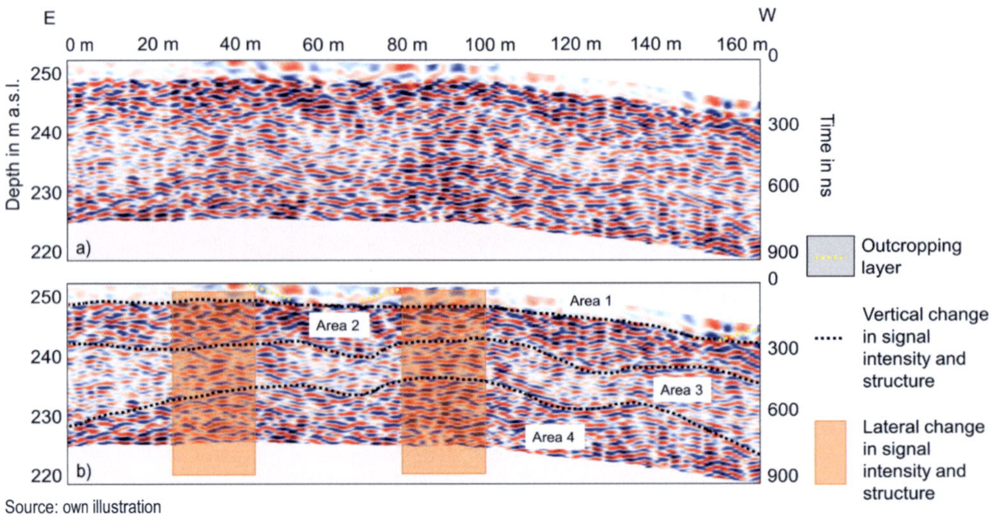

Source: own illustration

According to descriptions of the radargrams in Neal (2004), similar features to W1_GPR_a occur. The upper and lower boundaries are more or less concordant but are not well developed. Part-erosional truncation can is evident at the upper boundary, which is most recognizable in W3_GPR_d and -e at the outcropping layers. A well-developed shape of the reflections is not visible. However, they are most likely planar/even orientated with a more or less horizontal dipping. The relationship between the reflections is subparallel, and the reflection continuity is discontinuous; the northern part of W3_GPR_e is moderately continuous.

SRT

W3_SRT_a is parallel to the slope located in the northern meadow. W3_SRT_a is located almost in the same position as W3_ERT_a, intersecting W3_GPR_b and -c and missing W3_GPR_a by a few meters.

W3_SRT_a

In contrast to W3_ERT_a and the GPR measurements, W3_SRT_a (Figure 66) is not split into two sections within the seismic waves' velocity value distribution.

Figure 66: Seismic refraction tomography – W3_SRT_a

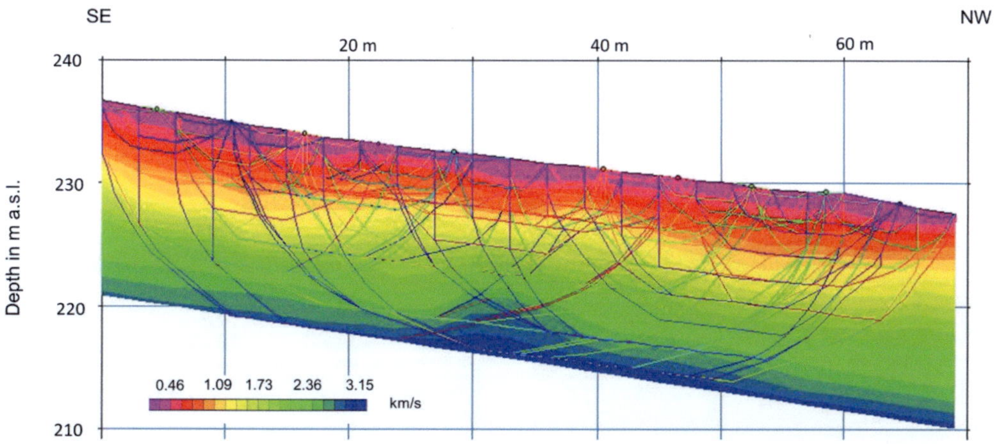

Source: own illustration

However, almost at the position of the increase of resistivity values within the central part of the ERT_a (Figure 57), at meter 30 of the SRT profile, at depths over 10 m, the velocity of the radar wave increased much faster than in other parts of the SRT.

W3_SRT_b and -c

Analog to W3_SRT_a, W3_SRT_b is almost at the position where W3_ERT_b is located, farther downslope. In contrast to W3_SRT_a and W3_SRT_b, for W3_SRT_c (cf. Figure 54), there is no parallel ERT profile. However, GPR_f and -g are located in the same meadow, more or less subparallel to this profile.

Although the structures displayed in the resistivity distribution are less visible in W3_SRT_a, W3_SRT_b (Figure 67) displays most of the structures of W3_ERT_b (Figure 68) within the velocity distribution of the seismic waves. Between meters 30 and 50, a layer of velocity values between 2.0 and 2.5 km/s narrows. In comparison with W3_ERT_b, these velocity values could classify the layer of low resistivity values between meters 40 and 75 of the ERT tomogram. Underneath this layer, faster velocity values occur stepwise. Some of these steps are visible northward, too.

Figure 67: Seismic refraction tomography – W3_SRT_b

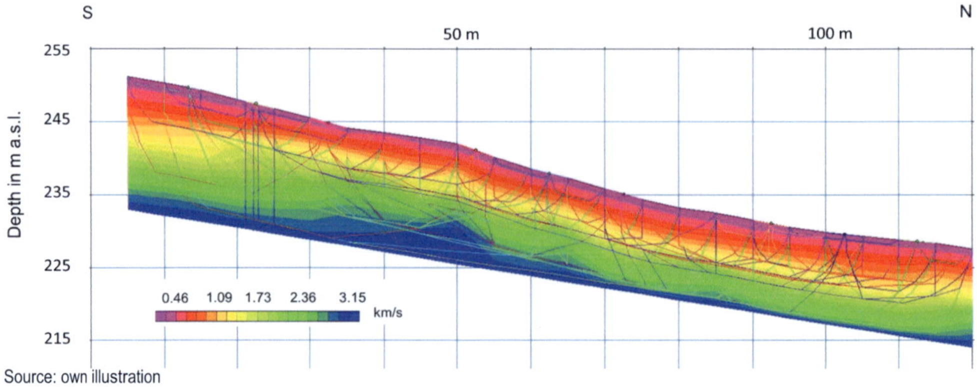

Source: own illustration

Figure 68: Seismic refraction tomography – W3_SRT_c

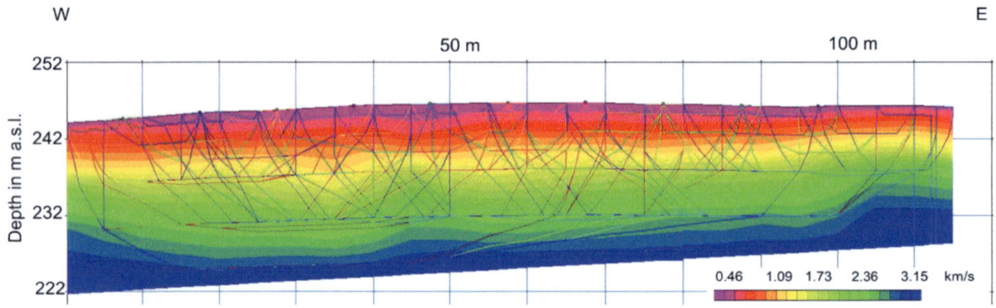

Source: own illustration

Whereas only small differences are visible within the uppermost layers of W3_SRT_c, the bottommost layer's size and form greatly vary.

4.1.6.1 Invasive methods and outcrops

The results within this subsection invasive methods and outcrops are associated with workflow Step 4 (S4) – *Ground-truth*.

Borehole

The soil columns as results from the boreholes are generally described and simplified according to the main lithological units to focus on the proposed scientific questions.

While W3_BH_a is located within an orchard, W3_BH_b and W3_BH_c are located within the northern and southern meadows, respectively.

W3_BH_a (Figure 69a) reveals over 100 cm of different PCBs on top of weathered Bröckelschiefer. Within the Bröckelschiefer layer, mainly claystone or its weathering products occur, but also some sandstone and siltstone, as well as their weathering products. Underneath, first evidence of (grano-)diorite occurs in the form of the basal-breccia.

Figure 69: Boreholes – W3: a) W3_BH_a; b) W3_BH_b; c) W3_BH_c

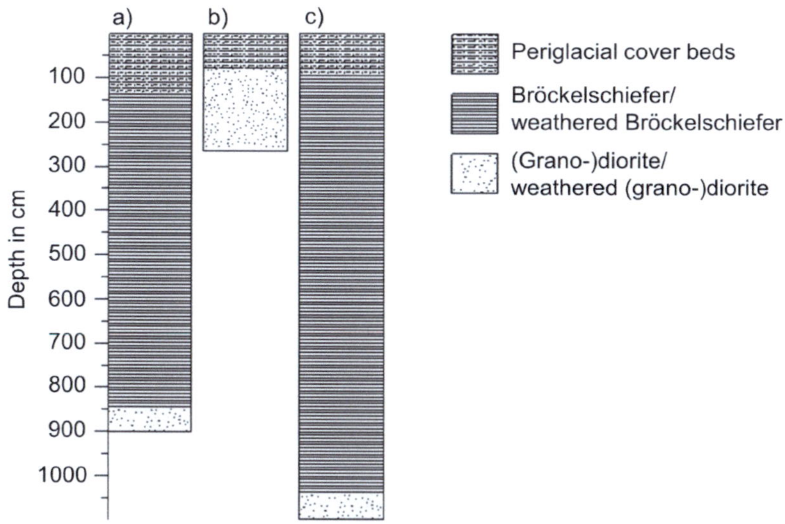

Source: own illustration

Within profile W3_BH_b (Figure 69b), the uppermost layer of PCBs is better subdivided than in W3_BH_a. Higher silt and fine sand content indicate the existence of the PCBs to a depth of over 100 cm. In the bottom part of this layer, the clay content increases, indicating a displacement of the materials from the upslope Bröckelschiefer layers. Sliding structures within this layer affirm this theory. However, this layer is highly weathered. Below the Bröckelschiefer, the weathered (grano-)diorite crops out.

Underneath the first meter of PCBs, which are well segmented, almost 10 m of Bröckelschiefer or its weathering products were found in W3_BH_c (Figure 69c). This layer mainly consists of claystone and its weathering products, but also sandstone and siltstone and their weathering products occur. At the bottom of this layer, the described basal-breccia of the Bröckelschiefer is best exposed among all the Waldaschaff profiles and turns over to pure (grano-)diorite at the bottom. The (grano-)diorite is only slightly weathered and almost impossible to penetrate with the used method.

4.1.7 W3 – Fault system: interpretation

The interpretation of the data was performed according to the workflow Step 5 (S5) – *Interpretation of the data.*

Synthesis

With the combination of all the data, three to four lithological units can be distinguished at the meadows surrounding the rough relief of W1. Within the uppermost area, parts of PCBs can be identified; those are small-sized and not resolved well in the ERT or SRT. Only the GPR revealed indications of them as small layers narrowing at the surface. In contrast, they were visible within all the soil columns of the invasive data.

Underneath this layer, there are several meters of low-resistive Bröckelschiefer. Due to the difference in weathering and alternating sand and siltstone layers within the claystone, the intensity of GPR reflections was high. With the combination of the second GPR area and the bottom of the SRT data, as well as the boreholes and the ERT data, the size of this layer varies a great deal.

At the bottom of the profiles, the (grano-)diorite crops out. The weathered (grano-)diorite has resistivity values between about 100 and 250 Ωm and is marked by a low number of reflectors in the GPR data, probably due to its more even weathering. However, underneath this saprolitized (grano-)diorite, the (grano-)diorite bedrock crops out. The resistivity values of the bedrock increase with depth and the GPR reflections are more intense. Only in the eastern parts is Bröckelschiefer missing and weathered (grano-)diorite is located near the surface, underneath the PCBs.

Interpretation

How do the mapped faults affect the development of relief?

Parts of the faulting system, important for the relief development surrounding *Wahlmich Castle*, can be displayed. The combination of the borehole data and ERT especially reveals different lithological units, which must have been shifted toward each other to create their current position. However, SRT and GPR delivered information on the intense variation between the boundaries of these lithological units and the expected faulting zones in between. Additionally, the shifting in the GPR data and the wavering bottom of the seismic velocities below 2.5 km/s indicate the existence of several more faults, as resolved in the ERT and mapped in the geological map. It is estimated that the same lithological units are staggered to each other and caused the refraction of the GPR wavetrain due to the offset, whereas the SRT revealed the bottom edge of the layers. Therefore, punctual boreholes and ERT measurements cannot resolve this issue.

When converting this information into conceptual models by examining the northern meadow (Figure 70a), it is clear that two lithological units are located underneath the PCBs. BH_a, located in the northwestern part of the meadow, has meters of Bröckelschiefer, whereas the material of BH_b, located within the southeastern parts of the meadow, consists of (grano)diorite. ERT_a indicates a horizontal change at meter 16. Combining the soil columns of the boreholes and the SRT and GPR results, which show changes at this position, a fault is located almost exactly at this position. However, the GPR data especially allows for the assumption that faulting did not stop after the accumulation of the Bröckelschiefer (e.g., W1), even if the geological map lets one assume this. Nevertheless, some areas with intense faulting are visible within the Bröckelschiefer and the underlying (grano)diorite in this area.

In a north-south direction (Figure 70b), the same lithological units are visible. In the central part of ERT_b, the high-resistive (grano-)diorite is visible. North and south of the (grano-)diorite, low-resistive Bröckelschiefer layers are located. In BH_c at about 10 m of depth, is the basal-breccia, underlaid by the (grano-)diorite at about

Figure 70: Conceptual models – W3: a) CM_a in northwest-southeast direction; b) CM_b in north-south direction

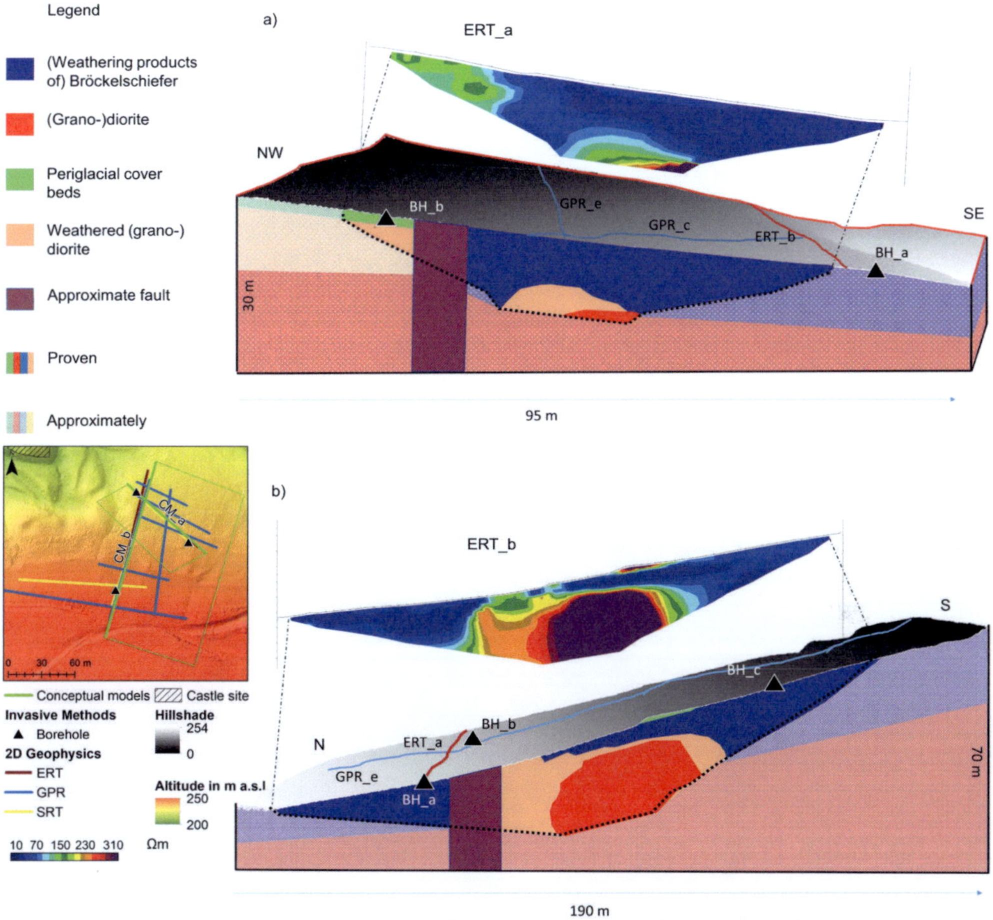

Sources: own illustration, based on Bavarian Agency for Digitalisation, High-Speed Internet and Surveying, 2014

11 m of depth. This size is comparable to literature values (Scheinpflug, 1992). However, at about meter 50 of ERT_b, a horizontal change is visible, presumably associated with another fault. Whereas SRT_b indicates a faulting process of different fragments between meters 35 and 60 of the SRT profile, GPR_d identifies other faulting processes within the meadow's northern part. These faults are detached by the present-day relief, including a recognizable step at meter 96 of ERT_b. Eastward, GPR_e indicates a further expansion of the faulting system within the bedrock of (grano-)diorite at this position.

To answer the question, many more faults and a more complex faulting system are resolved and are displayed in Chapter 5 to discuss the relief development of the study site in Waldaschaff. However, in brief, the faulting system affected the entire southern and southeastern surroundings of the castle site and was one of the main driving forces for the relative distribution of the lithological units within this area.

4.1.8 W4.1 – Building structures on-site: results

Do unknown building structures exist at the castle site?

To generate data at different resolutions, on the one hand, a cross-profile ERT was utilized for an overview of the structures within the subsurface; on the other hand, 3D ERT, 3D GPR and magnetic surveying were performed to create detailed subsurface information at spots where wall structures were suggested (Figure 71).

Figure 71: Location of measurement – W4.1

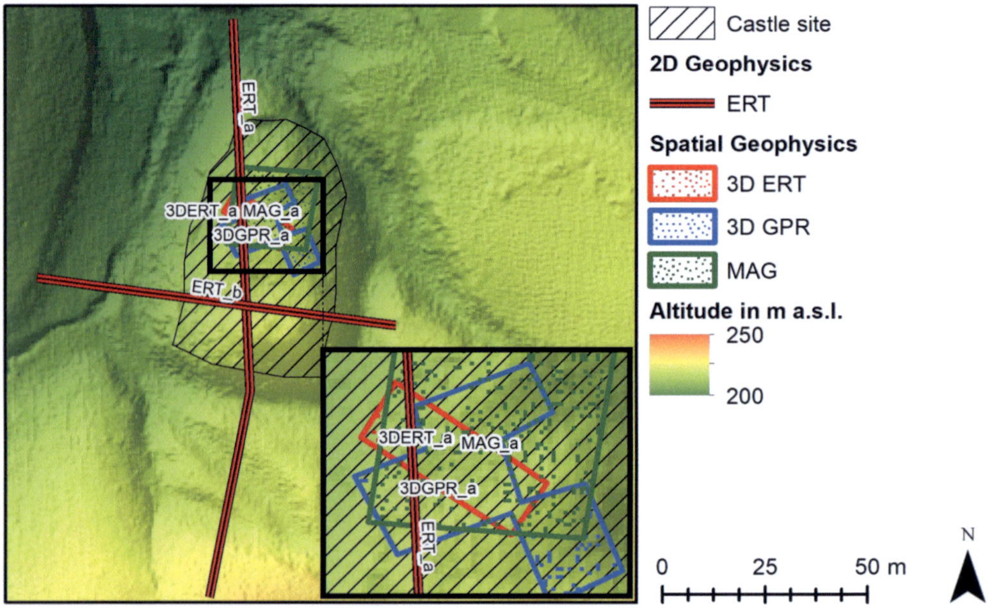

Source: Bavarian Agency for Digitalisation, High-Speed Internet and Surveying, 2014
Map creation: J. Trappe

4.1.8.1 Geophysical data

The geophysical data of W4.1 are associated with workflow Step 3 (S3) – *Geophysical methods.*

ERT

First, the inverted resistivity distribution of the profiles is described. Then, their relative distribution is discussed according to the guidelines for ERT description.

W4.1_ERT_a and W4.1_ERT_b resistivity distribution

W4.1_ERT_a and W4.1_ERT_b cross the castle hill and intersect almost orthogonally. W4.1_ERT_a starts upward at the breaking-off edge, passes some of the small, elevated areas, passes the castle hill with the different parts of the castle and finishes downhill from the castle hill within the lowland. W4.1_ERT_b starts at the floodplain west of the castle hill, crosses it and finishes at the most dominant of the small, elevated areas.

Figure 72: Electrical resistivity tomography – W4.1_ERT_a

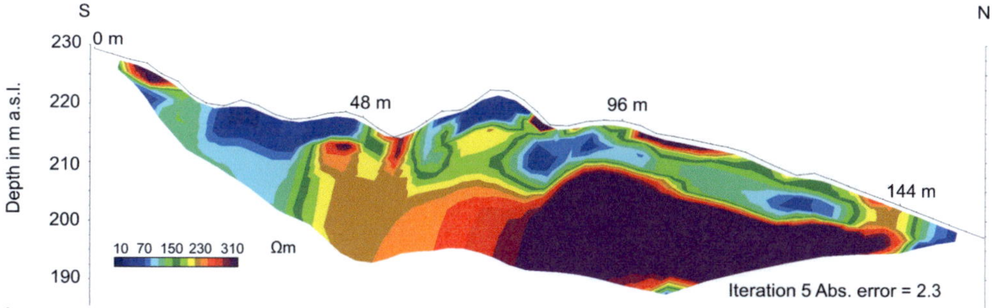

Source: own illustration

W4.1_ERT_a (Figure 72) and W4.1_ERT_b (Figure 73) represent cross-sections of the castle hill along two axes. At a depth of 5 to 10 m, high-resistive areas are visible, with resistivity values between 200 and over 300 Ωm. In W4.1_ERT_a, the location of this area starts around meter 48 of the profile and ends almost at meter 144. Within the northern part of the profile, resistivity values are higher.

Figure 73: Electrical resistivity tomography – W4.1_ERT_b

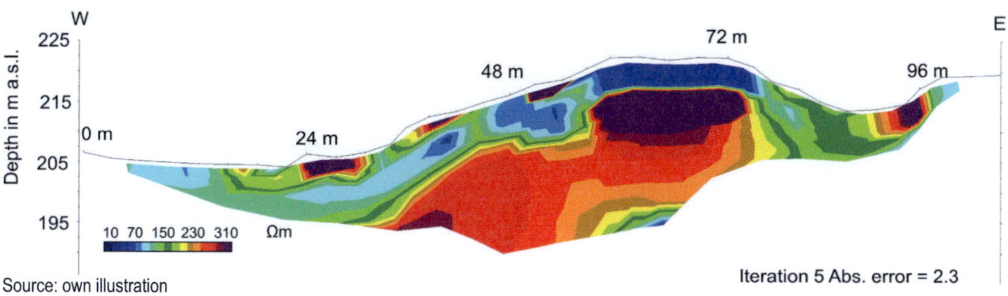

Source: own illustration

The high-resistive area within profile W4.1_ERT_b is located between meters 27 and 75 of the profile. This area's highest resistivity values are located between meters 54 and 75 of the profile at a depth of 7 to 15 m. The matrix of the remaining areas within both profiles has different resistivity values between 10 and 150 Ωm. Within this matrix, different high-resistive lenses are located mostly in the uppermost parts of the profile but also partly connected to the high-resistive area at the bottom. These lenses are usually larger in the surface-parallel direction. The resistivity values increase to over 300 Ωm.

W4.1_ERT_a and W4.1_ERT_b description after the guidelines

With the addition of the information of the boundary analysis (Figures 74 and 75) to the resistivity distribution, two r-layers are visible in the two profiles (Table 17).

The boundary analysis of W4.1_ERT_a and W4.1_ERT_b reveals a sharp boundary between the top (1) and bottom (2) r-layers within the eastern and northern

Figure 74: Boundary analysis – W4.1_ERT_a: a) sharpness of r-layer boundary including the position of r-layers and r-anomalies; b) orientation of the boundary gradient

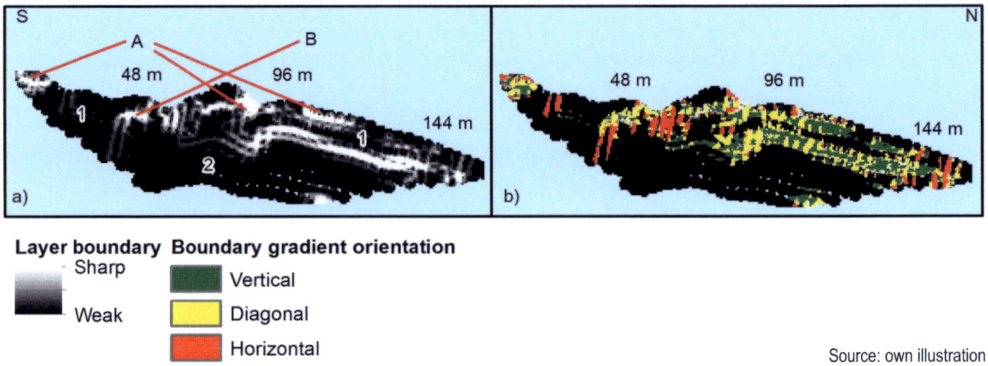

Source: own illustration

Figure 75: Boundary analysis – W4.1_ERT_b: a) sharpness of r-layer boundary including the position of r-layers and r-anomalies; b) orientation of the boundary gradient

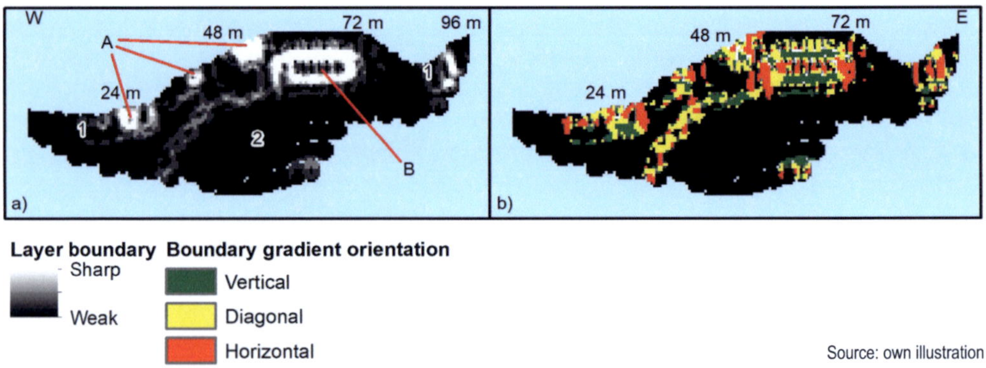

Source: own illustration

Table 17: ERT guidelines – W4.2_ERT_a and -b

R-layer	*Homogeneity*	*R-layer thickness*	*Layering*	*Horizontal changes*	*Boundary of r-layers*	*R-layer orientation*
R-layer 1	1	2	3	1	2–5	7
R-layer 2	5	3		0		
R-anomaly	*Number of r-anomalies*	*R-anomaly location*	*R-anomaly form*	*R-anomaly size*	*R-anomaly boundary*	*Boundary orientation of r-anomalies*
R-anomaly Type A	3	1	1, 3	1–2	5	2
R-anomaly Type B	1	2–3	1	3	5	1

Source: own research

parts. In the western and southern parts, this boundary is far weaker. The boundary orientation is mostly vertical-to-diagonal and is only horizontally located at the lateral parts in W4.1_ERT_b. In W4.1_ERT_a, there are a couple of high-resistive r-anomalies (A) with sharp boundaries. Even with the exclusion of the top r-layer's

Figure 76: 3D electrical resistivity tomography – W4.1_3DERT_a including anomalies a and b: a) resistivity scale 0.05–0.8 kΩm; b) resistivity scale 0.2–3.2 kΩm

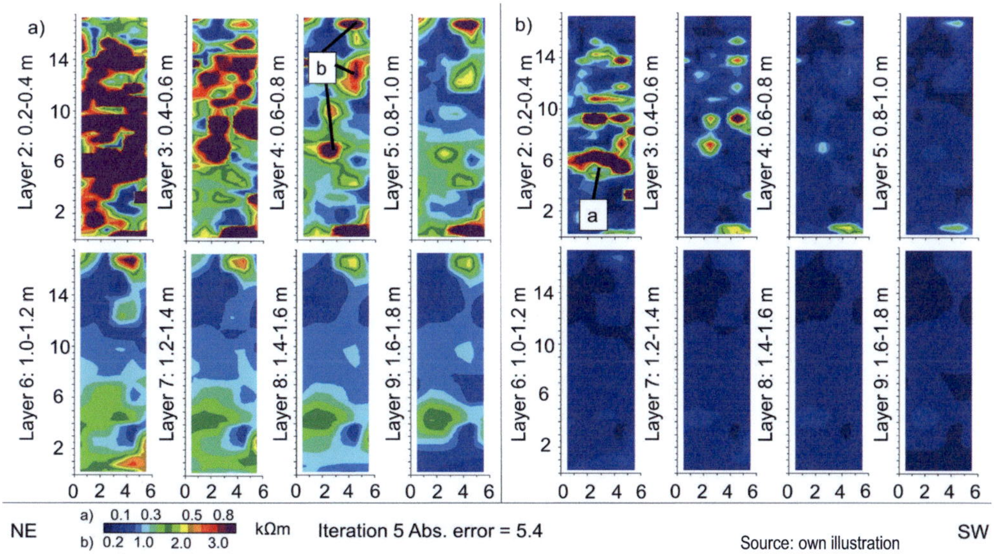

Source: own illustration

r-anomalies, the resistivity distribution is still heterogeneous, with a medium-developed layering to the bottom r-layer, which is more heterogenous, and which only indicates some boundaries at the bottom. Several r-anomalies have sharp boundaries with varying sizes without any definable form in the top r-layer. The upper boundary of the anomalies is not visible at the surface. However, some weaker parts are located within it. At the bottom of the first r-layer, another type of r-anomaly (B) occurs, with a sharp boundary surrounding it fully.

3D ERT

W4.1_3DERT_a is located on top of the castle hill at a small, leveled area in the northern part. For a better visual representation of all features within W4.1_3DERT_a (Figure 76), it is displayed at two different scales.

In the second layer, different anomalies are located. These anomalies have resistivity values of more than 3,000 Ωm (a). One of these anomalies is located almost in a north-south direction and is visible in all lines in Layer 2 of this quasi-3D measurement. The other anomalies are smaller and have no significant orientation. Most parts of the matrix have lower resistivity values, between 50 and around 500 Ωm, except for some smaller anomalies (b) in Layers 4 and 5. Beneath a depth of 1 m, the matrix in the eastern part has resistivity values around 200 Ωm, and in the western part, the values are up to 400 Ωm.

3D GPR

Partly intersecting with W4.1_3DERT_a on top of the castle hill, W4.1_3DGPR_a (Figure 77) is located on top of the castle hill, too, but also covers other flattened areas on top of the hill.

Figure 77: 3D ground-penetrating radar – W4.1_3DGPR_a

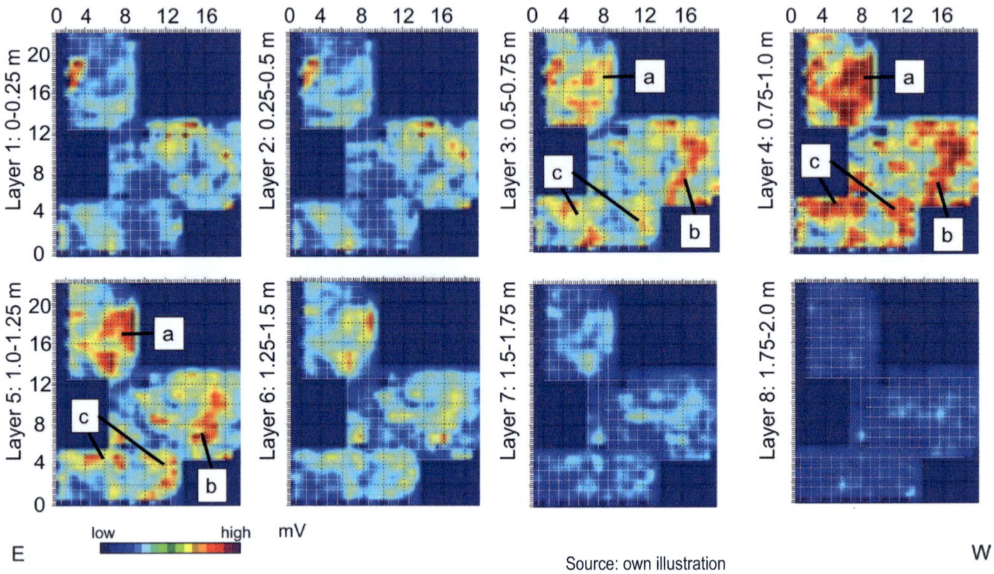

Source: own illustration

Within all the layers, differences within the intensity of the radar waves are documented. Most of these anomalies lack a clear boundary and vary in size. Larger anomalies are only visible in Layers 3 to 5. One anomaly (a) is located at 4–8 m along the x-axis and 12–20 m along the y-axis, more or less oval and hollow. The other (b) is located between 8–16 m along the x-axis and 0–12 m along the y-axis. This anomaly is linear, about 1 m thick and 8 m long. Other anomalies (c) are located at 0–8 m along the x-axis and 4 m along the y-axis, and at 12 m along the x-axis and at 8 m along the y-axis, respectively.

Figure 78: Magnetic survey – W4.1_MAG_a

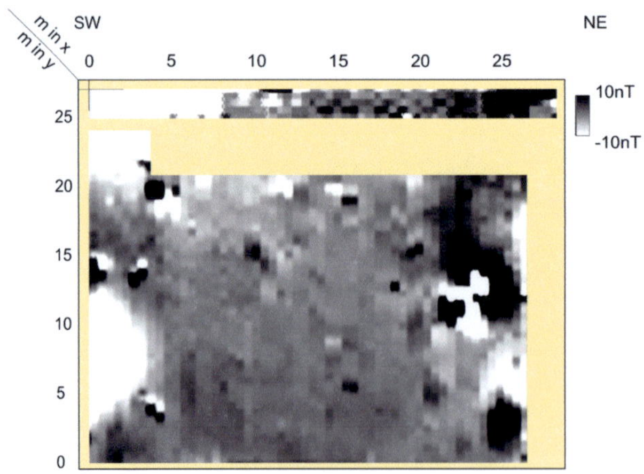

Source: own illustration

Magnetic survey

W4.1_MAG_a (Figure 78) is also located on top of the castle hill and intersects with W4.1_3DERT_a and W4.1_3DGPR_a.

Except for the northeastern part of the magnetogram, several dipole anomalies are visible across the entire grid. These anomalies mask all possible small shades within the data. However, they are particularly grouped in the southwestern part.

4.1.8.2 Invasive methods and outcrops

The study site was excavated during and after the measurements. Therefore, to answer the question of W4.1, no invasive methods were used, and no outcrops are described in this section. Nevertheless, the excavated wall structures were documented for archaeological reasons and were interpreted in combination with the collected geophysical data. Therefore, these data can be associated with workflow Step 4 (S4) – *Ground-truth* – for this question.

4.1.9 W4.1 – Building structures on-site: interpretation

The interpretation of the data was performed according to workflow Step 5 (S5) – *Interpretation of the data.*

Synthesis

Within the ERT measurements, high-resistive r-anomalies occur near the surface, surrounded by sharp boundaries at or in the surroundings of the castle hill. These r-anomalies are correlated with wall structures made of sandstone in line with the literature's resistivity values of them (Reynolds, 2011). The wall structures are partly visible or excavated, others can be expected. The GPR reveals many anomalies within the top 1.5 m, which can be correlated with those structures as well. Additionally, the homogeneity of the 2D ERT data is low, indicating additional, smaller anomalies. However, the magnetic survey indicates an intense usage of the hill and many ferromagnetic objects on the hill without the possibility of a more precise determination.

Additionally, the abovementioned low-resistive Bröckelschiefer and high-resistive (grano-)diorite occur, as well as stratigraphical units (the latter at least in the 2D ERT). The center of the hill is split by a sharp boundary between the Bröckelschiefer and, most likely, intensely weathered parts of the (grano-)diorite. In contrast, these weathered parts only show medium-to-weak boundaries between the layers above.

Interpretation

Do unknown building structures exist at the castle site?

All the geophysical data suggest different building structures on top of the castle hill. With the help of these data, a targeted excavation was conducted in 2018, which also provides the ground-truth for the geophysical models.

In both the 3D GPR and the 3D ERT data, a linear structure is visible in the western part of the castle hill. The excavation revealed this structure to be a wall (Figure 79). Upon closer inspection of W4.1_ERT_a at this position, a linear r-anomaly of 18 m in length was located, too. Therefore, it is expected that the wall continues further northward. Additional other reflectors are visible within the GPR. The 3D ERT displays anomalies with various resistivity values, and the magnetic survey suggests that there are other structures east of it. Unfortunately, this area is covered with anthropogenic debris as well; therefore, those anomalies can hardly be classified as in-situ structures.

The contamination with this debris covers a further differentiation of different parts within the occupation layer. However, W4.1_ERT_a and -b display the high-resistive (grano-)diorite underneath low-resistive clay-rich material, in which the occupation layer seems to be resolved. Up to 5 m of low-resistive material, mixed with the anthropogenic debris, occurs on top of the (grano-)diorite. This material underlines a theory that the removed materials consisting of PCBs and Bröckelschiefer were put on top of the castle hill to flatten and raise the topography (Rosmanitz & Bachmann, 2017; Rosmanitz *et al.*, 2019b). However, the 2D ERT displays that the prong consists of high-resistive dioritic bedrock.

4.1.10 W4.2 – Outer bailey: results

Where is the location of the outer bailey?

Even though the outer bailey of *Wahlmich Castle* is commonly suspected to be within the valley farther downslope, and most parts of the castle surroundings are unfavorable for building such structures, some parts of their functions could have been taken over from the buildings in the surroundings of the castle. Therefore, at some places in the surroundings of the castle site, spatial magnetic surveys were conducted to confirm or deny the existence of the outer bailey in the valley (Figure 80).

4.1.10.1 Geophysical data

The geophysical data of W4.2 are associated with workflow Step 3 (S3) – *Geophysical methods.*

Magnetic survey

W4.2_MAG_a is located on top of the most prominent of the elevated areas. W4.2_MAG_b and -c are southeast and upslope of the area surrounding the castle, at the southern meadows.

W4.2_MAG_a

In contrast to W1_ERT_a, located in the same position, W4.2_MAG_a prospected possible manmade structures rather than sedimentological prospection.

In the magnetogram (Figure 81), several small dipole anomalies are visible. At y-meter 5, they form a linear structure (a) almost along the x-axis. Between meters

Figure 79: Interpretation – W4.1: a) combination of the 3D-GPR-results and the excavated walls; b) combination of the 3D-ERT results and the excavated walls

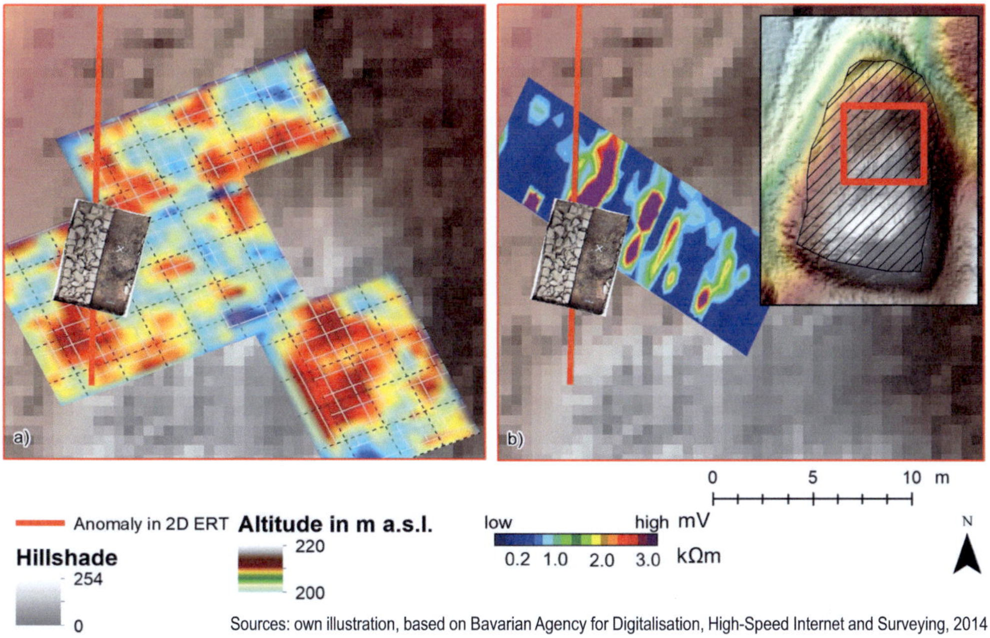

Sources: own illustration, based on Bavarian Agency for Digitalisation, High-Speed Internet and Surveying, 2014

Figure 80: Location of measurements – W4.2

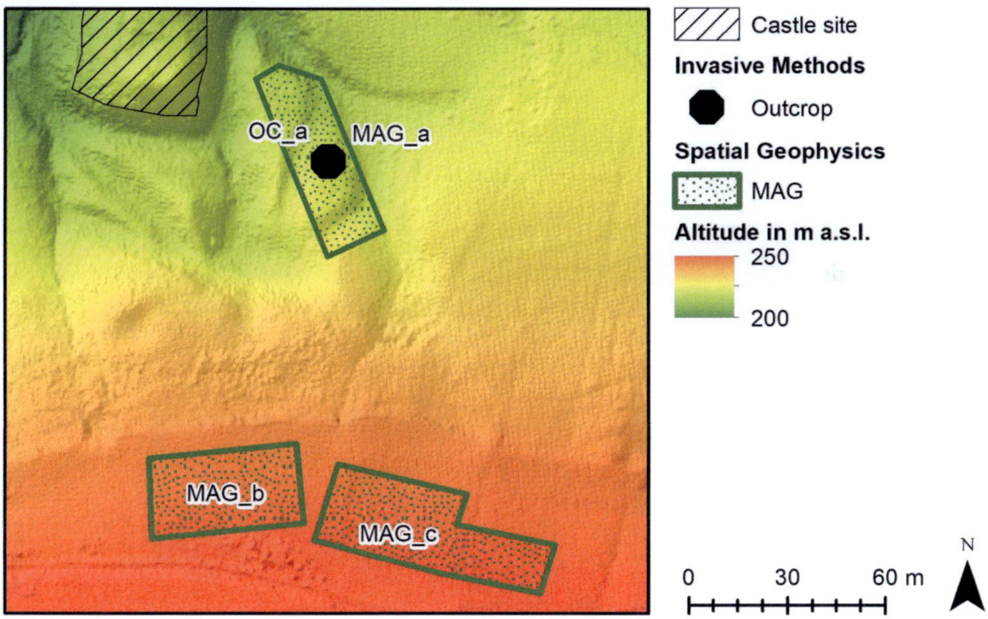

Source: Bavarian Agency for Digitalisation, High-Speed Internet and Surveying, 2014
Map creation: J. Trappe

Figure 81: Magnetic survey – W4.2_MAG_a

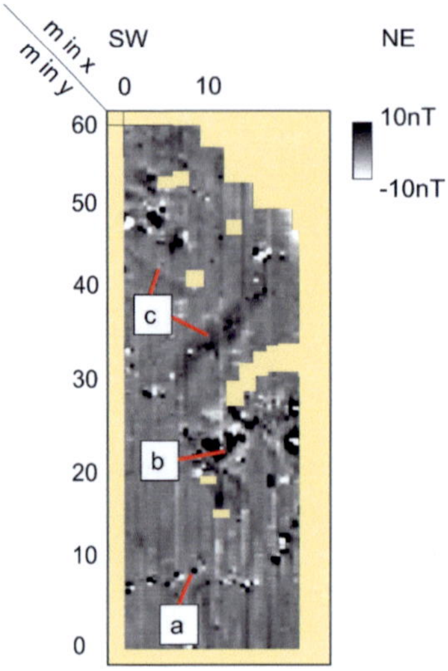

Source: own illustration

Figure 82: Magnetic surveys – W4.2: a) W4.2_MAG_b; b) W4.2_MAG_c

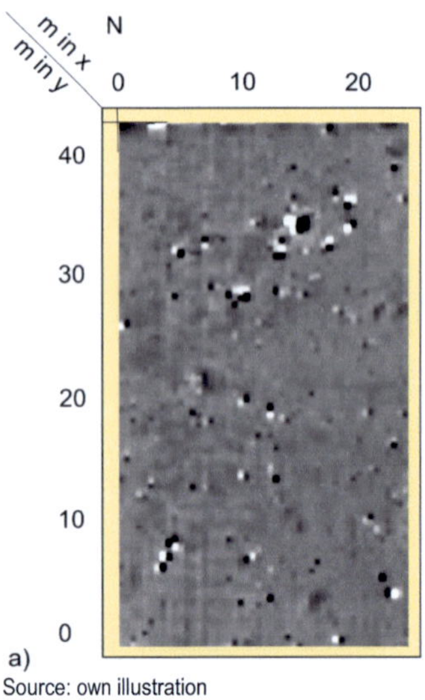

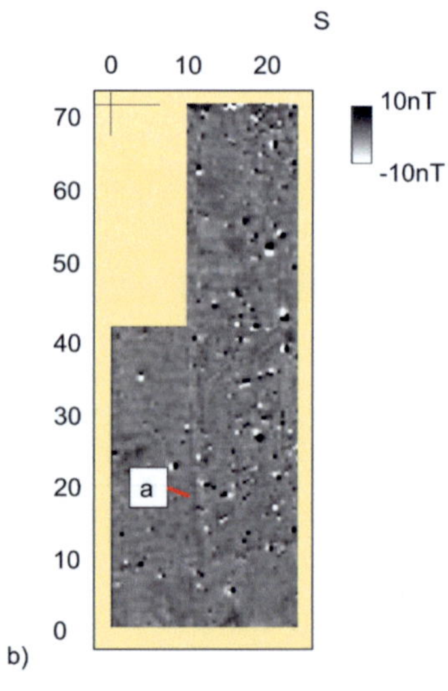

Source: own illustration

20 and 30 along the y-axis, dipole anomalies (b) cluster between x meters 10 and 20. However, nT values are more positive than negative in this area.

Two shades (c) with positive nT values are located between y-meters 30 and 50. They face almost north-south and east-west, respectively. The north-south facing one is stronger and has a length of about 15 m, whereas the west-east-facing one is lighter and has a length of about 10 m. It seems as if the lighter one leads to the stronger one orthogonally.

W4.2_MAG_b and -c

In W4.2_MAG_b (Figure 82a), only dipole magnetic anomalies are visible. All these anomalies are poorly sorted, and no further structures are visible.

In W4.2_MAG_c (Figure 82b), there are only some unsorted, dipole, magnetic anomalies analog to W_4.2_MAG_b. In contrast to W_4.2_MAG_b, some shades with nT values only slightly above 0 are visible (a). These linearly formed shades are visible at x-meter 10 and y-meters 10 to 30.

4.1.10.2 Invasive methods and outcrops

The results within this subsection invasive methods and outcrops are associated with workflow Step 4 (S4) – *Ground-truth.*

Outcrop

W4.2_OC_a is located at the most prominent of the small, elevated area, directly in the position of the most intense anomaly of 4.2_MAG_a.

Figure 83: Outcrop – W4.2_OC_a

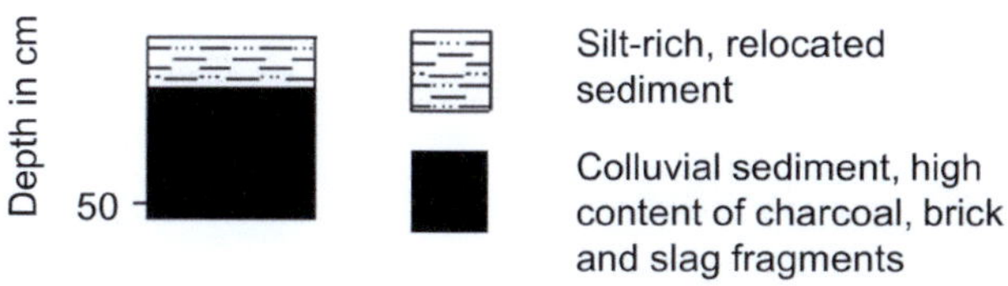

Source: own illustration

W4.2_OC_a (Figure 83) was previously designated for researching the layering of the bedrock; however, different relocated and colluvial layers appeared. Within the uppermost layer, silt-rich material has accumulated. Here, no hints of anthropogenic use are visible, but nor is there natural layering. Therefore, this is probably a relocated layer that was not in intense usage.

Underneath this layer is a colluvial layer. This layer is massively affected by anthropogenic usage as it contains: charcoal, brick and slag fragments. The material between 15 and 30 cm of depth is silt-rich, and beneath this depth, the clay content increases, and even claystone occurs. In the top part of the second layer, the material is dark, whereas the color of the bottom part is reddish. However, in the bottom part, still fragments from anthropogenic usage appear, such as charcoal and slag.

4.1.11 W4.2 – Outer bailey: interpretation

The interpretation of the data was performed according to workflow Step 5 (S5) – *Interpretation of the data.*

Synthesis

The different anomaly types indicate different causes. Dipole dots (high and low nT-values next to each other) are associated most likely with ferromagnetic materials in the subsurface. Larger negative or positive nTs probably hint at anthropogenic use or location of ferrimagnetic material, as 4.2_OC_a shows for the Anomaly b. The remnant magnetization allows for the detection of moved and relocated material, comparable to the Anomaly c.

Interpretation

Where is the location of the outer bailey?

W4.2_MAG_a reveals some structures with slightly higher nT values are associated partly with the depressions (Figure 84) within the elevated area (c).

Figure 84: Interpretation – northern part of W4.2

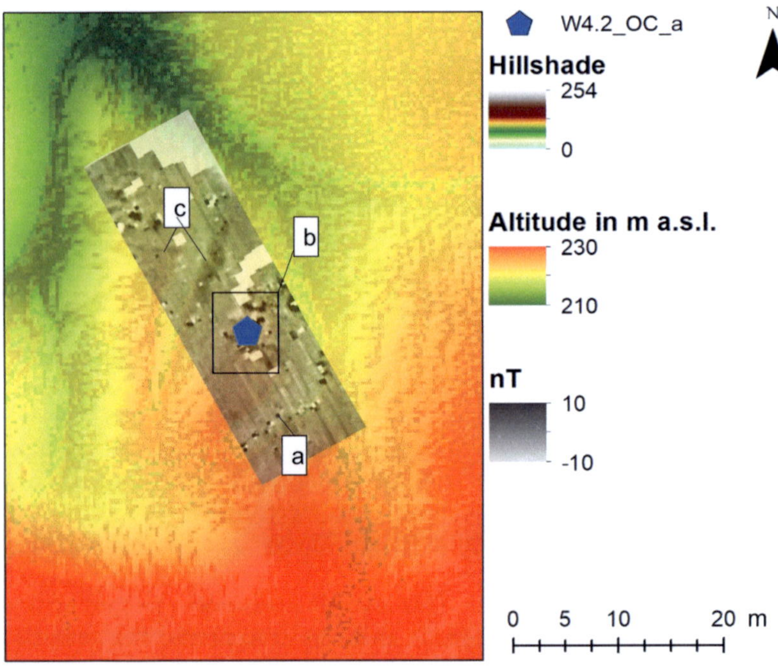

Sources: own illustration, based on Bavarian Agency for Digitalisation, High-Speed Internet and Surveying, 2014

Additionally, a dotted line of dipole anomalies is visible (a). Field observations (not shown) revealed the remnants of a 20th-century fence at this position. Only the higher nT values in the central section are associated with anthropogenic activity during the period of interest (b). The outcrop was previously intended to see the layering of the Bröckelschiefer to answer the question of W1. However, underneath a few centimeters of culminated culture-free material at the top of a massively affected and colluvial soil. Remarkably, the position is exposed and includes a portion of slope to the creek nearby; therefore, the position is not predestined to be an archive at the elevated area. However, charcoal, bricks and slag fragments were accumulated within the second layer, suggesting it was an occupation layer. Therefore, anthropogenic use, probably in the medieval period, is assumed. Nevertheless, the extent of the nT anomaly is only a couple of square meters in size. Therefore, most likely, the area is not the outer bailey itself, just a small, used area.

In the meadows, the data from the magnetic surveying provide no indication of any anthropogenic building structures, but they do indicate several small ferromagnetic objects, which can be linked most likely to the agricultural usage of the area. However, the described structure in W4.2_MAG_c could still be manmade and needs to be further investigated.

Nevertheless, a further investigation of these areas is currently prohibited. Therefore, this will stay state of the art until the next excavation.

4.1.12 Quantification of the anthropogenic, medieval influence in Waldaschaff

This Subsection can be associated with workflow Step 6 (S6) – *Quantification of the anthropogenic influence.*

First, all results and interpretations at Waldaschaff are briefly summarized, and then the medieval, anthropogenic influence on the relief development is estimated according to Section 3.5.

W1 – Origin of the rough relief

Detected using geophysical and invasive methods, the origin of the rough relief south of the castle hill in Waldaschaff can be classified as predominantly of natural origin (Category 1). No evidence of intense earthworks or massive building activity was found despite numerous invasive methods, which were also extrapolated with the help of the geophysical methods. These results indicate that paleo mass movements once created this landscape, and these are discussed in W3 (cf. Figure 85).

W2 – Neck ditch

Only in the northern part of the rough relief, directly at the castle site, do neck ditches incise deeply through the sediments, as mentioned first by the 2016 excavation (Rosmanitz & Bachmann, 2017). However, this was proven by geophysical results and indicated a preferred removal of specific materials, but also that this privilege can be deferred if necessary. The neck ditches can be classified as a massive anthropogenic effect on the relief (Category 4) directly at the castle site (cf. Figure 85).

W3 – Fault system

Using different geophysical and invasive methods, various faults were detected in the area of investigation. It is assumed that this faulting system was the driving factor that created the entire relief in the surroundings of the castle hill and was, thus, an important factor for the choice of the castle site. Even though it was not the aim of this specific question, no evidence of human effect (Category 1) on relief development was found (cf. Figure 85).

W4.1 – Building structures – on-site

With the help of 2D and 3D geophysics, different building structures were detected at the castle hill. Partly, these walls could be proven as in-situ structures belonging to the castle by the excavation in 2018 (Rosmanitz *et al.*, 2019b). Additionally, the material removed from the neck ditches was probably put on top of the prong for leveling or raising the altitude. Therefore, the entire prong can be described as massively anthropogenically (Category 4) influenced by building activity (cf. Figure 85).

W4.2 – Outer bailey

In combination with a prospection pit, magnetic surveying was conducted to detect the outer bailey. Evidence for an occupation layer and anthropogenic use could only be found on top of the largest ridge, southeast of the castle hill. However, south and southeast of the rough relief, no structures that could be linked to buildings were found (Category 1). Therefore, only the relief of the most prominent ridge can be classified as slightly anthropogenically influenced (Category 2), but a linkage to medieval building activity is still missing (cf. Figure 85).

Figure 85: Intensity and type of medieval, anthropogenic influence – Waldaschaff

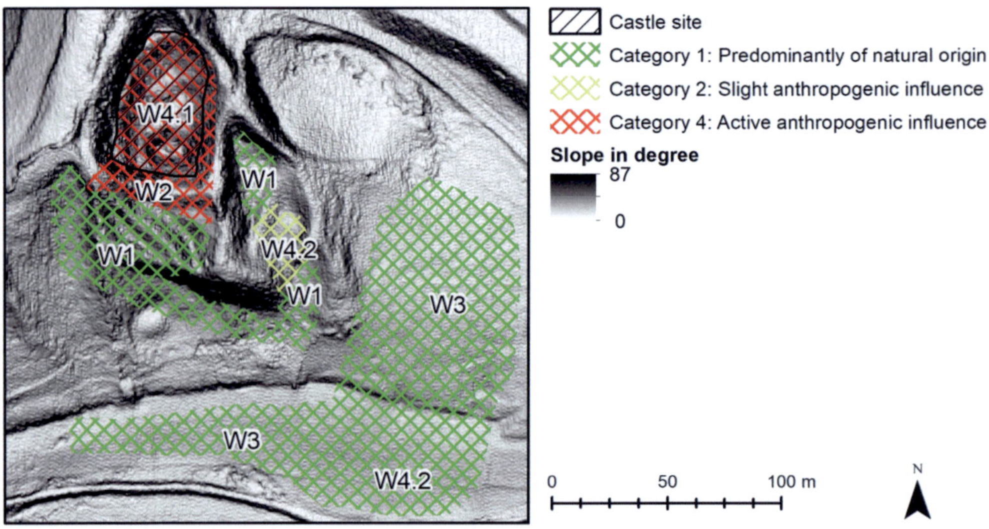

Source: Bavarian Agency for Digitalisation, High-Speed Internet and Surveying, 2014
Map creation: J. Trappe

4.2 Hauenstein

First, site-specific questions were derived with the help of the GIS-based terrain analyses at Hauenstein. The results for each question are provided and interpreted. Then, an estimation of quantification of the anthropogenic influence is made. A technical overview of the used application of the geophysical methods is given in Appendix 4 to Appendix 6.

4.2.1 Morphometric analysis for the creation of site-specific questions

The morphometric analyses for creating site-specific questions are associated with workflow Step 2 (S2) – *GIS-based terrain analyses*.

The TPI landform classification (Figure 86a) describes the castle hill as a local elevation in lowlands, surrounded by lowlands, small depressions in the deepest parts of the valleys and lowlands, large depressions at the bottom slopes and slightly elevated valley grounds. The slope analysis illustrates that the southeast-orientated slopes are terraced, whereas, in the northern creek valley, the talweg is leveled. Small ridge-like wall structures intersect this plain (not resolved well in the TPI and slope analysis) and separate several leveled areas associated with former ponds. Opposite to that, there is only direct at the castle hill, a small leveled area developed at the eastern creek valley. Southwest of the court and southwest of the castle hill on the northwest-orientated slope, meadows are located. Other parts of the castle's surroundings are forests. At the described meadows, southwestern of the present-day court, the slope and hillshade (Figure 7) displays round structures. The TWI (Figure 86b) indicates an expectable accumulation of the topographic culminated water into the fluvial system.

Figure 86: Morphometric analyses – Hauenstein: a) landform classification and slope; b) topographic wetness index

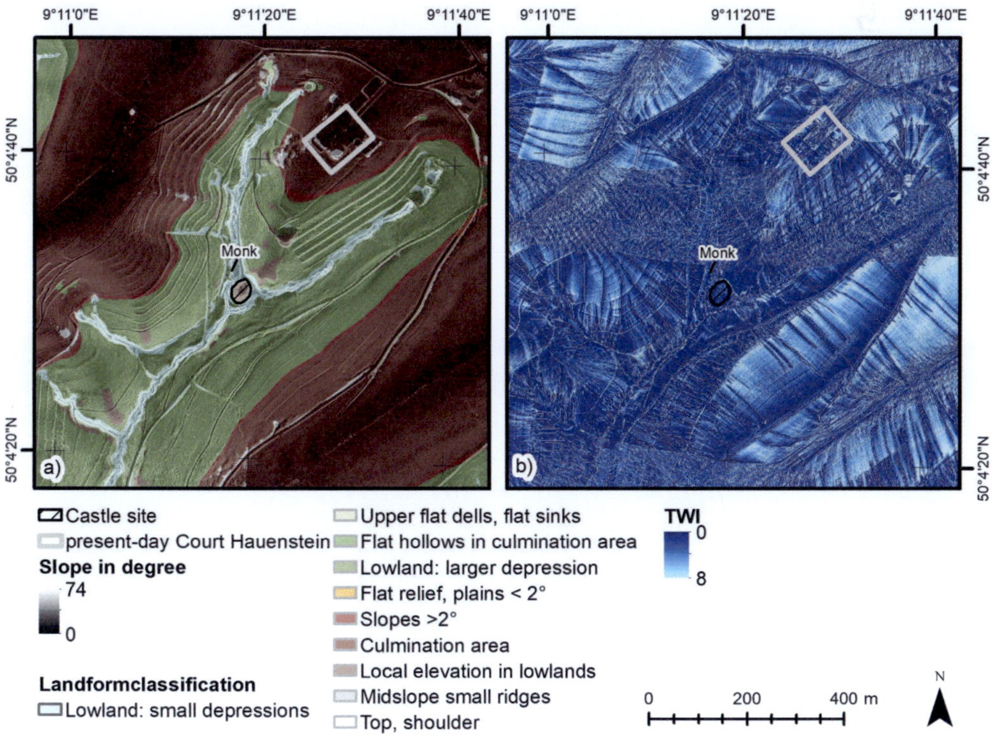

Source: Bavarian Agency for Digitalisation, High-Speed Internet and Surveying, 2014
Map creation: J. Trappe

With respect to the specific position of *Hauenstein Castle* within a valley by assessing its local importance, the following questions are derived (an overview is in Table 18):

Question H2 – surficial, anthropogenic features:

How has the northern talweg been developed?

The castle hill of *Hauenstein Castle* is located at the bottom of a valley, between two creeks. Additionally, some features belonging to a water mill were found, including a monk of mill ponds (Figure 86; Rosmanitz *et al.*, 2019a). Therefore, question H2 tries to figure out the origin and age of sediments accumulated in natural lakes or anthropogenic ponds for the mill. Is it possible to figure out whether the mill's creation had triggered the development of the sedimentation in this talweg, or was the talweg already developed it its present extent and the castle was founded there because of that?

Question H3 – buried, natural features:

Is the center of the castle hill build of bedrock?

The origin of the interior of the castle hill is still not proven. Even though a natural origin of it as a spur, like in Waldaschaff, seems likely, regional legends describe an up-to-30-m-high tower (Rosmanitz *et al.*, 2019a), which helped build the hill with its debris after destruction.

Question H4.1 – buried, anthropogenic features – on-site:

Do buried and as-yet-unknown building structures exist on the castle site?

Only parts of the castle hill were excavated in 2017 (Rosmanitz *et al.*, 2019a). However, further excavations are planned, and therefore, further building structures needed to be prospected. On top of the castle hill in *Hauenstein Castle*, several new-aged wall structures are visible. Underneath these structures, parts of the castle or damaged wall structures are expected and need further localization for these subsequent excavations.

Question H4.2 – buried, anthropogenic features – off-site:

Where is the location of the outer bailey?

Analog to Waldaschaff, an outer bailey has not been found yet in the Hauenstein study site. The round structures, only visible on the slope and the hillshade, are suspected to be manmade and need to be further investigated.

Table 18: Site-specific questions in Hauenstein

Category	*Questions*
H2	How has the northern talweg been developed?
H3	Is the center of the castle hill made of bedrock?
H4.1	Do buried and as-yet-unknown building structures exist on the castle site?
H4.2	Where is the location of the outer bailey?

Source: own research

4.2.2 H2 – Development of the talweg: results

How has the northern talweg been developed?

ERT profiles were positioned as cross-profiles at the bottom of the talweg to display the spatial distribution of the fluvial or limnic sediments. Additionally, in deep loose sediment, percussing core probing was performed, and samples were taken for ^{14}C-dating. In areas where smaller sediment bodies were expected, Pürckhauer soundings were performed (Figure 87).

Figure 87: Location of measurements – H2

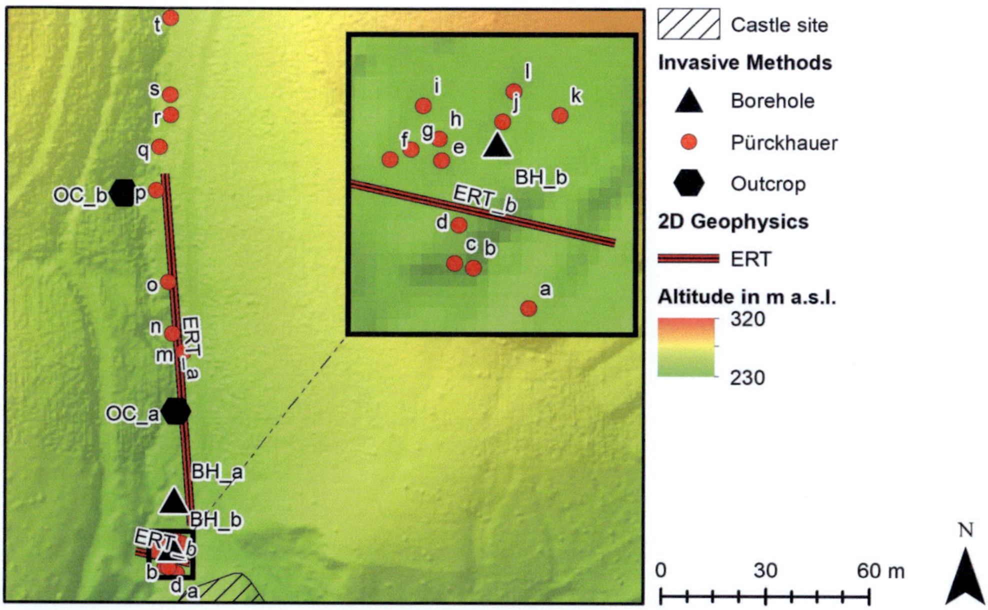

Source: Bavarian Agency for Digitalisation, High-Speed Internet and Surveying, 2014
Map creation: J. Trappe

4.2.2.1 Geophysical data

The geophysical data of H2 are associated with workflow Step 3 (S3) – *Geophysical methods*.

ERT

First, the inverted resistivity distribution of the profiles is described. Then, their relative distribution is discussed according to the guidelines for ERT description.

H2_ERT_a resistivity distribution

H2_ERT_a is located within a small valley in which a creek flows. Starting at a small plain, the profile crosses an earth wall structure at meter 55 and, after meter 60, the meandering creek itself.

Figure 88: Electrical resistivity tomography – H2_ERT_a

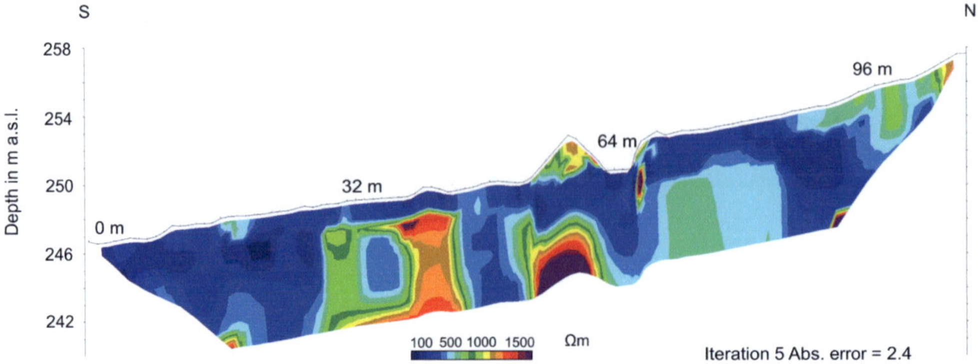

Source: own illustration

Within H2_ERT_a (Figure 88), several areas of low resistivity values of a few hundred Ωm are visible. The low-resistive area between meters 0 and 28 and from meter 44 to meter 52 is more than 5 m thick. However, this area has a smaller thickness of about 1.5 m northward until meter 44. The other parts have higher resistivity values up to more than 1.5 kΩm. At meter 60, located close to the surface, resistivity values over 1.0 kΩm occur.

H2_ERT_a description after the guidelines

With the information of the boundary analysis added (Figure 89) to the resistivity distribution, two layers appear (Table 19).

Figure 89: Boundary analysis – H2_ERT_a: a) sharpness of r-layer boundary including the position of r-layers and r-anomalies; b) orientation of the boundary gradient

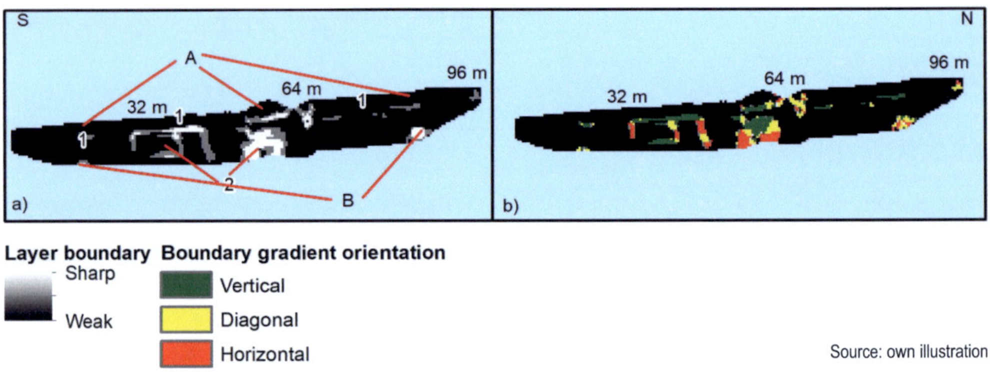

Source: own illustration

The top layer's (1) thickness varies massively, and the bottom layer (2) only occurs in some positions. Gaps are visible between different sections of this layer. The boundary orientation between these two layers varies and is mostly vertical or horizontal. The internal homogeneity of the two layers varies and a few boundaries are visible. The layering between these two layers is developed to a medium extent. A few anomalies are located in the top layer. On the one hand, exposed anomalies (A) in the wall structures are separated with a sharp boundary. On the other hand, small blocky and quasi-layer-formed anomalies with weak boundaries occur at the bottom. Additional parts of the approximated bottom layer need to be classified as an anomaly (B) surrounded by sharp boundaries.

Table 19: ERT guidelines – H2_ERT_a

R-layer	*Homogeneity*	*R-layer thickness*	*Layering*	*Horizontal changes*	*Boundary of r-layers*	*R-layer orientation*
R-layer 1	3	Mostly 1	3	2	2	Varying
R-layer 2	3	1–2				
R-anomaly	*Number of r-anomalies*	*R-anomaly location*	*R-anomaly form*	*R-anomaly size*	*R-anomaly boundary*	*Boundary orientation of r-anomalies*
R-anomaly Type A	2–3	1	1–2, 4	2	2–5	2
R-anomaly Type B	1	2–3	1	1–3	5	1

Source: own research

H2_ERT_b resistivity distribution

H2_ERT_b is located at a leveled area in the southern part of the small creek. It intersects with the creek and starts and finishes at the valley slopes.

Figure 90: Electrical resistivity tomography – H2_ERT_b

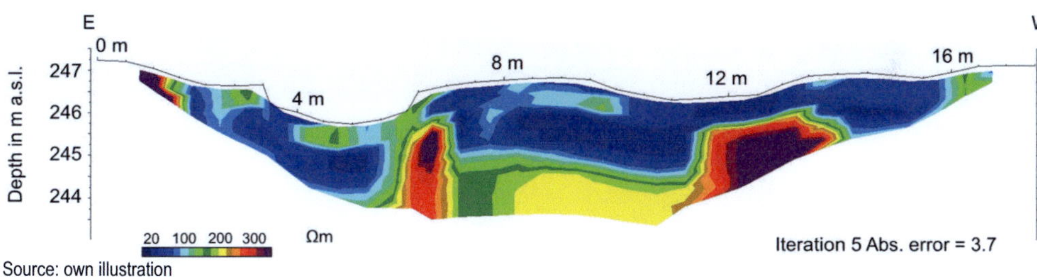

Source: own illustration

H2_ERT_b (Figure 90) displays a large low-resistive area with resistivity values mainly under 100 Ωm and thicknesses between 0.5 and 2 m. It covers the top of an area with varying but generally higher resistivity values in various forms. Within the top part, numerous lenses with higher resistivity values occur. Especially within the western part of the profile, the resistivity values of the bottom layer increase up to over 300 Ωm.

H2_ERT_b description after the guidelines

Combining the variable resistivity distribution with the boundary analysis (Figure 91), two r-layers are distinguished (Table 20).

Figure 91: Boundary analysis – H2_ERT_b: a) sharpness of r-layer boundary including the position of r-layers and r-anomalies; b) orientation of the boundary gradient

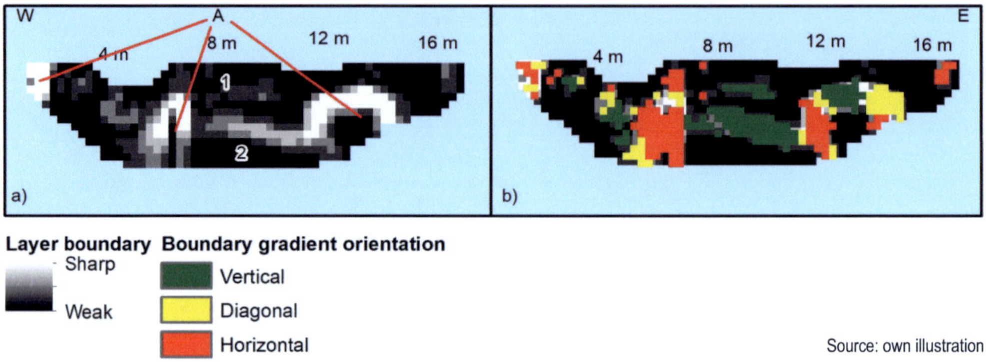

Source: own illustration

The uppermost r-layer's (1) thickness varies, and the boundary is medium-to-sharp. At the boundary between the top and the bottom r-layer (2), high-resistive lenses at 7 and 13 meters are classified as r-anomalies (A). The boundary between these two r-layers is mainly vertical; only at the two r-anomalies does this varies. The top r-layer's homogeneity varies, and many boundaries appear without being precisely classified as r-anomalies.

Table 20: ERT guidelines – H2_ERT_b

R-layer	*Homogeneity*	*R-layer thickness*	*Layering*	*Horizontal changes*	*Boundary of r-layers*	*R-layer orientation*
R-layer 1	3	Mostly 1	3	2	2	Varying
R-layer 2	3	1–2				
R-anomaly	*Number of r-anomalies*	*R-anomaly location*	*R-anomaly form*	*R-anomaly size*	*R-anomaly boundary*	*Boundary orientation of r-anomalies*
R-anomaly Type A	2–3	1	1–2, 4	2	2–5	2
R-anomaly Type B	1	2–3	1	1–3	5	1

Source: own research

4.2.2.2 Invasive methods and outcrops

The results within this subsection invasive methods and outcrops are associated with workflow Step 4 (S4) – *Ground-truth*.

Borehole and AMS ^{14}C-datings

The borehole H2_BH_a and -b are located at the leveled area north of the castle hill, near the creek. The laboratory results of the AMS ^{14}C-datings are attached in Appendix 7.

Figure 92: Boreholes and ^{14}C-ages – H2: a) H2_BH_a, H2_AGE_a, H2_AGE_b; b) H2_BH_b, H2_AGE_c for details see also Appendix 7)

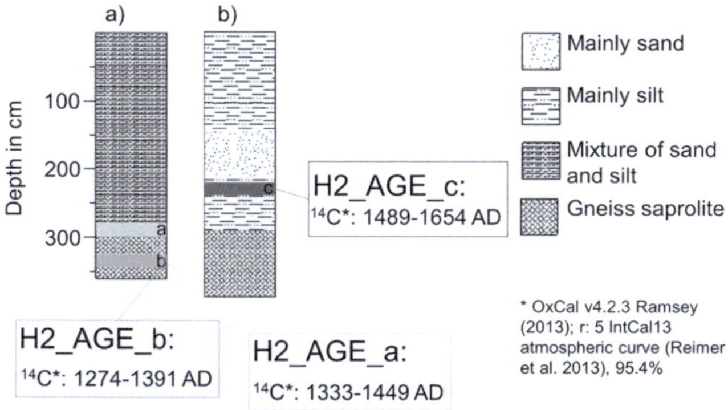

Source: own illustration

Up to a depth of 295 cm, the material within H2_BH_a (Figure 92a) consists of sandy loam, without any significant clay content. At the uppermost and the bottom of this layer, the silt content increases. To a depth of 115 cm, the oxidation features dominate the reduction ones. Underneath this depth, the reduction increases while the oxidation decreases. At 160 cm, the coarse particles > 2 mm increase by 50% while the coarse particles. However, at the bottom of this layer the number of coarse particles decreases to about 5%. In this area, the root content rises. One of these roots delivers the sample for the age-analysis H2_AGE_a (Figure 92a) at a depth of about 295 cm. The sample can be dated to 1333-1449 AD, which is the late period of the castle's use or shortly thereafter (Rosmanitz *et al.*, 2019a).

Whereby the first few centimeters of the part with high root content are located within the silty and sandy layer, the main parts are located underneath this layer, where at 295 cm the saprolite of the gneiss crops out. Within this layer, there are a few less-weathered parts as well, and still, there is a high root content. The last roots are located at a depth of about 333 cm. At this depth, the sample for H2_AGE_b (Figure 92a) was collected. This sample can be dated to the time of castle use, with 1274-1392 AD as well.

Even though H2_BH_b (Figure 92b) was located close to the position of H2_BH_a to prove its results, H2_BH_b differs a great deal from the other profile. While there was a more or less unsorted mixture of sand and silt in the upper layer of H2_BH_a, the single layers of H2_BH_b are far more sorted and segmented. The first 145 cm consists of almost pure silt. Within the silt, a layering is visible due to different colors, especially with varying darkness. The single bands or packages differ a great deal in size, but only solely in texture. In this layer, oxidation and reduction features alternate. This implies not only the influence of the groundwater surrounding the creek; it also implies that a couple of dense layers cause stagnic conditions above this layer. Underneath this layer, the silt content decreases during the first 15 cm, and after this, pure sand occurs. This layer ends at a depth of 210 cm, whereabouts another pure silt layer occurs. This layer is split into two parts. The upper 30 cm are dominated by reduction, whereby the bottom 50 cm of the layer is medium-heavy oxidized. In the reduced part of this layer, a few roots are located, which deliver the sample for H2_AGE_c (Figure 92b). This sample is dated after the use of the castle from 1489-1654 AD. At the bottom, almost 1 m of saprolite of weathered gneiss crops out.

Pürckhauer

Within the small valley of the described creek, several smooth leveled areas of different sizes are located. Some walls or heights of few meters in altitude subdivide these depressions. H2_PUE_a to H2_PUE_l are located in a small area within the first of these leveled areas, north of the castle hill. H2_PUE_m is located at the edge of the first leveled area close to the first height, while H2_PUE_n is directly located on top of it. H2_PUE_o, H2_PUE_q and H2_PUE_s are located on three different smaller leveled areas northward, whereby H2_PUE_p and H2_PUE_r are at the heights in between. H2_PUE_t is outside of the area specified with the heights and leveled areas, nearby the creek at a slope (for details, see Figure 87).

To a depth of 60 cm, within H2_PUE_a (Figure 93a), several different fluvial or limnic layers, including a high content of silt, are accumulated. Additionally, there are several roots and other organic material inside this layer. Underneath this layer, the position close to the creek causes gleyic conditions within a silt-rich layer. Within this layer a recognizable content of medium-sized sand occurs.

While the texture of the first layer of H2_PUE_b (Figure 93b) consists of coarse sand, the central layer is made of pure silt. The bottom layer consists of slightly loamy medium-sized sand. The whole profile comprises gleyic features.

The uppermost and bottom layers of H2_PUE_c (Figure 93c) are of loamy texture. Opposite to that, the other two layers are well-separated. The second layer consists of coarse sand which includes gravels. Underneath 30 cm of depth, pure silt occurs. While there is no indication of gleyic conditions within the first 19 cm, the other layers are influenced by changing groundwater levels and show reduction features underneath 30 cm.

In H2_PUE_d (Figure 93d), mostly well-segmented layers of different textures alternate. The first layer's texture consists of slightly loamy silt. Contrary, the second layer, to a depth of 37 cm, consists of fine sand. To a depth of 70 cm, the main grain-size is silt, while in the intermediate part, a recognizable amount of sand is visible. Underneath a small-sized, fine sand layer a pure silt layer is located.

Underneath the lost core in H2_PUE_e (Figure 93e), the main texture consists of silt. However, between 50 and 80 cm, a recognizable sand content exists, which increases bottomward. The entire profile shows gleyic conditions, starting at 50 cm, previously with reduction features.

To a depth of 74 cm, H2_PUE_f (Figure 93f) consists of sand-rich silt. Underneath, the texture consists of pure silt. The bottom layer shows reduction features.

Until a depth of 74 cm, sand is the main grain-size within H2_PUE_g (Figure 93g). These layers partly have high silt or gravel content, and the grain-size in between the sand fraction also varies.

Up to a depth of 70 cm, underneath the lost core in H2_PUE_h (Figure 93h), the texture consists of silt, including recognizable sand content. This layer is well-segmented to a well-sorted coarse sand layer on top of a well-sorted pure silt layer.

The top layer of H2_PUE_i (Figure 93i) consists of sandy silt. Underneath this layer, there are well-segmented and sorted sandy and silty layers.

Figure 93: Pürckhauer soundings – H2: a) H2_PUE_a; b) H2_PUE_b; c) H2_PUE_c; d) H2_PUE_d; e) H2_PUE_e; f) H2_PUE_f; g) H2_PUE_g; h) H2_PUE_h; i) H2_PUE_i; j) H2_PUE_j; k) H2_PUE_k; l) H2_PUE_l; m) H2_PUE_m; n) H2_PUE_n; o) H2_PUE_o; p) H2_PUE_p; q) H2_PUE_q; r) H2_PUE_r; s) H2_PUE_s; t) H2_PUE_t

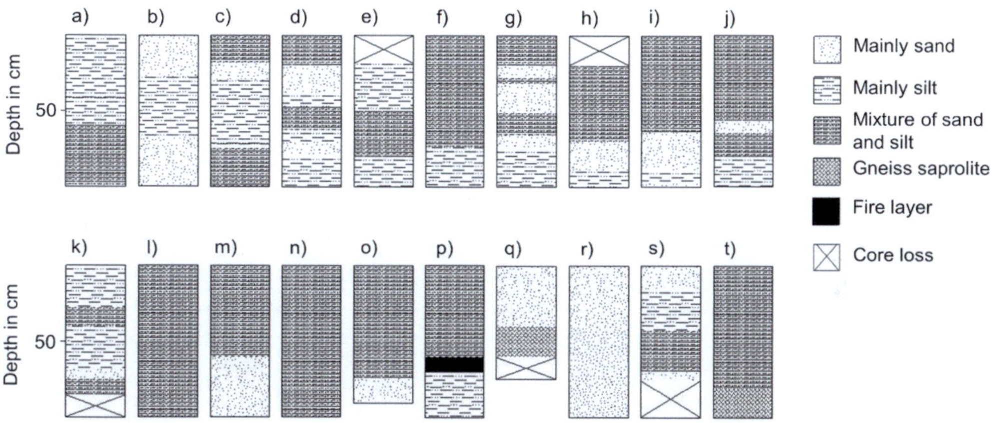

Source: own illustration

Within H2_PUE_k (Figure 93k), layers of different silt and sand content alternate. These layers are partly well-segmented, with mainly sharp but also diffuse boundaries. Due to the influence of the groundwater, the bottom sample was lost. In the remaining bottom layers, the changing groundwater level causes oxidation and partial reduction of the material.

H2_PUE_l (Figure 93l) mainly consists of sandy loam or silt loam with different sand or silt content, respectively. However, in all layers, significant clay content could be found.

To a depth of 75 cm within H2_PUE_m (Figure 93m), silt and sand-rich layers alternate. These layers have a higher silt content above 40 cm and a higher sand content at the bottom part. Underneath 75 cm, the grain-size is solely sand with small gravels. Underneath a depth of 40 cm, oxidation and reduction features alternate.

Due to its location on top of a wall, H2_PUE_n (Figure 93n) is not influenced by groundwater and has no spatially distributed oxidation or reduction features. The substrate is mixed, and no clear layering is visible. Nevertheless, the silt and sand content varies within the profile. No gravel or grus can be found in the entire profile.

H2_PUE_o (Figure 93o) has more silt than sand within the first layer. The sand content increases to a depth of 75 cm. Here, there are several lenses of sand, which originate from weathering grus in the matrix. Underneath, up to a depth of 90 cm, a sand layer occurs, which entirely consists of the weathering product of grus. Underneath this layer, the sample is lost.

Comparable to H2_PUE_n, H2_PUE_p (Figure 93p) is also located on top of a wall. Opposite to the other profile, the layers are very well-segmented. Above a depth of 30 cm, silt dominates the sand; however, the content of sand increases to a depth of 60 cm. In this part, the number of grus of gravel size rises. Between 60 and 70 cm, a black-colored layer with a high amount of charcoal within the matrix appears and can be classified as a fire layer. This fire layer has a texture made of loamy sand. The bottom layer has a sharp boundary with the fire layer and consists of pure silt. A lamination within the silt layers is visible.

Until a depth of 40 cm, in H2_PUE_q (Figure 93q), the matrix is loamy sand, without clay content. Underneath this layer, a pure sand layer of weathered saprolite occurs. The bottom of this layer is lost. Below a depth of 75 cm, no further penetration into the saprolite was possible.

Within the last wall structure, at H2_PUE_r (Figure 93r), the entire texture consists of loamy sand without any clay content. The content of sand and silt varies slightly but without any visible systematic variation.

In H2_PUE_s (Figure 93s), some alternating layers of sand and silt in the top and weathered grus in the bottom occur. In the first 15 cm, sand is the dominant grain-size, while the second layer, up to a depth of 42 cm, consists of almost pure silt. Underneath this layer, the weathering products and the remains of the grus are mixed in a silty matrix among this layer. The bottom layer consists of sand and gravel according to grus or saprolite of the gneiss. Due to the high water content, the bottom of the profile was lost.

H2_PUE_t (Figure 93t) is located outside the leveled areas and the wall structures. Underneath the topsoil made of silt loam to a depth of 15 cm, hillside loam is located. This loam usually has no further layering and is a mixture of loess deposits and the bedrock's weathering products. Underneath this layer, the saprolite of gneiss is located.

After the WRB classification system, the developed soils in the leveled areas are classified as fluvisols, partly with gleyic features in the bottom part. Outside the leveled areas and on top of the wall structures, the developed soils are classified as regosols.

In comparison, the BKA classification system would classify the soils as paternias or regosols, respectively.

Outcrop

The outcrops near *Hauenstein Castle* are located in the bed of the creek. Near the castle and further downslope, the creek bed consists of fluvial sediments in silt-, sand-, gravel- and stone-size, as H2_OC_a shows (Figure 94a). Starting at the location of H2_OC_b, the regime changes from accumulation to erosion and the gneiss bedrock crops out (Figure 94b). Here, sediments were accumulated only at positions of low fluvial energy. The position of photograph b is about two meters upslope from the position of H1_PUE_p.

Figure 94: Field pictures – outcrops H2: a) H2_OC_a Creek bed with fluvial deposits at the bottom b) H2_OC_b Creek bed with bedrock at the bottom

Source: own pictures

4.2.3 H2 – Development of the talweg: interpretation

The interpretation of the data was performed according to workflow Step 5 (S5) – *Interpretation of the data.*

Synthesis

Three different lithological units can be distinguished within the talweg. At the top, some r-anomalies occur with medium-high resistivity values, separated with various boundaries at exposed positions or as blocks, or "quasi-layer." These r-anomalies are associated with anthropogenic debris of various compositions illustrated in the Pürckhauer soundings.

Below these r-anomalies, a low-resistive r-layer of silt-rich deposits occur. The existence of this r-layer is affirmed by various invasive methods. The varying sharpness of boundaries in different orientations separates it from the bottom lithological unit. Caused by the admixture of different sand and grus content and different moisture, this r-layer's homogeneity is medium.

At the bottom, a high-resistive r-layer exists in fragments. Only small r-anomalies partly occur. This r-layer is medium-well separated from the one above. Due to its position, and the gneiss saprolite above, visible in some of the invasive methods, it is interpreted as the gneiss of the bedrock. Different weathering statuses cause high conductivity, especially of the mica (Reynolds, 2011). Therefore, parts of the saprolite might have resistivity values similar to those of the silt-rich deposit and difficult to be distinguished from them.

Interpretation

How has the northern talweg been developed?

The two ERT transects at H2 (Figures 88 and 90) display several low-resistive areas. Within these areas, especially the southern part, silt is accumulated. Within the northern part, sand is mixed into the silty matrix, and partly pure sand and even weathered grus layers are visible. In some of the boreholes (Figure 92) and Pürckhauer soundings (Figure 93), these layers are laminated. Most of the boundaries between the different substrates are very sharp; in other cases, the material of different textures was accumulated homogenously. All these layers could be interpreted as the sediments accumulated within several ponds. Due to its small sizes and an inflow and outflow, there was no opportunity to accumulate clay; additionally, the sorting of the material varies due to different intensities of water flow and frequency of changing the intensity of the flow. The archaeological interpretation of the castle as a mill (Rosmanitz *et al.*, 2019a) underlines this interpretation, caused by the small-sized creek mill ponds seem to be necessary. However, a pond monk was found in the archaeological record at the bottom pond's edge. Whenever the mill needed higher water flow, some water of the pond was drained, and the creek had, therefore, increased energy in the water at these times, which allows only coarser-sized grains to accumulate.

The slopes are terraced; therefore, a higher erosion rate due to agricultural usage can be expected. Because of that, a mixture of loamy slope deposits and the weathered gneiss was eroded at the slopes and accumulated within the ponds, too. These deposits were of different textures, which can explain the partly high content of sand or even grus within a silty matrix. Additional ponds were often used for fish farming carps (e.g. Schneider, 2014). These animals dig over the grounds of silent waters for foraging and cause bioturbation. While the boreholes show a depth of up to 3 m of this low-energy fluvial system sediments at the southern pond, the northern ones had smaller depths, as described in the ERT data and indicated by the outcrops (Figure 94). The depth of the sediment bodies within the northern ones was partly underneath 1 m.

Whenever the maximum depth of the bodies was reached, saprolite could be found in the sample. Saprolite is usually in-situ weathered material in which the layering or foliation of the bedrock is still comprehensible. The Pürckhauer soundings -a to -l (Figure 93) show high variations of different sediments within a small area and indicate the complexity of this sedimentation regime.

AMS-^{14}C-dates (Figure 92) of wood and roots in the boreholes date back to the late phase of the mill usage in the Late Middle Ages (see also laboratory report in Appendix 7). This indicates that the ponds needed to be maintained and dug out. This work suddenly stops after the destruction of the mill in the early 15th century (Rosmanitz *et al.*, 2019a). Such necessary works could explain the walls within the area. Whenever it was necessary, the ponds were dug out, and these sediments were put at the end of the single ponds. Therefore, in most of these walls, the material is homogenous, and mainly no sediments greater than sand or gravel could be found. Only the second wall is better segmented and has a fire layer on top of limnic silty sediment. The material within these wall structures indicates that they acted as a dump for the silty and sandy material which accumulated within the ponds. This material is unlikely to be used to build wall structures without sediments of stone size or anything comparable with it.

From the position of the second wall within the creek bed, the bedrock is located. The bedrock is only slightly weathered. Due to its density, most likely, there were

Figure 95: Conceptual model – H2 (south-north orientated)

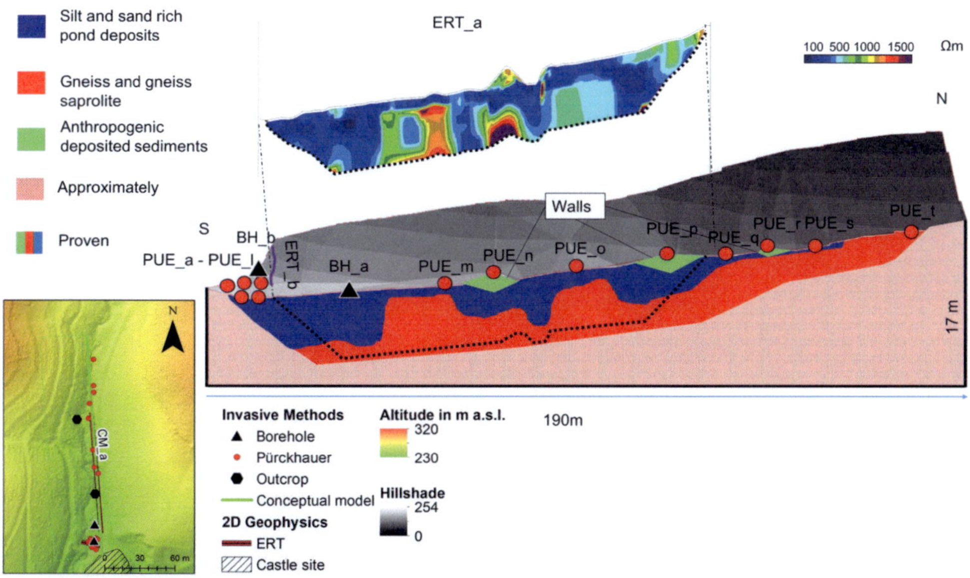

Sources: own illustration, based on Bavarian Agency for Digitalisation, High-Speed Internet and Surveying, 2014

no further occasions to build deeper ponds in this area without a labor-intensive deepening of the less-weathered bedrock.

As a combination of the described data, the conceptual model (Figure 95) illustrates differently sized sediment bodies belong to a variable sedimentation regime expectable in a pond. It seems that there were several ponds at different time spans, partly divided by wall structures between them; the wall structures were partly constructed subsequently. The grain-size distribution within the wall structures indicates a maintaining of the ponds and a relocation of the material in such structures. At the bottom of the ponds, the gneiss's saprolite was located, which is the bedrock within the investigation area.

According to the question, the combination of various-sized fluvial and limnic sediment bodies in a continuously flowing creek and the creek bed on top of the bedrock outside the anthropogenic usage indicates a massive anthropogenic impact on the creation of the sedimentation within the talweg. Due to the ^{14}C-dates, at least parts of the sediments accumulated within the talweg can clearly be associated with the mill's medieval usage.

4.2.4 H3 – Center of the castle hill: results

Is the center of the castle hill made of bedrock?

To investigate the origin of the castle hill, an ERT profile with 3 m spacing was conducted across the hill, and a borehole was drilled within the middle of the hill (cf. Figure 96).

Figure 96: Location of measurements – H3

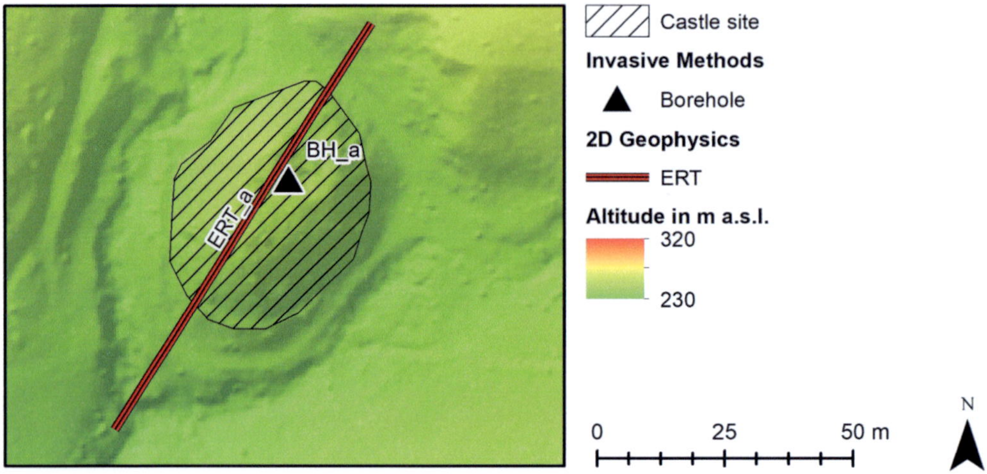

Source: Bavarian Agency for Digitalisation, High-Speed Internet and Surveying, 2014
Map creation: J. Trappe

4.2.4.1 Geophysical data

The geophysical data of H3 are associated with workflow Step 3 (S3) – *Geophysical methods.*

ERT

First, the inverted resistivity distribution of the profiles is described. Then, their relative distribution is discussed according to the guidelines for ERT description.

H3_ERT_a resistivity distribution

H3_ERT_a starts at the slope, northeast of the castle hill, and finishes at the creek, southwest of the hill.

Figure 97: Electrical resistivity tomography – H3_ERT_a

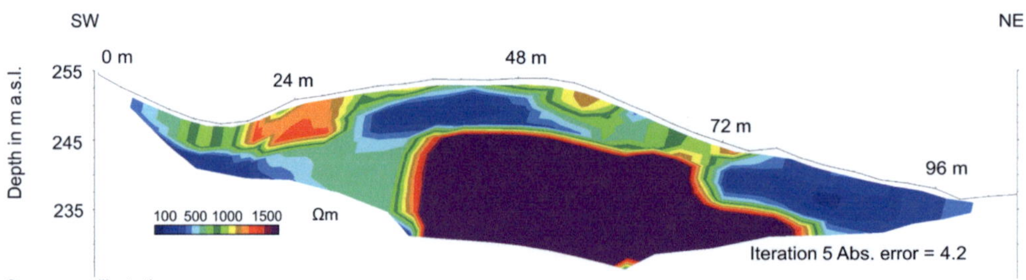

Source: own illustration

H3_ERT_a (Figure 97) illustrates a high-resistive area with resistivity values over 1.5 kΩm at the bottom, building the center of the castle hill. On top of this area, lower resistivity values between 100 and 500 Ωm occur. Within this part, the resistivity values vary strongly, and several lenses with high resistivity values can be found in H3_ERT_a.

H3_ERT_a description after the guidelines

With the addition of the information of the boundary analysis (Figure 98) to the resistivity distribution, three r-layers can be distinguished (Table 21).

Figure 98: Boundary analysis – H3_ERT_a: a) sharpness of r-layer boundary including the position of r-layers and r-anomalies; b) orientation of the boundary gradient

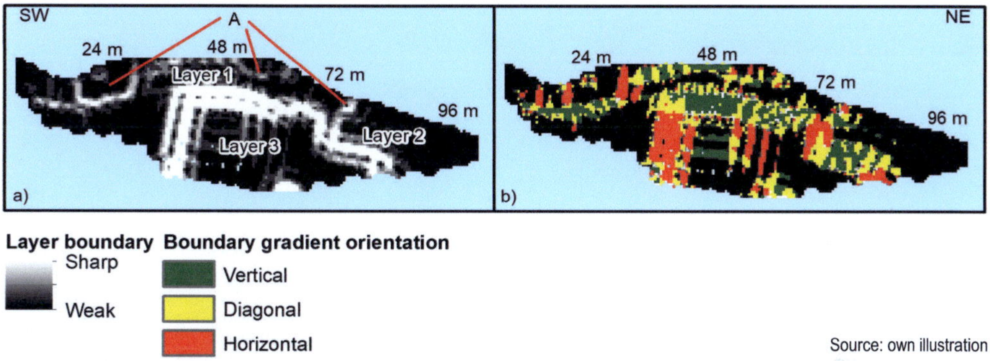

Source: own illustration

The top part shows two r-layers, separated at about meter 75. The top, southwestern (1) r-layer is mostly heterogeneous, with several boundaries within. In contrast, the northeastern r-layer (2) is mostly homogeneous, including only very few weak boundaries. The layering between the two top r-layers is not well-developed. The top r-layers are separated from the bottom r-layer (3) with a sharp, vertically orientated boundary at the southern part and are more or less surface-parallel at the northern part. The bottom part is medium homogeneous. The layering between the two top r-layers and the bottom r-layer is well-developed. The top r-layers are heterogeneous and include several r-anomalies (A), located directly below the ground surface and surrounded by a medium-weak boundary.

Table 21: ERT guidelines – H3_ERT_a

R-layer	*Homogeneity*	*R-layer thickness*	*Layering*	*Horizontal changes*	*Boundary of r-layers*	*R-layer orientation*
R-layer 1	2	2	2	2	3	5
R-layer 2 (split horizontal)	4	2	-	-	-	-
R-layer 3	3	3	5	2	5	8: 36: 6 37–80 m: 7
R-anomaly	*Number of r-anomalies*	*R-anomaly location*	*R-anomaly form*	*R-anomaly size*	*R-anomaly boundary*	*Boundary orientation of r-anomalies*
R-anomaly Type A	3	1	1	1–2	3	2

Source: own research

4.2.4.2 Invasive methods and outcrops

The results within this subsection invasive methods and outcrops are associated with workflow Step 4 (S4) – *Ground-truth.*

Borehole

H3_BH_a is located almost in the middle of the castle hill, at a position where penetration through the first meters was possible. In most other positions, modern walls (Rosamnitz et al. 2019) and remaining or destructed parts of the castle prevent penetration into the ground.

Figure 99: Borehole – H3_BH_a

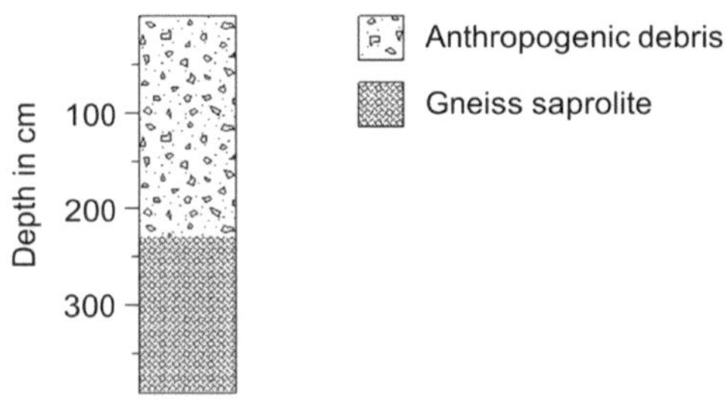

Source: own illustration

Within the uppermost 230 cm of H3_BH_a (Figure 99), anthropogenic debris is accumulated. The debris is loosely compacted and enriched with organic materials. The grain-size mainly consists of sand, but also, many fragments of grus are mixed in this matrix. The grus partly consists of the surrounding gneiss but is mainly made of Buntsandstein fragments. From a depth of 230 cm, the anthropogenic fragments disappear. Here, many mica and quartz grains are visible. At this position, in-situ weathering of the gneiss started isovolumetric, and it is described as saprolite. In deeper areas of the profile, partly hydromorphic features are visible within the saprolite. Within the bottom 50 cm, the foliation structure of the gneiss saprolite is still visible.

4.2.5 H3 – Center of the castle hill: interpretation

The interpretation of the data was performed according to workflow Step 5 (S5) – *Interpretation of the data.*

Synthesis

On top, different areas of various resistivity values occur (Figure 97). These areas correlate with anthropogenic debris and wall structures, visible in the borehole (Figure 99) and at the hill's surface (These wall structures are further investigated in H4.1). Further north, the low-resistive area is most likely divided into a top layer made of fluvial deposits and a bottom part made of saprolite.

The low-resistive areas in the first layer consist of gneiss saprolite, as proven in the borehole and indicated by a comparison with literature values (Reynolds, 2011), which describe a similar decrease by the weathering of comparable materials.

H3_ERT_a displays a high-resistive bottom part, with lower resistivity values on top. In combination with the saprolite within the borehole H3_BH_a, the high-resistive bottom part can be interpreted as gneiss.

Interpretation

Is the center of the castle hill made of bedrock?

By transferring this information into a conceptual model (Figure 100), it is displayed that the castle hill consists of gneiss. In several parts, the saprolite of the gneiss is visible on top of the bedrock, and much anthropogenic debris exists at the uppermost, including fragments of manmade building structures, which are investigated in the following subsection.

Figure 100: Conceptual model – H3

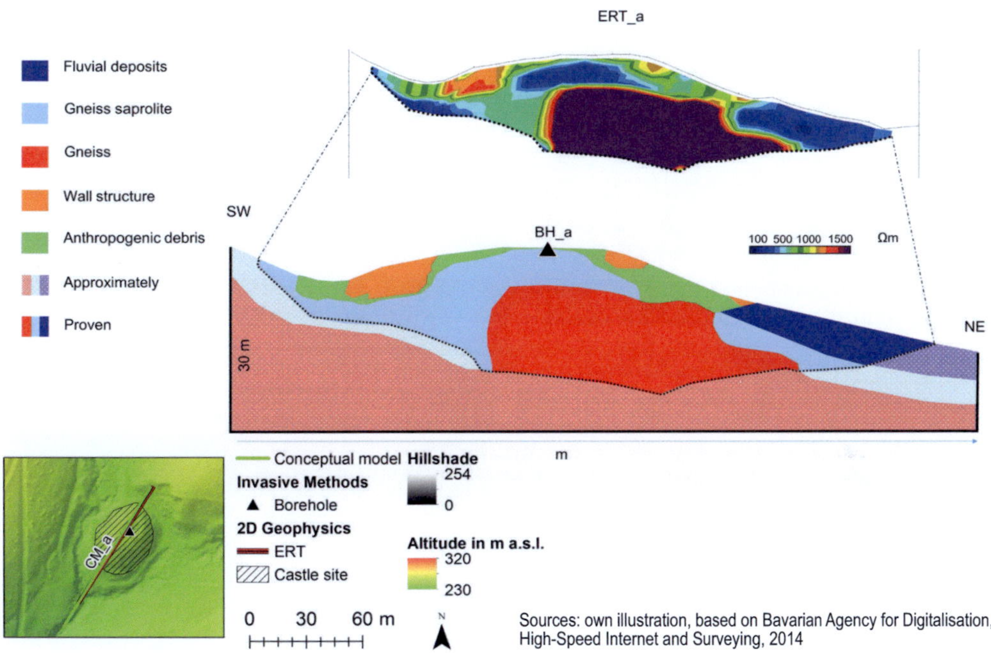

Sources: own illustration, based on Bavarian Agency for Digitalisation, High-Speed Internet and Surveying, 2014

4.2.6 H4.1 – Building structures on-site: results

Do buried and as-yet-unknown building structures exist on the castle site?

For the prospection of buried wall structures, three ERT profiles, with 0.5 m spacings, were positioned on top of the castle hill (cf. Figure 101). Spatial surveys or other geophysical techniques could not be performed at this position due to different new-aged wall structures and the rough topography on top of the castle hill. Information of H3_ERT_a and the archaeological excavation was combined with the results of the geophysical surveys.

Figure 101: Location of measurements – H4.1

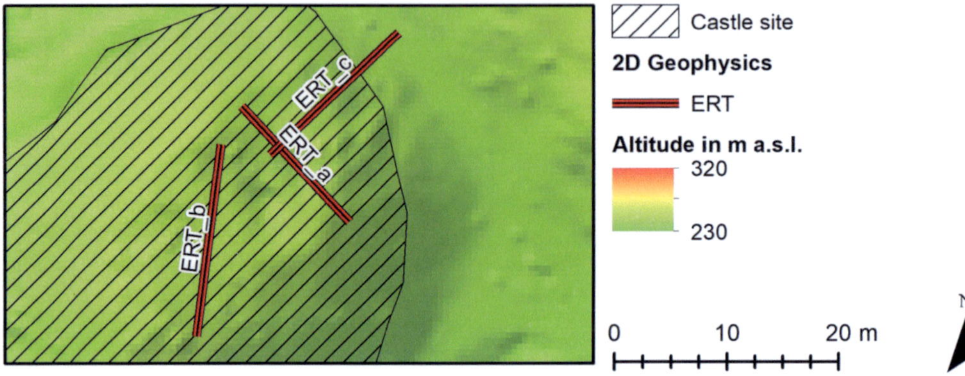

Source: Bavarian Agency for Digitalisation, High-Speed Internet and Surveying, 2014
Map creation: J. Trappe

4.2.1.1 Geophysical data

The geophysical data of H4.1 are associated with workflow Step 3 (S3) – *Geophysical methods.*

ERT

First, the inverted resistivity distribution of the profiles is described. Then, their relative distribution is discussed according to the guidelines for ERT description.

H4.1_ERT_a resistivity distribution

H4.1_ERT_a is located fully on top of the castle hill and crosses some depressional structures.

Underneath a low-resistive area of about 1 m depth with resistivity values of a few 100 Ωm, resistivity values increase to about 1.0 kΩm in H4.1_ERT_a (Figure 102). Within the matrix of these higher resistivity values, high-resistive lenses are located. Especially in the northwestern part, resistivity values increase to 1.5 kΩm.

Figure 102: Electrical resistivity tomography – H4.1_ERT_a

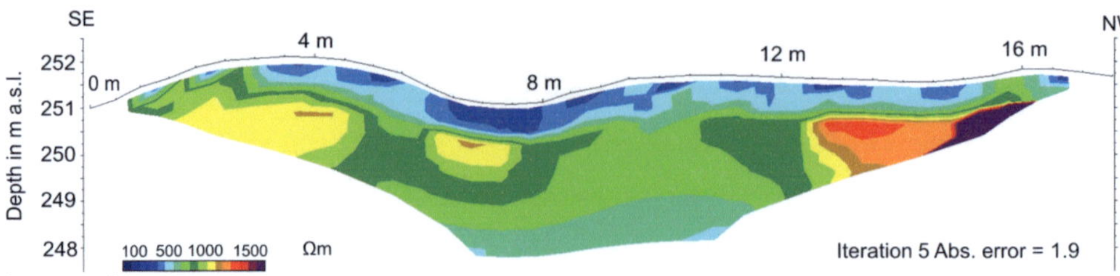

Source: own illustration

H4.1_ERT_a description after the guidelines

With the combination of the resistivity distribution and the boundary analysis (Figure 103), two r-layers are visible (Table 22).

Figure 103: Boundary analysis – H4.1_ERT_a: a) sharpness of r-layer boundary including the position of r-layers and r-anomalies; b) orientation of the boundary gradient

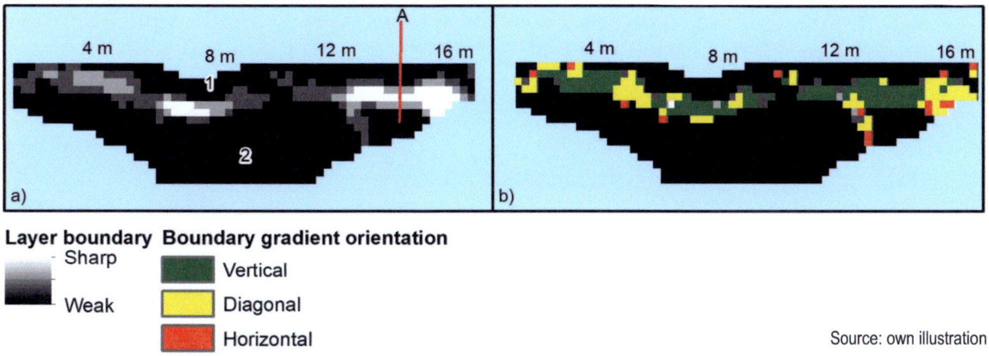

Source: own illustration

The boundary separating the top (1) from the bottom (2) r-layer follows the surface topography and varies in intensity. The top r-layer is medium homogeneous and the bottom r-layer can be classified as more homogeneous. An r-anomaly (A) is located at the northwestern part, having a sharp boundary to the top r-layer and a weak one to the bottom r-layer.

Table 22: ERT guidelines – H4.1_ERT_a

R-layer	*Homogeneity*	*R-layer thickness*	*Layering*	*Horizontal changes*	*Boundary of r-layers*	*R-layer orientation*
R-layer 1	3	1	3	1	2	7
R-layer 2	4	1	-	-	-	-
R-anomaly	*Number of r-anomalies*	*R-anomaly location*	*R-anomaly form*	*R-anomaly size*	*R-anomaly boundary*	*Boundary orientation of r-anomalies*
R-anomaly Type A	1	1	1	1	3	1

Source: own research

Figure 104: Electric resistivity tomography – H4.1_ERT_b

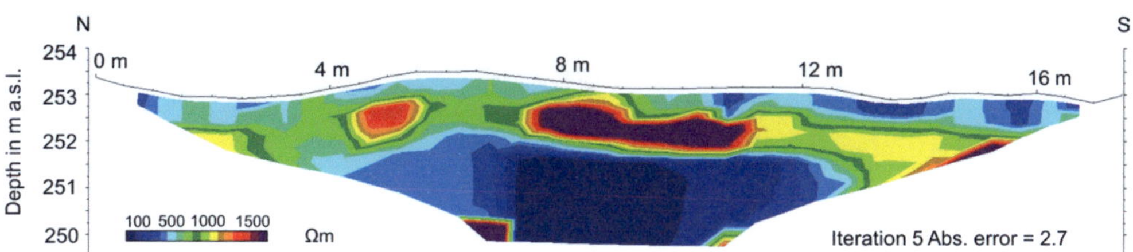

Source: own illustration

H4.1_ERT_b resistivity distribution

H4.1_ERT_b is located entirely on top of the castle hill as well. Its position is further southeast, compared to H4.1_ERT_a.

In the top part of H4.1_ERT_b (Figure 104), low resistivity values occur. Underneath this part, a higher-resistive area with values around 1.0 kΩm is located. Within this area, high-resistive lenses of over 1.5 kΩm are embedded. In contrast to H4.1_ERT_a, underneath a depth of about 2 m, the resistivity decreases to values between 100 and 300 Ωm. Here, some high-resistive lenses occur at the bottom.

H4.1_ERT_b description after the guidelines

With the information of the boundary analysis (Figure 105) added to the resistivity distribution, three r-layers become visible (Table 23).

Figure 105: Boundary analysis – H4.1_ERT_b: a) sharpness of r-layer boundary including the position of r-layers and r-anomalies; b) orientation of the boundary gradient

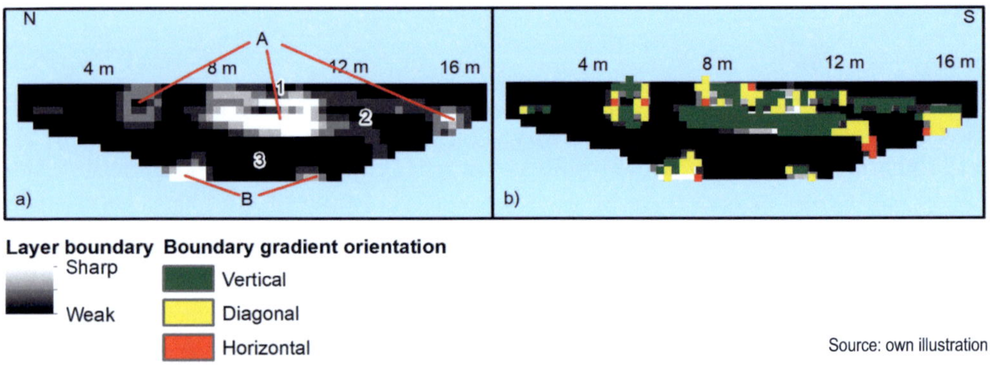

Source: own illustration

These r-layers are separated by weak-to-medium-weak boundaries. The uppermost r-layer (1) and the central (2) r-layer have thicknesses of about 1 m, whereas the thickness of the bottom r-layer (3) is larger. The layering is developed to a medium extent, and small boundaries are visible in all r-layers. The orientation of the boundaries broadly follows the surface topography. However, especially in the southern part of the profile, horizontal changes are visible. Within the second r-layer, several r-anomalies (A) of different sizes occur. These r-anomalies are surrounded by medium-to-sharp boundaries. The outlines of these anomalies are classified as blocky to linear. At the bottom of the profile, a few high-resistive r-anomalies (B) with a sharp boundary at the top exist as well. Those can be most likely associated with inversion artifacts.

H4.1_ERT_c resistivity distribution

Contrary to H4.1_ERT_a and -b, H4.1_ERT_c starts in the neck ditch and only finishes on top of the castle hill.

In H4.1_ERT_c (Figure 106), a low-resistive area covers an area with higher resistivity values analog to H4.1_ERT_a and H4.1_ERT_b. At the southwestern part, the resistivity values of the bottom area increase substantially.

Table 23: ERT guidelines – H4.1_ERT_b

R-layer	*Homogeneity*	*R-layer thickness*	*Layering*	*Horizontal changes*	*Boundary of r-layers*	*R-layer orientation*
R-layer 1	3	1	4	1	1–2	7
R-layer 2	3	1	4	1	2	7
R-layer 3	5	1	-	-	-	-
R-anomaly	*Number of r-anomalies*	*R-anomaly location*	*R-anomaly form*	*R-anomaly size*	*R-anomaly boundary*	*Boundary orientation of r-anomalies*
R-anomaly Type A	2	1	1, 3	1–2	3, 5	1
R-anomaly Type B	2	2	2	1	5	2

Source: own research

Figure 106: Electrical resistivity tomography – H4.1_ERT_c

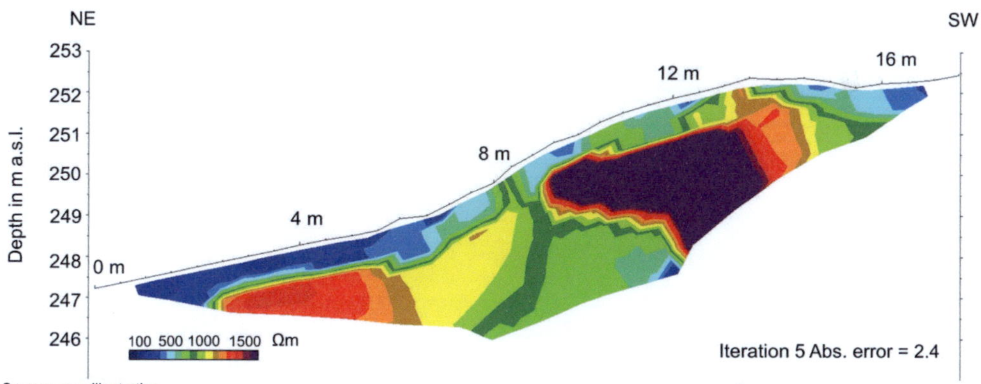

Source: own illustration

Figure 107: Boundary analysis – H4.1_ERT_c: a) sharpness of r-layer boundary including the position of r-layers and r-anomalies; b) orientation of the boundary gradient

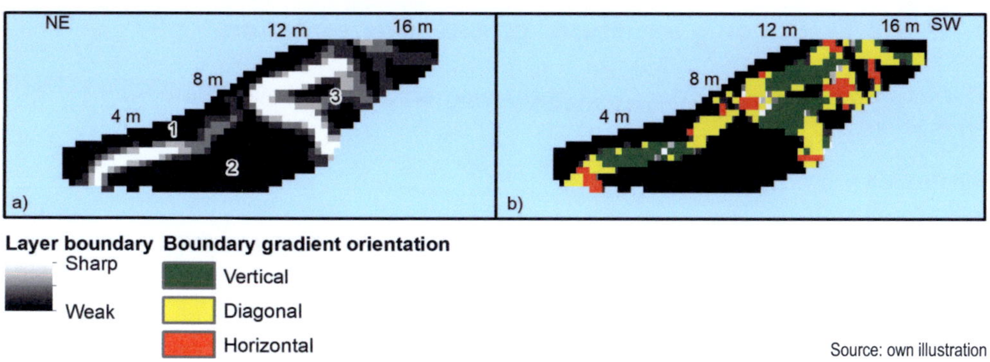

Source: own illustration

H4.1_ERT_c description after the guidelines

With a combination of the resistivity distribution and the boundary analysis (Figure 107), three r-layers become visible (Table 24).

The low-resistive top (1) r-layer has a sharp boundary with the bottom two. The course of the boundary follows the surface. The top r-layer is more or less homogeneous; only a few boundaries occur. Horizontal resistivity changes split the bottom r-layers. The southwestern (2) r-layer is characterized by a sharp boundary at the top and in the north, by having a medium homogeneity and by including some boundaries. The other bottom (3) r-layer has lower resistivity values and is homogeneous. No r-anomalies are visible within the entire profile.

Table 24: ERT guidelines – H4.1_ERT_c

R-layer	*Homogeneity*	*R-layer thickness*	*Layering*	*Horizontal changes*	*Boundary of r-layers*	*R-layer orientation*
R-layer 1	4	1	4	0	5	7
R-layer 2	5	1	5	2	5	3
R-layer 3 (split horizontal)	4	1	-	-	-	-
R-anomaly	*Number of r-anomalies*	*R-anomaly location*	*R-anomaly form*	*R-anomaly size*	*R-anomaly boundary*	*Boundary orientation of r-anomalies*
R-anomaly Type A	0	-	-	-	-	-

Source: own research

4.2.1.1 Invasive methods and outcrops

The study site was excavated during the measurements. Therefore, no invasive methods could be performed to address the question of W4.1, and no outcrops can be described for this sub-subsection. Nevertheless, the excavated wall structures were documented for archaeological reasons and can be interpreted in combination with the collected geophysical data. Therefore, this data can be associated with workflow Step 4 (S4) – *Ground-truth*.

4.2.7 H4.1 – Building structures on-site: interpretation

The interpretation of the data was performed according to workflow Step 5 (S5) – *Interpretation of the data.*

Synthesis

In all ERT profiles of H4.1, high-resistive r-anomalies and small r-layers are visible. These high-resistive parts are often surrounded by sharp boundaries. Besides these r-anomalies, the homogeneity of the surrounding r-layer is medium. With a combination of the position of these r-anomalies in a few meters of depth and the depth information of H3, it is visible that these high-resistive areas are located above the bedrock within a layer of cultural debris. Therefore, it can be assumed that the high-resistive parts represent wall fragments, which consist of the local gneiss (Rosmanitz *et al.*, 2019a).

The resistivity values of this debris are comparable to those of H3. However, the low-resistive bottom r-layers of H4.1_ERT_b (Figure 104) and those in -c (Figure 106) are associated with the saprolite of the gneiss, displayed in H3_ERT_a.

Interpretation

Do buried and as-yet-unknown building structures exist on the castle site?

The high-resistive areas interpreted as wall structures are mainly located at the edge of the hill, but also at some areas in the center. The largest of these areas, visible in H3_ERT_a (Figure 97), is at the northeastern edge of the hill, above the neck ditch. Additional parts of the walls are collapsed into the neck ditch, visible in H4.1_ERT_c (Figure 106). In extension to the most prominent excavated wall structures southwest of the castle hill, H3_ERT_a indicates the existence of an additional part of this structure. Nevertheless, it seems to be smaller than the other parts.

Figure 108: Interpretation – H4.1

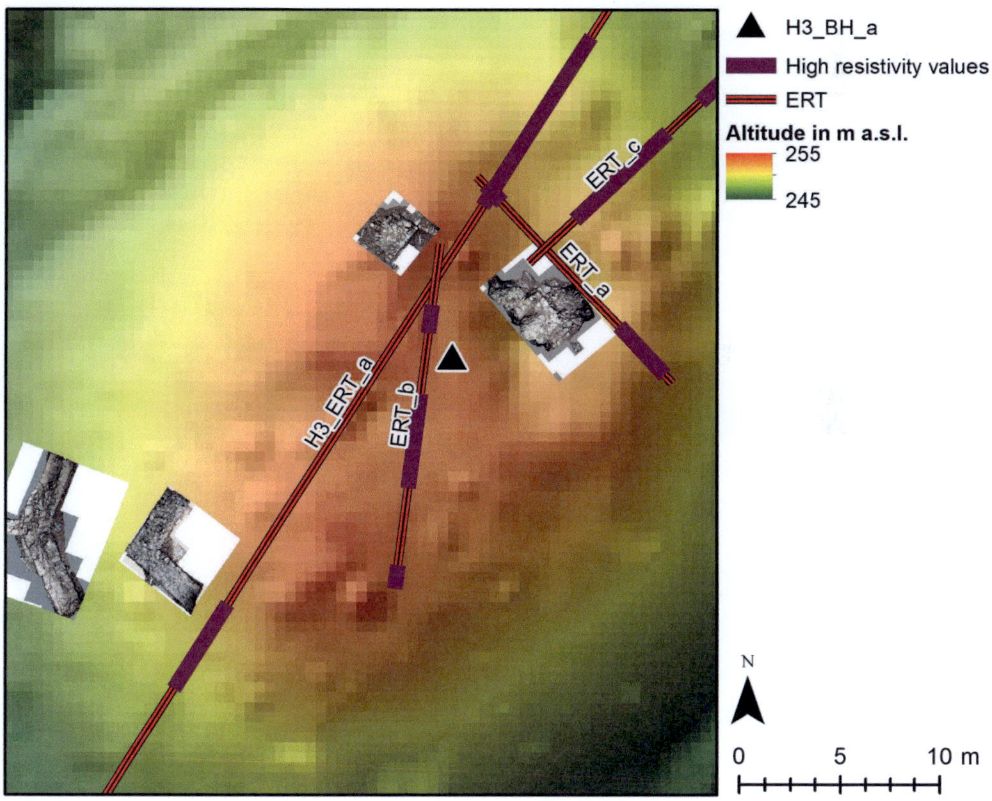

Source: Bavarian Agency for Digitalisation, High-Speed Internet and Surveying, 2014
Map creation: J. Trappe

According to the question of H4.1, several more wall structures are expected on top of the castle hill (Figure 108). In particular, the high-resistive resistivity values highlighted in Figure 108 may be such wall structures. After the destruction of the castle, several fragments were removed and are not located in-situ anymore. Thus, it is difficult to distinguish which parts are in-situ structures and which parts are relocated fragments. However, results indicate promising positions for further excavations at the described positions.

4.2.8 H4.2 – Outer bailey: results

Where is the location of the outer bailey?

On a meadow northeast of the castle site, 3D GPR and magnetic surveying were performed to find the outer bailey (cf. Figure 109). Additionally, a borehole was drilled for ground-truth information.

Figure 109: Location of measurements – H4.2

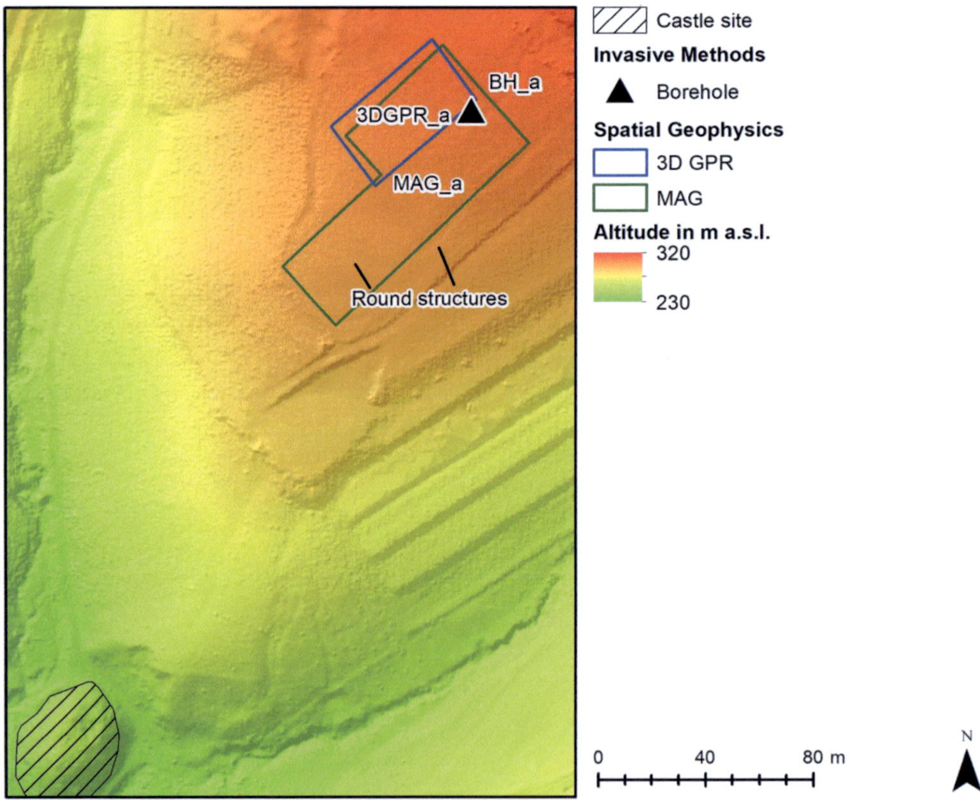

Source: Bavarian Agency for Digitalisation, High-Speed Internet and Surveying, 2014
Map creation: J. Trappe

4.2.3.1 Geophysical data

The geophysical data of H4.2 are associated with workflow Step 3 (S3) – *Geophysical methods.*

3D GPR

The 3D GPR grid was performed on the northwestern part of the meadow.

Figure 110: 3D ground-penetrating radar measurement – H4.2_3DGPR_a

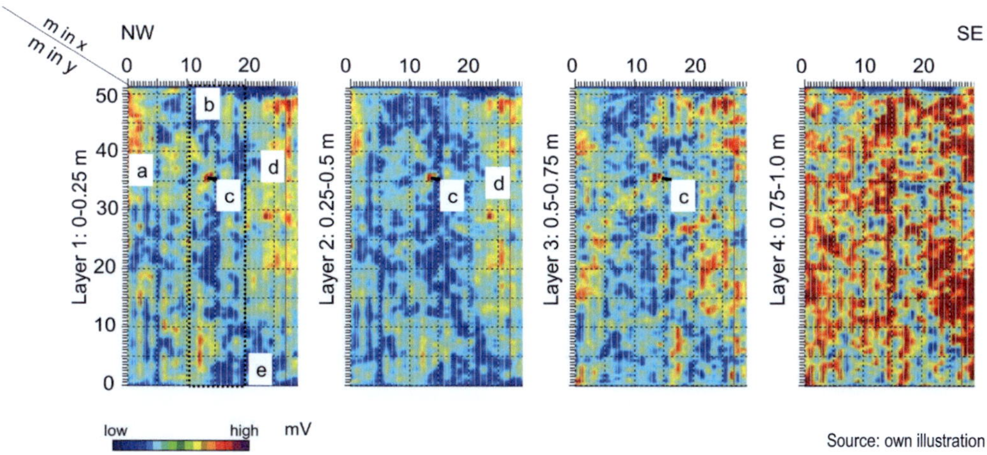

Source: own illustration

Within H4.2_3DGPR_a (Figure 110), several changes in the dielectric constant indicate human activity. In the uppermost layer (0-0.25 m depth) within the first 10 m along the x-axis, several small reflectors are visible without a distinct cluster. Only in the first meters along the x-axis and between 40 and 50 m along the y-axis, one large or a cluster of reflectors is visible (a). Between 10 and 20 m along the x-axis, reflection amplitude diminishes within this layer (b). However, at 35 m along the y-axis and 15 m along the x-axis between 0 and 0.75 m depth, a strong and constant reflector appears (c). This reflector passes through all layers. Further southeast, starting at 20 m along the x-axis, linear reflectors are visible without forming a clear structure (d). Solely in the first 10 m along the y-axis, no reflections are visible (e). The southeastern reflections (d) endure into the second layer (0.25-0.5 m depth), and in the northwestern part (a), reflections get weaker. Some of these reflections endure even further into the third layer (0.5-0.75 m depth). Here, other reflections occur diffusely, and no clear clustering is visible anymore. An area-wide reflector is visible, starting at Layer 4 (0.75-1.0 m depth).

Magnetic survey

The magnetic survey at H4.2 also covered the northwestern part of the meadow but comprised further areas in the southeastern area as well.

Magnetogram H4.2_MAG_a (Figure 111) illustrates an almost east-west striking structure. This structure is most visible in the eastern and northwestern parts,

Figure 111: Magnetic survey – H4.2_MAG_a

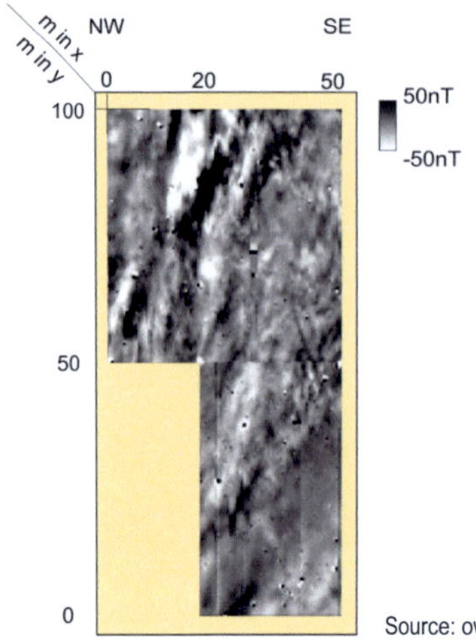

Source: own illustration

whereas it is less visible in the southeastern and southern parts. However, the highest dipole intensity is at around 20 m along the x-axis and 80 m along the y-axis. Further small dipole anomalies are visible in the entire magnetogram.

4.2.3.2 Invasive methods and outcrops

The results within this subsection invasive methods and outcrops are associated with workflow Step 4 (S4) – *Ground-truth.*

Borehole

The borehole H4.2_BH_a (Figure 112), located in the middle of the meadow, displays only a thin layer of soil development. Typical structures of land use appear only in the uppermost 20 cm. Underneath this depth, the material consists of in-situ weathered saprolite of gneiss.

Figure 112: Borehole – H4.2_BH_a

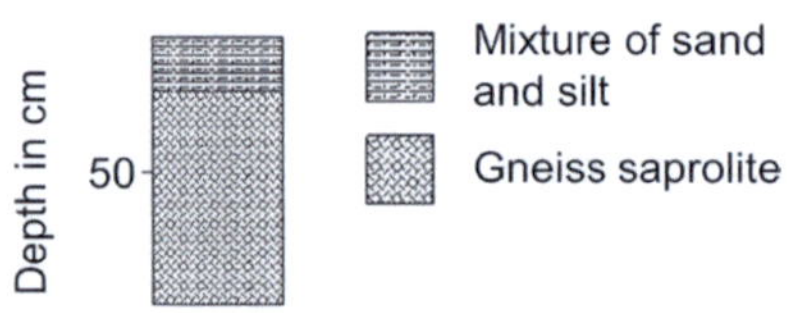

Source: own illustration

4.2.9 H4.2 – Outer bailey: interpretation

The interpretation of the data was performed according to workflow Step 5 (S5) – *Interpretation of the data.*

Synthesis and interpretation

Where is the location of the outer bailey?

The round structures, visible within the slope and the hillshade (Figure 7), point to anthropogenic building structures or at least a usage of this area. Additionally, magnetic surveying (Figure 111) shows structures, which indicate earthworks, indicated by the remnant magnetization, and the GPR (Figure 110) results display many reflections of different sizes and shapes, indicating a possible anthropogenic usage of the area. However, as the borehole displays (Figure 112), only the uppermost centimeters provide an indication of anthropogenic usage of this area. In deeper areas, a natural development of the subsurface is expected. The course of the boundary between soil and bedrock can be detected with GPR, and results indicate a location between 0.5 and 1 m in the third and fourth depth-layer. However, compared to the borehole, this might be slightly too deep. Most likely, this position is already within the saprolite layer, showing a decrease in weathering, causing the change in dielectric constant and thus the reflection of the signal.

As the meadows are used as couples at the moment, the magnetic anomalies are most likely caused by recent agricultural usage and result from a mixture of the developed soils, not by medieval building structures. The data lacks any indication of an outer bailey.

4.2.10 Quantification of the medieval, anthropogenic influence in Hauenstein

This subsection is associated with workflow Step 6 (S6) – *Quantification of the anthropogenic influence.*

First, all results and interpretations for Hauenstein are briefly summarized, and the medieval, anthropogenic influence on the relief development is estimated according to Section 3.5.

H2 – Development of the talweg

ERT measurements and invasive methods display several former ponds located in the northern creek valley at Hauenstein. In combination with ^{14}C-datings, the sediments within the ponds can clearly be associated with the period of castle use, including a mill at Hauenstein. Therefore, the sediments' entire accumulation regime in the talweg can be classified as (Category 3) massively, passively affected (cf. Figure 113). The construction of the ponds and the construction of the wall structures in between them have also affected the development of the talweg actively too. However, some of this construction had not yet been dated, so this work focuses on the passive influence.

H3 – Center of the castle hill

The ERT measurement, combined with a borehole, displays that the castle hill is made of local bedrock. Of course, this bedrock was prepared, as, for example, neck ditches were dug or building structures were constructed on top. However, local legend suggested that the hill is made of a collapsed tower (Rosmanitz *et al.*, 2019a). In contrast, the bedrock indicates an important impact of the relief for the construction of the castle. Therefore, the center of the castle hill can be classified as Category 1, but, of course, the uppermost 2 m can be classified as (Category 4) intensely active influenced by the building structures (cf. H4.1, Figure 113).

H4.1 – Building structures on-site

ERT measurements on top of the castle hill, in combination with the results of the excavation, illustrate various building structures on top of the castle hill. Clear differentiation in in-situ and collapsed structures cannot be performed with these results. Anyway, the structures are a massive impact (Category 4) on relief development (cf. Figure 113).

H4.2 – Outer bailey

Magnetic surveying, in combination with a GPR grid and a borehole, was performed to search for the outer bailey. Unfortunately, the geophysical data indicate hints of the usage of these meadows, but as the borehole indicates, this use is most likely linked to recent agricultural use. Therefore, the area was most likely of natural origin and not in usage during the medieval period, thus classified as Category 1 (cf. Figure 113).

Figure 113: Intensity and type of medieval, anthropogenic influence – Hauenstein

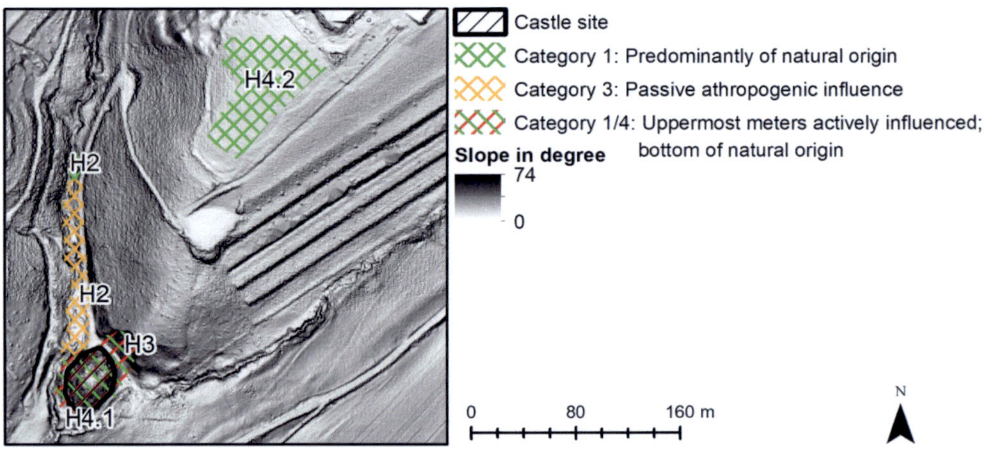

Source: Bavarian Agency for Digitalisation, High-Speed Internet and Surveying, 2014
Map creation: J. Trappe

4.3 Mömbris

First, a site-specific question was derived with the help of the GIS-based terrain analyses at Mömbris. The results for this question are provided and interpreted. Then, an estimation of quantification of the anthropogenic influence is made. A technical overview of the used application of the geophysical methods is given in Appendix 8 to Appendix 10.

4.3.1 Morphometric analysis for the creation of site-specific questions

The morphometric analyses for creating site-specific questions are associated with workflow Step 2 (S2) – *GIS-based terrain analyses*.

Within the expected castle site of Mömbris, a plateau of higher altitude values is located (Figure 114). The plateau is surrounded by a steep slope of about 2 to 3 m of altitude. This plateau is surrounded by a ditch, followed by an embankment, ending in an old wall of the graveyard in the southwestern direction. In the northwestern area, the castle site is limited by a leveled area, less than 10 m in size, finishing at a graveyard's current fence. In the northeastern area, a building of the 20th century CE is located, and in the southeastern direction nowadays, there is a garden of a modern building. Because of the heavily embossed site at Mömbris, only the building structures are investigated.

Figure 114: DEM of the castle site in Mömbris

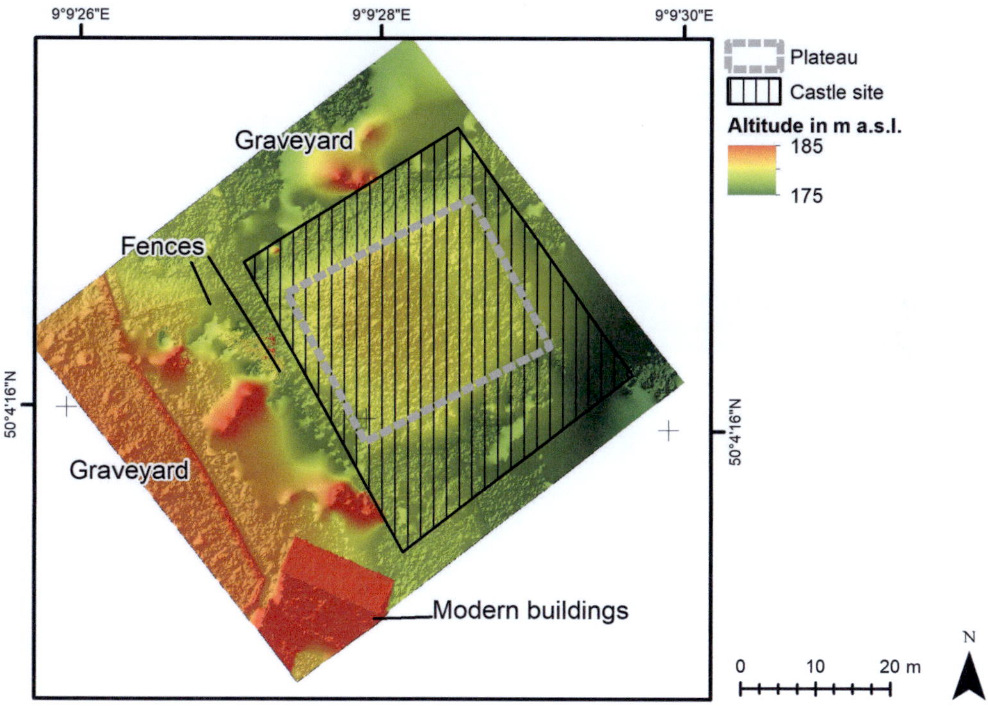

Source: DEM, calculated via structure from motion: C. Büdel, 2019
Map creation: J. Trappe

With respect to the plateau at the study site, the following question is derived.

Question M4.1 – buried, anthropogenic features – on-site:

Where are the building structures of the castle?

A rectangle plateau near the graveyard at Mömbris is generally associated with the former castle hill's position. Combined analyses of different geophysical data should help to identify wall structures within the plateau and its surroundings.

4.3.2 M4.1 – Building structures: results

Where are the building structures of the castle?

According to the expected building structures on top and beside the castle hill, 3D GPR, 3D ERT and magnetic surveying were performed on the top and beside the castle hill to answer the question. Additionally, a borehole was drilled on top of the castle hill (Figure 115).

Figure 115: Location of measurements – M4.1

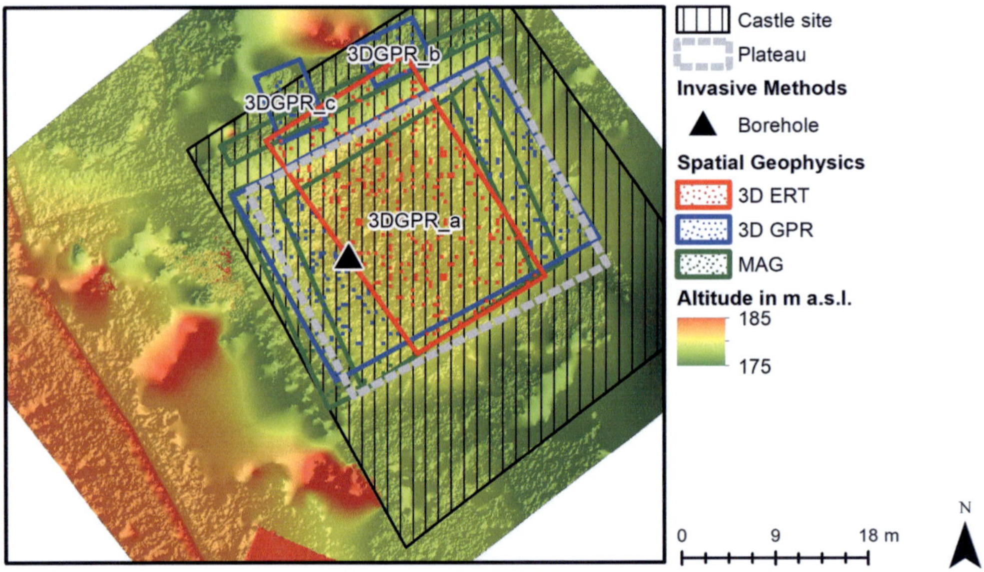

Source: DEM, calculated via structure from motion: C. Büdel, 2019
Map creation: J. Trappe

4.3.2.1 Geophysical data

The geophysical data of M4.1 are associated with workflow Step 3 (S3) – *Geophysical methods*.

3D ERT

M4.1_3DERT_a is located on top and in the northwestern direction of the plateau.

Figure 116: 3D electrical resistivity tomography – M4.1_ERT3D_a

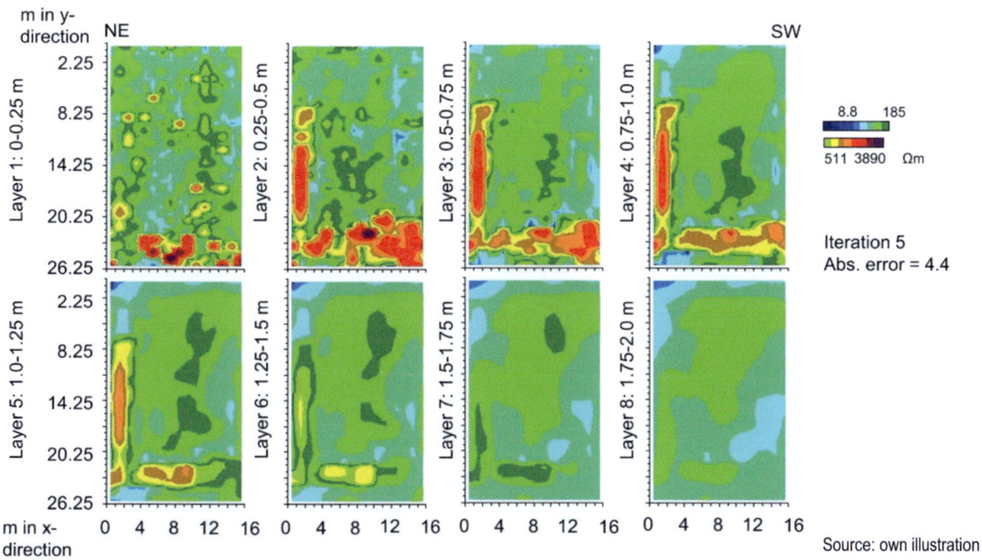

Source: own illustration

Figure 117: 3D model – M4.1_ERT3D_a

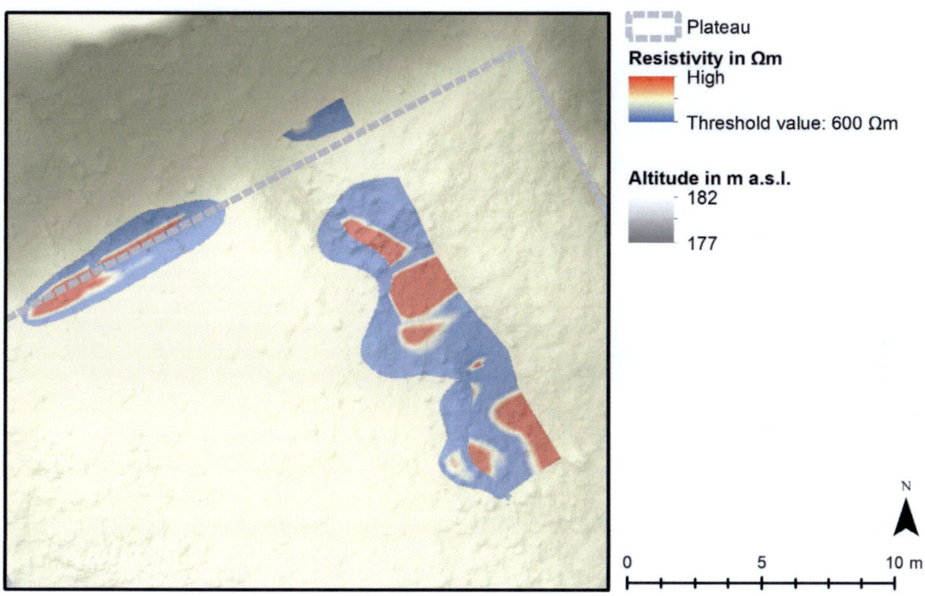

Source: DEM, calculated via structure from motion: C. Büdel, 2019
Map creation: J. Trappe

3D ERT resistivity distribution

The results of M4.1_3DERT_a (Figure 116) illustrate, in the first two layers, different high-resistive anomalies with resistivity values of up to 4,000 Ωm diffusely spread over a matrix with about 100 Ωm.

However, starting in the second layer, the anomalies get more and more sorted and build an L-shaped structure. The extent of the structure is through the entire length along the x-axis and between 8 and 22 m along the y-axis. Underneath 50 cm, the structure is entirely visible and disappears between 1.5 and 2 m of depth.

3D display ERT

With the transfer of the data into three dimensions, the use of a threshold value of 600 Ωm and the combination of it with the DEM, the positions of the L-shaped structure becomes visible (Figure 117). This threshold value was chosen to show those values that differ from the surrounding matrix. Whereas the northern anomaly is located precisely at the plateau's sharp edge, the southern one is located within the plateau orthogonally to it.

3D GPR

M4.1_3DGPR_a has its location completely on top of the plateau, which was initially suggested as being the castle hill. M4.1_3DGPR_b and -c are located underneath the plateau, close to the border of the graveyard.

Figure 118: 3D ground-penetrating radar measurement – M4.1_3DGPR_a

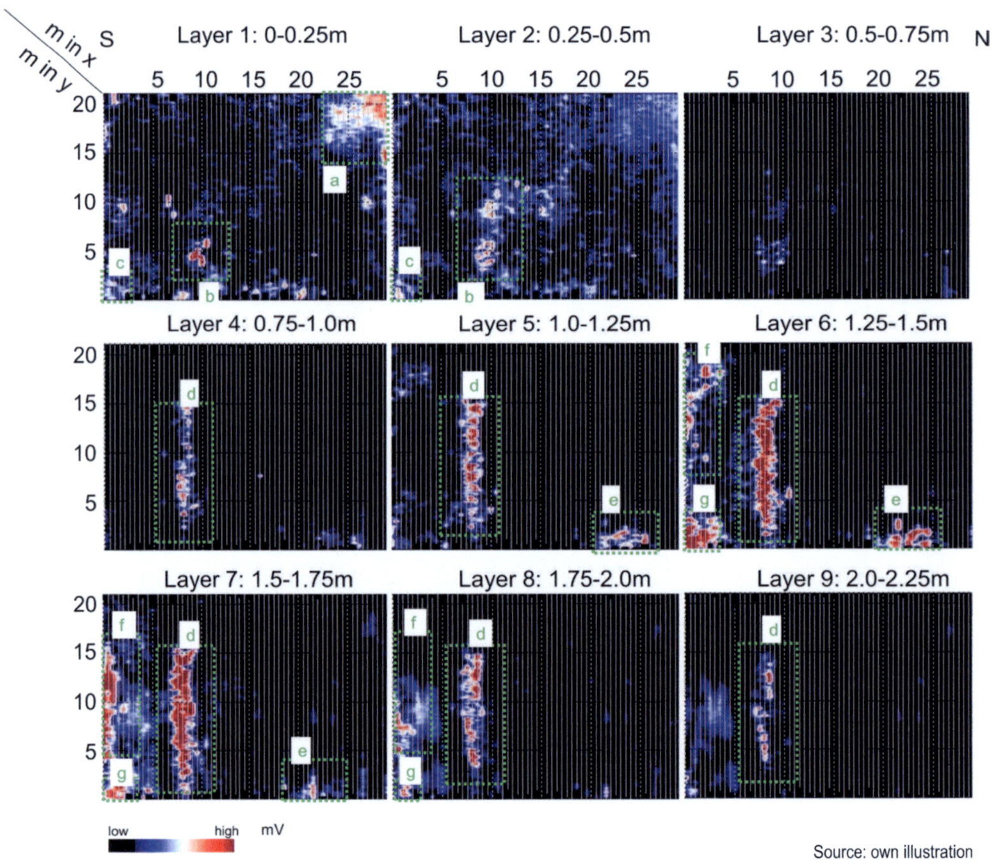

Source: own illustration

M4.1_3DGPR_a

Within the first 50 cm of M4.1_3DGPR_a (Figure 118), several small reflectors are visible. Additionally, within the northwestern corner, a strong polygonal-shaped bunch with high intensity of reflections (a) is located.

Starting in a depth of about 20 cm at 8–10 m along the x-axis and 2–12 m along the y-axis, other more or less continuous strong reflections (b) appear. This high intensity of reflections stops at 60–70 cm of depth. Another bunch of such reflections (c) occurs; it starts at a depth of 20 cm beginning at x and y 0 m and ending at x and y 2 m, at a depth of 60 cm. At the same position at a depth of 110 cm, another bunch with high intensity of reflections (g) appears with a maximum extent of up to 4 m in both directions. This anomaly starts to disappear at a depth of 190 cm and entirely disappears underneath 230 cm. West of this anomaly, several small and large reflections (f) appear and disappear at about a depth of 110 cm. This area of appearing and disappearing reflections ends at a depth of 230 cm. Starting at a depth of 100 cm, between meters 21 and 26 along the x-axis and between meters 0 and 3 along the y-axis, a strong signal (e) appears at a depth of almost 200 cm. The intensest bunch of reflection (d) of the radargram is located between 7 and 9 m along the x-axis. It starts almost exactly at 14.5 m at a depth of 80 cm and disappears at depths over 230 cm.

Figure 119: Lines of 3D ground-penetrating radar – M4.1: a) M4.1_3DGPR_a: Line 16; b) M4.1_3DGPR_a: Line 2; c) M4.1_3DGPR_a: Line 48

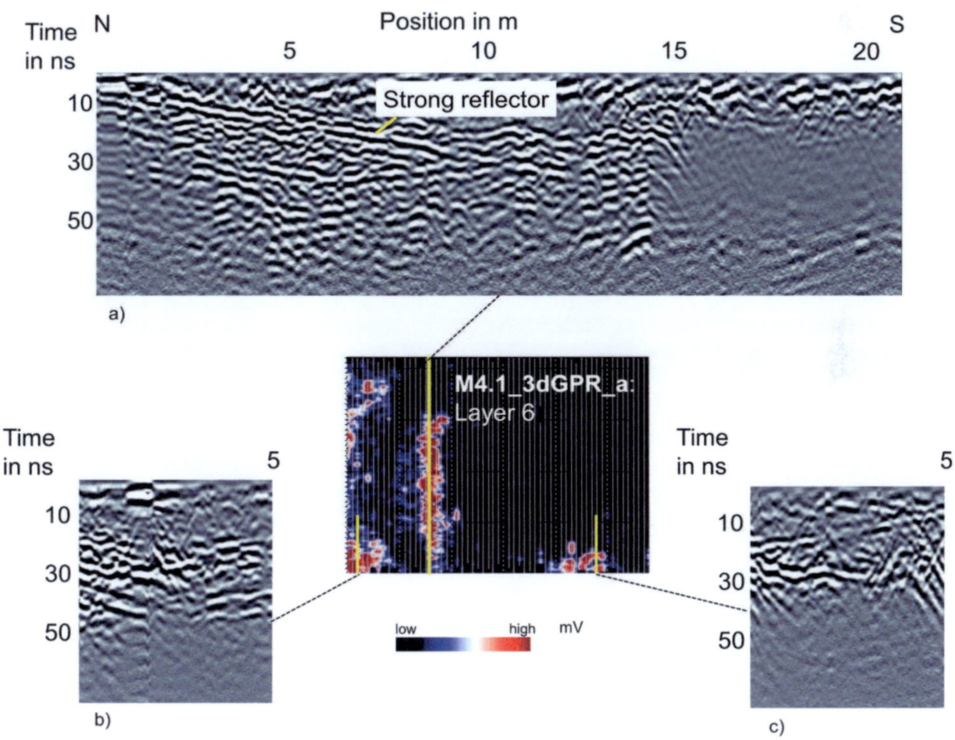

Source: own illustration

Within line 16 (Figure 119a), there is a strong reflector visible with several other reflectors diffusely underneath it. This reflector starts at meter 0 close to the surface and ends between 14 and 15 meters sharply. The inner structure of the reflector within anomaly d and f is comparable with the one of g (Figures 119b and c), but is less intensely developed.

However, with a description of the radargrams in line with Neal (2004), the upper and lower boundaries are classified more or less as concordant without being well-developed. The reflectors in line 16 are some kind of curved, sigmoidal, but also planar/even, such as the ones on line 2 and line 48. On all lines, a dipping in different angles can be seen. The relationship between the reflections is subparallel and partly oblique, and the reflection continuity is discontinuous in the southern part of line 16. Some of the reflectors in the northern part of line 16, line 2 and line 48 are continuous, others moderately continuous.

M4.1_3DGPR_b and -c

Underneath a couple of reflections within the first 60 cm of M4.1_3DGPR_b (Figure 120a), a continuous reflector between 2 and 4 m along the x-axis and through the entire grid along the y-axis appears and starts disappearing at a depth of 190 cm.

A strong, profile-long reflector is continuously detectable within the single lines of the first 3 m along the x-axis at a depth of 150 cm. Within M4.1_3DGPR_c (Figure 120b), there is as well a linear reflector visible at 2.5 m along the x-axis.

Figure 120: 3D ground-penetrating radar: a) M4.1_3DGPR_b; b) M4.1_3DGPR_c

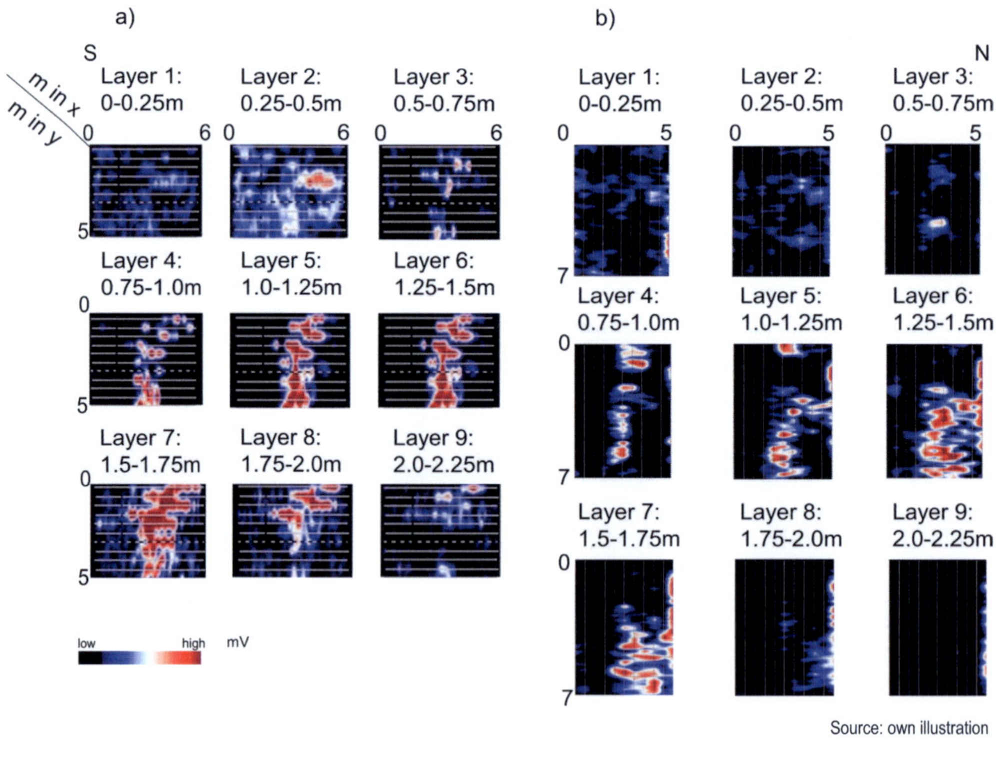

Source: own illustration

Figure 121: Lines of 3D ground-penetrating radar: a) M4.1_3DGPR_b: Line 10; b) M4.1_3DGPR_c: Line 5

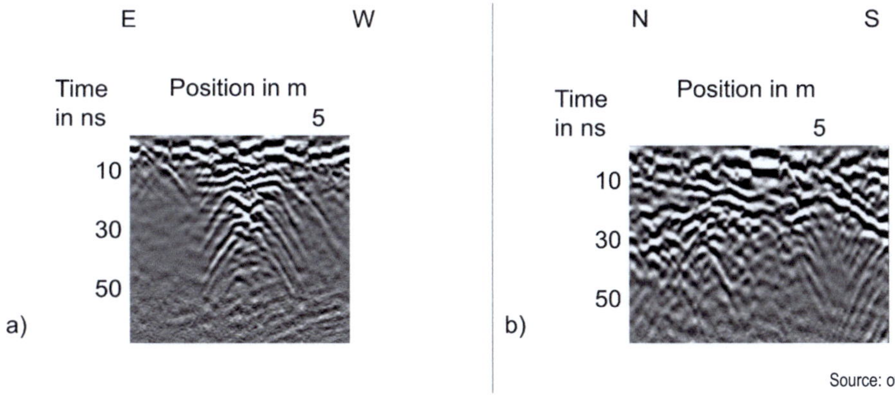

Source: own illustration

With descriptions of the radargrams in line with Neal (2004), the upper and lower boundaries are classified more or less as concordant, without being well-developed. The shape of reflectors in 3DGPR_b are planar/even with a horizontal dip, whereby the shape of -c cannot be classified but is dipping. The relationship between the reflections of -a is subparallel to parallel but in -c oblique. Both reflection continuities are classified as moderately continuous.

The strong reflector within M4.1_3DGPR_b crosses all lines of the profile. The inner structure looks rather the same as the one in M4.1_3DGPR_a line 16 does (Figure 121a), even though it detects the reflections in the orthogonal direction. In contrast, the inner structure of M4.1_3DGPR_c (Figure 121b) varies from the ones of M4.1_3DGPR_b and M4.1_3DGPR_a.

3D display GPR

With the transfer of this data into a 3D model (Figure 122), combined with the DEM, it becomes visible that some of the reflectors are located at unexpected positions on top of the plateau (a) or beside the plateau (b). The reflector d of M4.1_3DGPR_a especially is arranged orthogonally to the edges of the plateau and placed in the middle of it. However, the southern end is still located about 5 m away from the next edge (not shown in Figure 122). In contrast, the other described reflectors within M4.1_3DGPR_a are located at the edges or corners of the plateau. Furthermore, the reflector within M4.1_3DGPR_b is located down the plateau, orientated about orthogonally away from it. Remarkably, this reflector is not in the extension of the reflector g of M4.1_3D_GPR_b.

Magnetic survey

M4.1_MAG_a is subdivided into several parts. Whereas there are small sections surrounding the plateau to the north- and southeast, the most extended part is located on top of the plateau.

The result of the fluxgate magnetic (Figure 123) displays huge negative anomalies at the western part of the magnetogram. On the eastern part of the magneto-

Figure 122: 3D model –: M4.1_3DGPR_a and -b

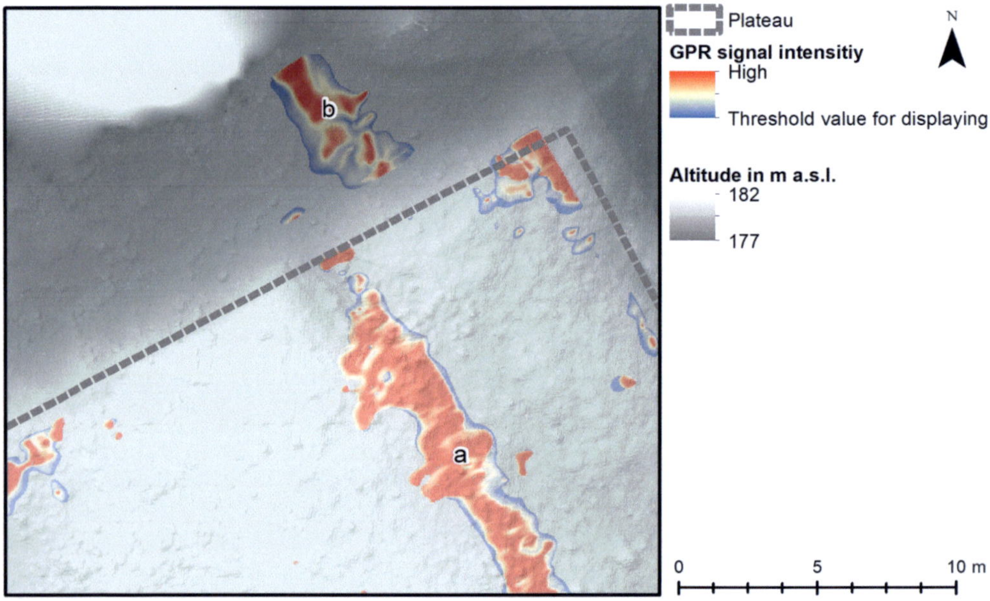

Source: DEM, calculated via structure from motion: C. Büdel, 2019
Map creation: J. Trappe

Figure 123: Magnetic survey – M4.1_MAG_a

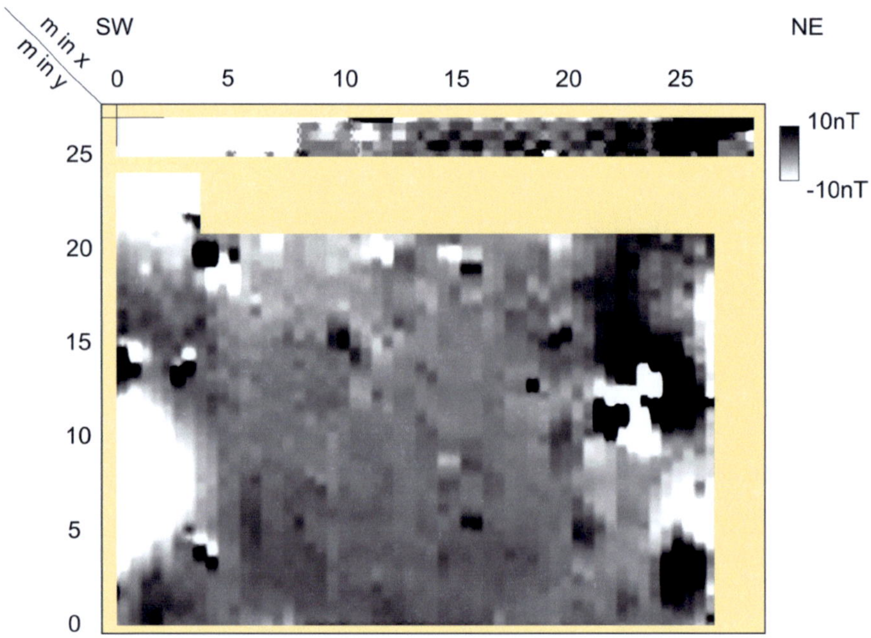

Source: own illustration

gram, several positives anomalies are visible as well, but also two massive dipoles. In the central part, only a few small dipoles are visible.

4.3.2.2 Invasive methods and outcrops

The results within this subsection invasive methods and outcrops are associated with workflow Step 4 (S4) – *Ground-truth.*

Borehole

The borehole M4.1_BH_a is located in the western part of the described plateau. The detailed grain-size distribution is given in Table 25.

Figure 124: Borehole – M4.1_BH_a

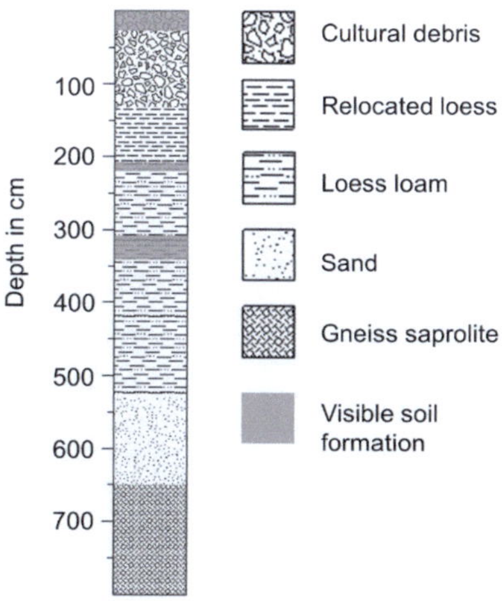

Source: own illustration

Until a depth of 130 cm, Borehole M4.1_BH_a (Figure 124) shows cultural debris. This layer consists of silt loam with low sand content and higher clay content. However, the grain-size distribution varies a great deal within this layer. Until a depth of 210 cm, the material is classified as a colluvial layer, including clay enrichment and less anthropogenic features. Nevertheless, it still needs to be described as relocated.

Underneath this layer, a huge loess loam layer is located with a size of about 320 cm until a depth of 530 cm. The loess loam layer has a high silt content and varying sand and clay contents and is classified as silt loam or silty clay loam. Within this layer, hints of soil formation are visible, at least in two sections. Here the initial building of a fossil A-horizon is detectable, and clay mineral building and clay cutaneous, as a first clue of the creation of a cambic and luvic horizon, are recognizable. Additionally, redoximorphic features in several horizons indicate the impact of the aquiferous slope water. Other horizons within this layer are laminated and show loose sediments without any indication for soil development processes. The laminated layering gets askew at the bottom of the layer as a hint of solifluction.

Below the loess loam layer, pure sand layers occur. The border between this layer and the one above is diffuse. The impact of aquiferous layers is visible in redoximorphic features in this layer.

The sandy layer turns over gradually into the saprolite of the gneiss as bedrock. The typical lamination and foliation of the gneiss are visible within this layer, which indicates the in-situ weathering of it.

Table 25: Grain-size distribution – M4.1_BH_a pipetting

Depth in cm	*S*	*cU*	*mU*	*fU*	*C*	*Grain-size BKA*	*Grain-size WRB*
110	8.74	36.41	22.46	6.35	26.04	Lu	Silt loam
210	7.84	52.91	16.20	4.57	18.49	Ut4	Silt loam
230	9.33	79.27	4.76	0.97	5.68	Uu	Silt
270	16.71	38.87	13.15	5.12	26.15	Lu	Silt loam
290	27.88	46.41	8.16	3.06	14.50	Ul3	Silt loam
310	3.49	44.52	24.60	4.82	22.56	Ul4	Silt loam
320	3.44	46.96	13.79	6.51	29.30	Ul4	Silty clay loam
345	23.37	54.63	5.66	2.17	14.17	Ul3	Silt loam
435	6.94	58.30	14.56	3.87	16.33	Ul3	Silt loam
530	4.87	59.58	17.23	4.15	14.17	Ul3	Silt loam
542	12.43	49.95	16.05	3.59	17.97	Ul3	Silt loam

Source: own research

Additionally, the study site was excavated after these measurements. Therefore, for M4.1, the excavated wall structures were documented for archaeological reasons and can be interpreted in combination with the geophysical data and can be associated with workflow Step 4 (S4) – *Ground-truth.*

4.3.3 M4.1 – Building structures: interpretation

The interpretation of the data was performed according to workflow Step 5 (S5) – *Interpretation of the data.*

Synthesis

Loess loam-rich sediments with a high silt content are very conductive compared to wall structures made of gneiss. Additionally, the permittivity of those materials differs. Because of that, the high-resistive structures and those showing an intense reflection in the GPR are most likely anthropogenic wall structures. In the case of long linear features, in-situ wall structures seem to be a realistic possibility to create such anomalies.

Interpretation

Where are the building structures of the castle?

With the combination of the geophysical results and the subsequently performed excavation, most of the anomalies and reflectors appear to be wall structures or collapsed wall structures.

Whereas the archaeological Section 1 and 3 were planned regardless, Sections 2 and 2 a) were made following the prospection data (e.g., Figure 125, Figure 126, Rosmanitz *et al.*, 2020).

Figure 125: Interpretation – M4.1_3DERT_a in combination with photographs of the excavation

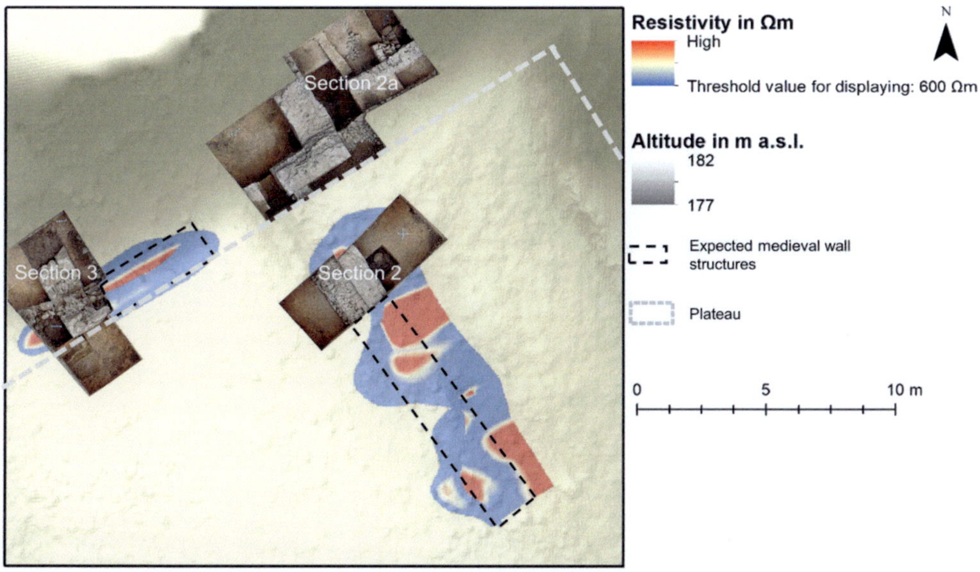

Source: DEM, calculated via structure from motion: C. Büdel, 2019; wall pictures S. Bachmann, 2019
Map creation: J. Trappe

The anomaly orientated in the southeastern direction (cf. Figure 125 at section 2) and reflector g (cf. Figure 126 at section 2) appear as a double wall consisting of sections of two different periods. At least one of these construction periods dates to the 14th century CE (Rosmanitz *et al.*, 2020). These walls end suddenly in the middle of the plateau, proven with a Pürckhauer (not shown).

The wall structure resolved in Section 3 can be correlated with the high-resistive anomaly orientated in a southwestern direction (Figure 125). The location of the wall structure is at the steepest parts of the edge of the plateau and, therefore, not measured with the GPR.

The ERT displays only a small anomaly at the position of wall structures within Section 2a), whereas especially M4.1_3DGPR_b shows the wall structures extending in a northwestern direction. Here the edge of the ERT profiles is located. Most likely, the penetration depth of the ERT was too shallow to display it at this position. As illustrated in the photographs, these and the surrounding wall structures appear detached from the one in the middle of the plateau (e.g., Figure 125 Figure 126). At the northeastern part of the plateau in Section 1, a small dry stone wall from about the 17th century CE appears near the surface. Due to its position, the GPR was not able to resolve this wall. However, about 1 m away, some reflectors suggest a further extent of this structure in a southwestern direction.

Figure 126: Interpretation – M4.1_3DGPR_a and -b in combination with photographs of the excavation

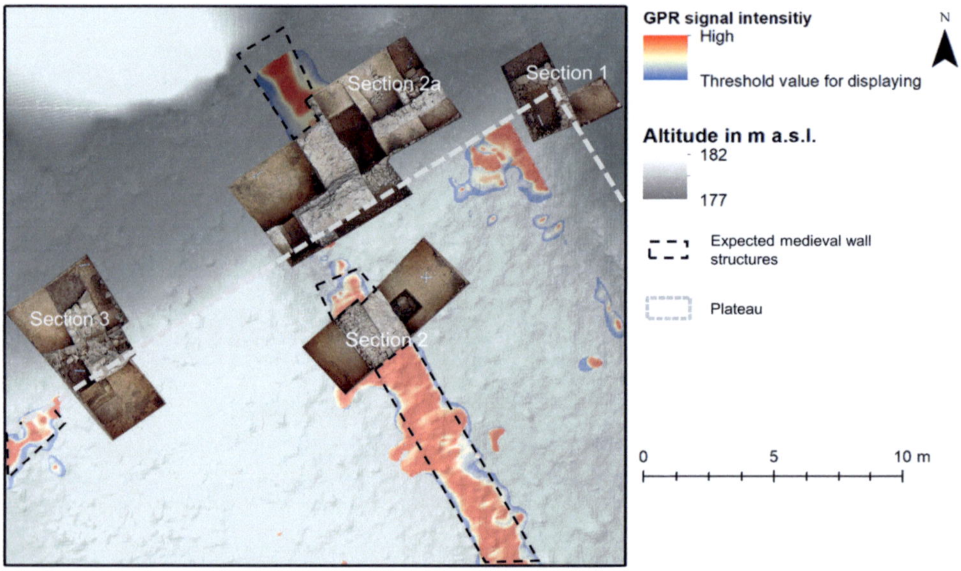

Source: DEM, calculated via structure from motion: C. Büdel, 2019; wall pictures S. Bachmann, 2019
Map creation: J. Trappe

With the combination of this information and the findings, it becomes clear that the extent of the plateau, generally associated with *Mömbris Castle's* extent (Gröber & Karlinger, 1916), needs to be interpreted as detached from the castle itself. Instead of belonging to the medieval castle, the wall structures at the corners of the plateau belong to a garden wall created in the 17th century CE. However, within Sections 3 and 2a), wall structures belonging to the castle itself could be found. These structures only occur at the northwestern edge of the plateau. Most of them are located underneath it, following in the direction of the graveyard and possibly underneath it.

By contrast, the wall structure in the middle of the plateau could be dated as well to the medieval period with two different building phases, but not to the castle itself. They are most likely made to beware of the upslope area by eroding the culminated loess loam sediments shown within the borehole (Figure 124). However, as the borehole displays, the upper 210 cm are not accumulated in-situ, and at least the uppermost 130 cm were raised manually.

4.3.4 Quantification of the medieval, anthropogenic influence in Mömbris

This chapter is associated with workflow Step 6 (S6) – *Quantification of the anthropogenic influence.*

M4.1 – Building structures

3D geophysics combined with a borehole and an archaeological excavation display various wall structures, partly belonging to the medieval castle of Mömbris. Additionally, massive earthworks could be detected with the borehole's help in the western part of the castle hill. Therefore, the site in Mömbris can be classified as Category 4. The entire study site was of active anthropogenic influence during the medieval period. Because of that, no map was created.

4.4 Alzenau

First, site-specific questions were derived with the help of the GIS-based terrain analyses at Alzenau. The results for each question are provided and interpreted. Then, an estimation of quantification of the anthropogenic influence is made. A technical overview of the used application of the geophysical methods is given in Appendix 11 to Appendix 13.

4.4.1 Morphometric analysis for the creation of site-specific questions

The morphometric analyses for creating site-specific questions are associated with workflow Step 2 (S2) – *GIS-based terrain analyses.*

The topographic position of the castle hill is at the end of a ridge surrounded by steep slopes. Opposite of the other castle hills, the castle hill of the *Rannenburg Castle* near Alzenau is a spur of a ridge, and the entire ridges are classified after TPI landform classification as top, shoulder (cf. Figure 127a). South and east of the ridge, creeks formed deep valleys with steep slopes. The TPI landform classification identifies these deep valleys as lowlands, large depressions, whereby the present-day creek beds are classified as lowlands, small depressions. Westwards the relief flattens and turns over to the Lower Main Valley.

The castle hill is intersected by three neck ditches, graduating the hill shoulder in different parts. Surrounding the castle, deep ditches were dug, culminating in a round depression in the north of the hill. The neck ditches and the other ditches are classified as either culmination areas or as flat hollows in culmination areas. A couple of hundred meters north of the castle area, gully erosion formed decameter-deep rills. Westward, several hollow ways are visible. In the surroundings of the castle site, few areas linked to other medieval buildings are visible: for example, the rectangle ditch about 200 m northeast of the castle site or the area southeast of the castle hill on the opposite side of the eastern valley (Bavarian Ministry of Finance and Regional Identity, 2020: Bodendenkmäler). The TWI (Figure 127b) illustrates a low amount of wetness at the castle site and the ridge. Conversely, at the upslope northern slopes, the surface drains further downslope into the gullies. Within this gully system, a large amount of water accumulates.

Figure 127: Morphometric analyses – Alzenau: a) landform classification and slope; b) topographic wetness index

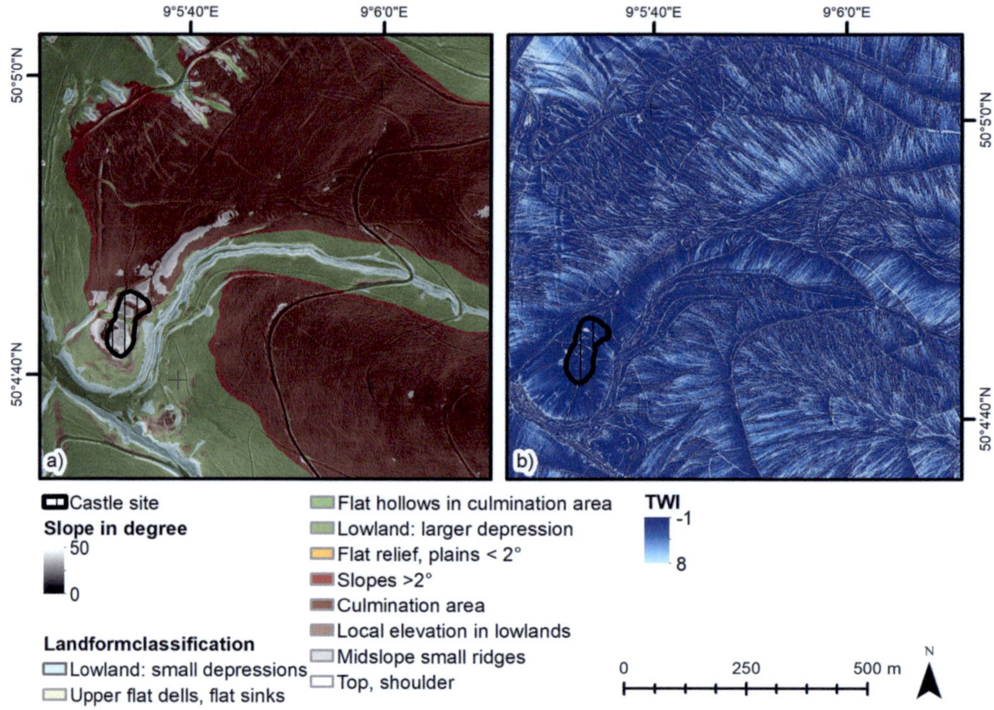

Sources: own illustration, based on Bavarian Agency for Digitalisation, High-Speed Internet and Surveying, 2014
Map creation: J. Trappe

With respect to its specific position on a spur and the intense gully erosion in its surrounding, the following questions are derived (an overview of is in Table 26).

Question A1 – surficial, natural features:

What is the extent of loess cover in the study area?

The intense gully erosion and the geological map (cf. Section 2.5; Okrusch *et al.*, 1967) indicate massive loess packages as a burden cap north of the castle. The extent of such deposits, unstable for buildings but fertile as agricultural land (cf. Section 2.1), is essential for medieval settlements and castles and need to be mapped and analyzed.

Question A4.1 – buried, anthropogenic features – on-site:

Are there buried building structures on the castle site, and if so, where can they be found?

Analog to the other study areas, in Alzenau, a geophysical prospection should help to identify manmade structures such as walls and buried basements for future excavations.

Table 26: Site-specific questions in Alzenau

Category	*Questions*
A1	What is the extent of loess cover in the study area?
A4.1	Where are buried building structures on the castle site?

Source: own research

4.4.2 A1 – Loess coverage: results

What is the extent of loess cover in the study area?

To generate information of the extent and size of the accumulated loess in the surroundings of *Rannenburg Castle*, some ERT profiles were collected and compared with boreholes (cf. Figure 128). Additionally, at several positions, outcrops can be described at uprooted trees. The ERT profiles were spread over an area of a couple of hundred meters to display the spatial distribution of the loess coverage. For a more detailed view, ERT_c was measured in a smaller spacing.

Figure 128: Location of measurements – A1

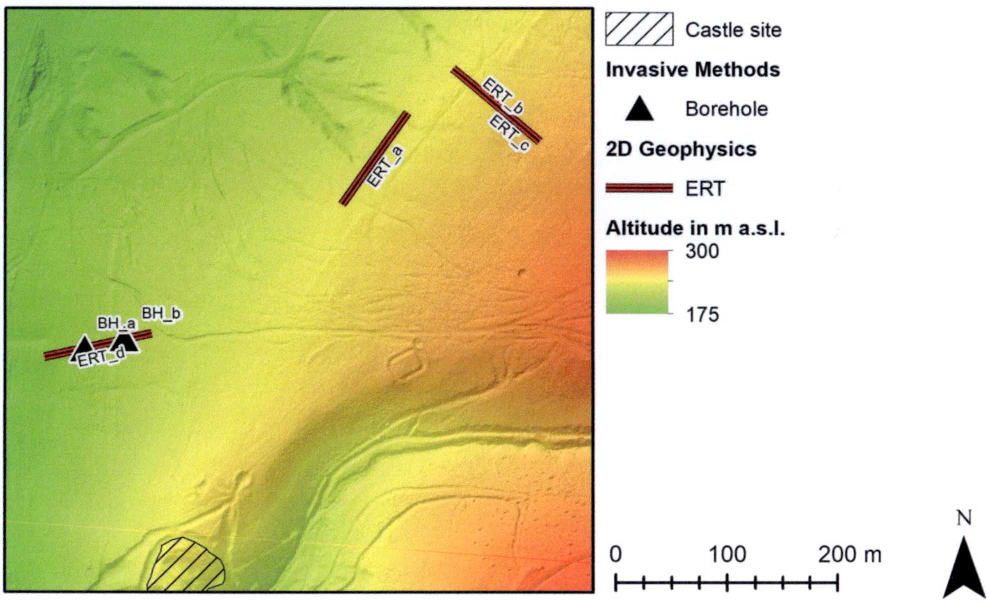

Sources: own illustration, based on Bavarian Agency for Digitalisation, High-Speed Internet and Surveying, 2014
Map creation: J. Trappe

4.4.2.1 Geophysical data

The geophysical data of A1 are associated with workflow Step 3 (S3) – *Geophysical methods.*

ERT

First, the inverted resistivity distribution of the profiles is described. Then, their relative distribution is discussed according to the guidelines for ERT description.

A1_ERT_a resistivity distribution

A1_ERT_a is located northeast of the castle hill, uphill, an area affected by gully erosion, close to the area of erosion in a slope-parallel orientation.

Figure 129: Electrical resistivity tomography – A1_ERT_a

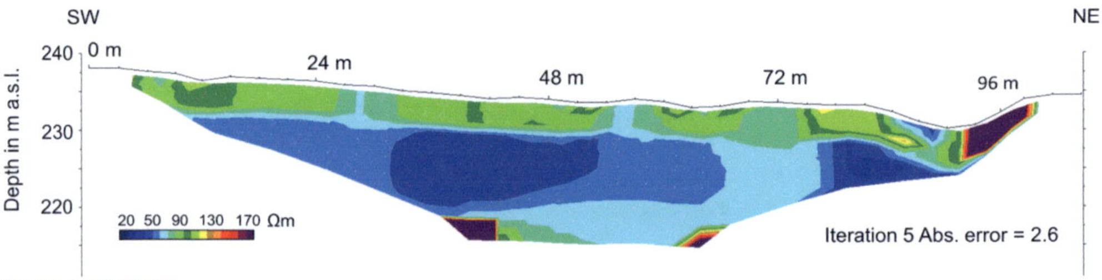

Source: own illustration

The uppermost part of A1_ERT_a (Figure 129) has a size between 6 and 8 m and is characterized by resistivity values around 80 to 100 Ωm. Starting at 93 m, a high-resistive lens, with resistivity values over 170 Ωm, occurs. The bottom part of the profile is classified by resistivity values slightly lower than those of the top part, showing resistivity values between 40 and 60 Ωm. At a depth of between 15 and 20 m, a few high-resistive lenses occur.

A1_ERT_a description after the guidelines

With the addition of the information of the boundary analysis (Figure 130) to the resistivity distribution (Table 27), two r-layers become visible: a top (1) and a bottom

Figure 130: Boundary analysis – A1_ERT_a: a) sharpness of r-layer boundary including the position of r-layers and r-anomalies; b) orientation of the boundary gradient

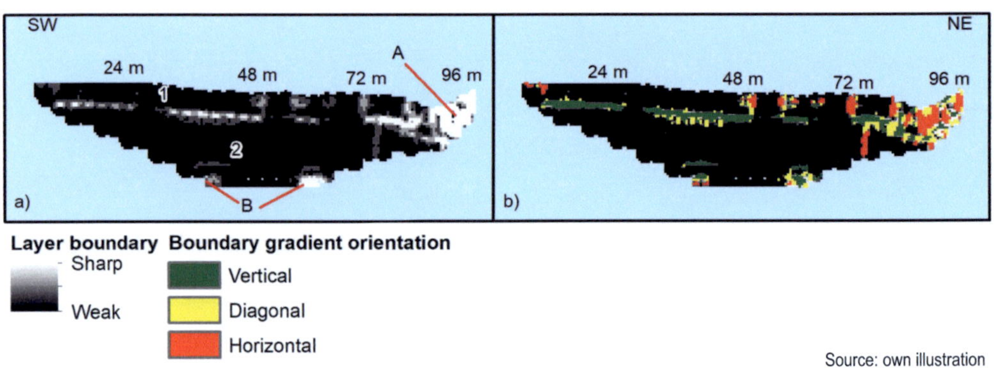

Source: own illustration

(2) r-layer. Although the difference in the first two r-layers' resistivity values seems not so significant, the vertical boundary could be classified as medium, including some weaker and sharper parts. The r-layers are of medium homogeneity with an intense layering between each other. The r-anomaly (A) within this r-layer has a sharp boundary surrounding it. Caused by the quality management during the standardization of the boundary analysis, some of the potential r-anomalies (B) at the bottom profile had to be deleted for the analysis.

Table 27: ERT guidelines – A1_ERT_a

R-layer	*Homogeneity*	*R-layer thickness*	*Layering*	*Horizontal changes*	*Boundary of layers*	*Layer orientation*
Layer 1	3	2	5	1	2	7
Layer 2	4	3	-	-	-	-
R-anomaly	*Number of r-anomalies*	*R-anomaly location*	*R-anomaly form*	*R-anomaly size*	*R-anomaly boundary*	*Boundary orientation of r-anomalies*
R-anomaly Type A	1	1	1	2	5	2
R-anomaly Type B	2	3	1	1	5	1

Source: own research

A1_ERT_b resistivity distribution

A1_ERT_b is located upslope from the gullies, in a downslope orientation.

Figure 131: Electrical resistivity tomography – A1_ERT_b

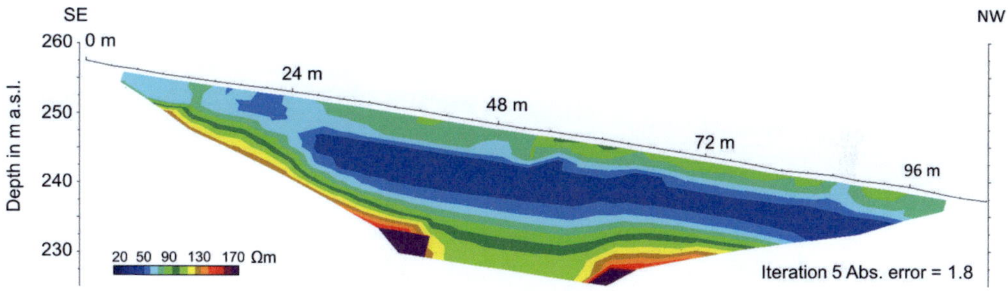

Source: own illustration

Opposite to A1_ERT_a, a segmentation of three different areas within A1_ERT_b (Figure 131) is classified. The top part has a size of about 5 m and resistivity values between 80 and 100 Ωm. There is an area about 10 m underneath this area, classified by resistivity values between 50 and 70 Ωm. Within the bottom section, the resistivity values increase again and rise partly up to over 170 Ωm. Within the southeastern part of meters 6 to 24, the top and central parts seem to intermix, and no clear differentiation can be performed anymore. At this part, the bottom area is located closer to the surface.

A1_ERT_b description after the guidelines

With the addition of the information of the boundary analysis (Figure 132) of A1_ERT_b to the resistivity distribution, three r-layers appear (Table 28).

Figure 132: Boundary analysis – A1_ERT_b: a) sharpness of r-layer boundary including the position of r-layers and r-anomalies; b) orientation of the boundary gradient

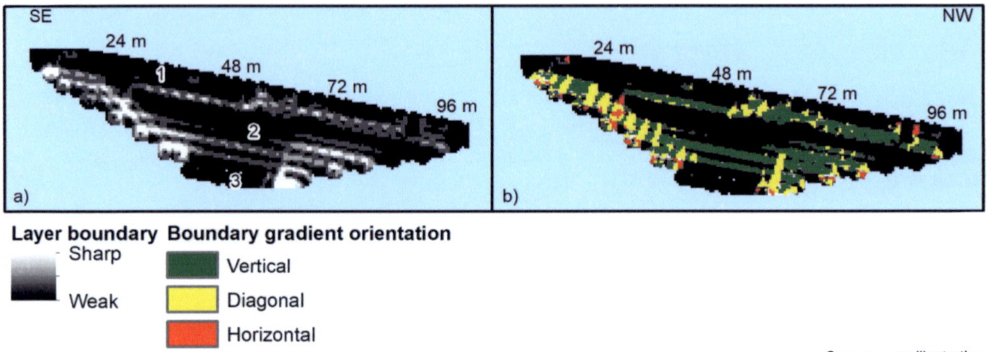

Source: own illustration

The first r-layer (1) is located at the top of the entire profile. A medium-sharp, surface-parallel boundary separates it from the r-layer (2) below. Although the resistivity distribution suggests an intermixing of the top two r-layers, a clear boundary is visible between them even in the southeastern parts of the profile. The bottom r-layer (3) is characterized by an almost surface-parallel boundary between meter 45 and the northwestern end of the bottom r-layer. Southeastward, the orientation is diagonal, and the central r-layer is cropping out. Especially within the diagonal part, the boundary is sharp, whereby in the other sections, it is classified as medium-sharp-to-sharp. All r-layers have a size of around 10 m, and the layering between these r-layers is well-developed. The top and bottom r-layers display a few boundaries within it and are, therefore, mostly homogeneous. In contrast, within the central r-layer, no boundaries are shown.

Table 28: ERT guidelines – A1_ERT_b

R-layer	*Homogeneity*	*R-layer thickness*	*Layering*	*Horizontal changes*	*Boundary of r-layers*	*R-layer orientation*
R-layer 1	3	1–2	4	0	3	7
R-layer 2	5	2	4	1	3–5	OI: 3 (24 m–45 m) OI: 7
R-layer 3	3	2	-	-	-	-
R-anomaly	*Number of r-anomalies*	*R-anomaly location*	*R-anomaly form*	*R-anomaly size*	*R-anomaly boundary*	*Boundary orientation of r-anomalies*
R-anomaly Type A	0	-	-	-	-	-

Source: own research

A1_ERT_c resistivity distribution

A1_ERT_c is located exactly at the top part of A1_ERT_b, only with a smaller spacing.

Figure 133: Electrical resistivity tomography – A1_ERT_c

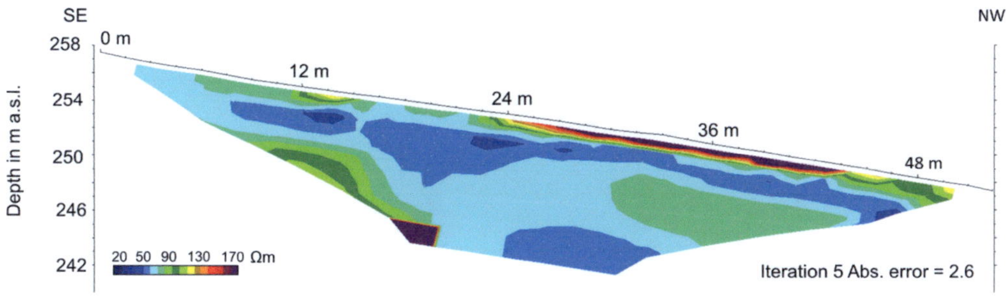

Source: own illustration

Due to its location at the southeastern part of A1_ERT_b, A1_ERT_c (Figure 133) illustrates the same features as A1_ERT_b does at a higher resolution. Starting at meter 24, the thin top part has a thickness of only 1 or 2 m and resistivity values between 100 and over 170 Ωm. Underneath, the resistivity values decrease to values between 50 and 70 Ωm. Within the southeastern part, the high-resistive bottom area, only visible at this part in A1_ERT_c, is coming closer to the surface. Opposite of A1_ERT_b, partly higher resistivity values occur in the top part, too.

A1_ERT_c description after the guidelines

With a combination of the resistivity distribution and the boundary analysis (Figure 134), two r-layers are visible (Table 29). A variable boundary separates the top r-layer (1) from the r-layer below (2).

This boundary is partly sharp; in other areas, it is not visible anymore. The layering is only well-developed at the sharp boundary. Within the top and bottom r-layer, a few weak boundaries are visible. The r-layers show a medium-to-high homogene-

Figure 134: Boundary analysis – A1_ERT_c: a) sharpness of r-layer boundary including the position of r-layers and r-anomalies; b) orientation of the boundary gradient

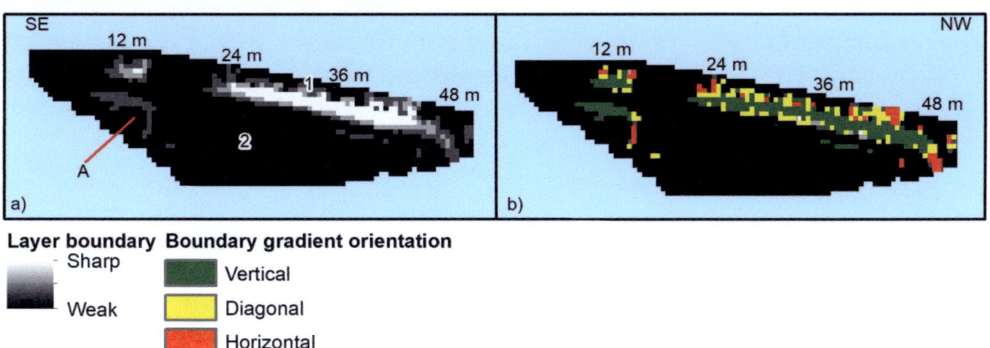

Source: own illustration

ity. At the bottom, an r-anomaly (A) appears with a weak boundary surrounding it. Other high resistivity values needed to be deleted for the boundary analysis due to quality management.

Table 29: ERT guidelines – A1_ERT_c

R-layer	*Homogeneity*	*R-layer thickness*	*Layering*	*Horizontal changes*	*Boundary of r-layers*	*R-layer orientation*
R-layer 1	4	1	2	0	0–5	7
R-layer 2	4	2	-	1	-	-
R-anomaly	*Number of r-anomalies*	*R-anomaly location*	*R-anomaly form*	*R-anomaly size*	*R-anomaly boundary*	*Boundary orientation of r-anomalies*
R-anomaly Type A	1	2	2	2	1	3

Source: own research

A1_ERT_d resistivity distribution

A1_ERT_d is located westwards of the area affected by gully erosion, almost north of the castle hill.

Figure 135: Electrical resistivity tomography – A1_ERT_d

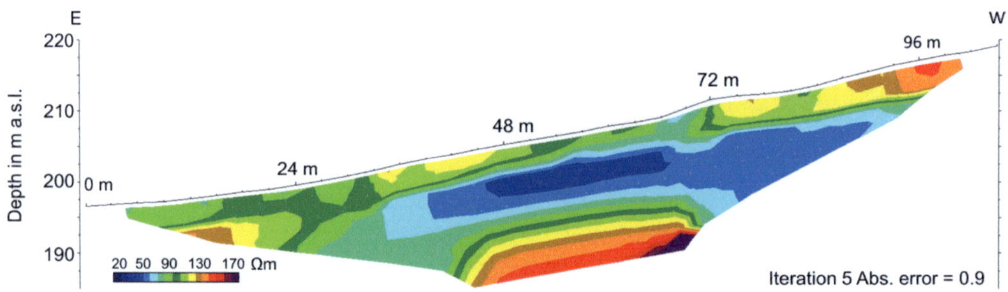

Source: own illustration

All three areas of different resistivity values of the other profiles of A1 can be observed in A1_ERT_d (Figure 135). The resistivity values within the uppermost part vary between 80 and 150 Ωm. These values decrease in the central area to values between 50 and 70 Ωm. In the bottom area, the values rise again to resistivity values of over 170 Ωm.

A1_ERT_d description after the guidelines

With the addition of the boundary analysis to the resistivity distribution (Figure 136), three r-layers become visible (Table 30).

The top r-layer (1) is separated with a varying boundary from the central one (2), and the central one is separated with a medium-sharp boundary from the bottom (3) r-layer. Due to a successive rise in resistivity, the bottom boundary's layering is less

Figure 136: Boundary analysis – A1_ERT_d: a) sharpness of r-layer boundary including the position of r-layers and r-anomalies; b) orientation of the boundary gradient

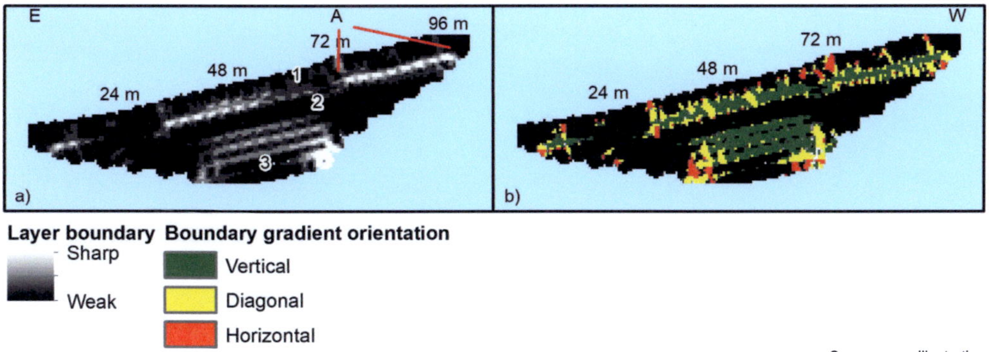

Source: own illustration

good developed than the upper boundary's layering is. The resistivity distribution is less homogeneous within the top and the bottom r-layer than in the intermediate r-layer. Only within the top r-layer, the varying resistivity distribution is classified as r-anomalies (A). Those are small-to-medium-sized and have a weak boundary except at the surface.

Table 30: ERT guidelines – A1_ERT_d

R-layer	*Homogeneity*	*R-layer thickness*	*Layering*	*Horizontal changes*	*Boundary of r-layers*	*R-layer orientation*
R-layer 1	2	1	4	0	0–5	7
R-layer 2	4	2	2	1	3	7
R-layer 3	2	1	-	-	-	
R-anomaly	*Number of r-anomalies*	*R-anomaly location*	*R-anomaly form*	*R-anomaly size*	*R-anomaly boundary*	*Boundary orientation of r-anomalies*
R-anomaly Type A	2	1	2	1–2	1	2

Source: own research

4.4.2.2 Invasive methods and outcrops

The results within this subsection invasive methods and outcrops are associated with workflow Step 4 (S4) – *Ground-truth*.

Borehole

Both boreholes are located among A1_ERT_d at a depth of about 50 and 76 m in the profile, upslope and downslope of a small geomorphological step.

Both boreholes of A1 consist entirely of loose sediments. In A1_BH_a (Figure 137a), sand is the dominant fraction of the uppermost 140 cm. Within the sand, mainly silt but also slightly clay is mixed. Charcoal fragments and a diffuse layering

Figure 137: Boreholes – A1: a) A1_BH_a; b) A1_BH_b

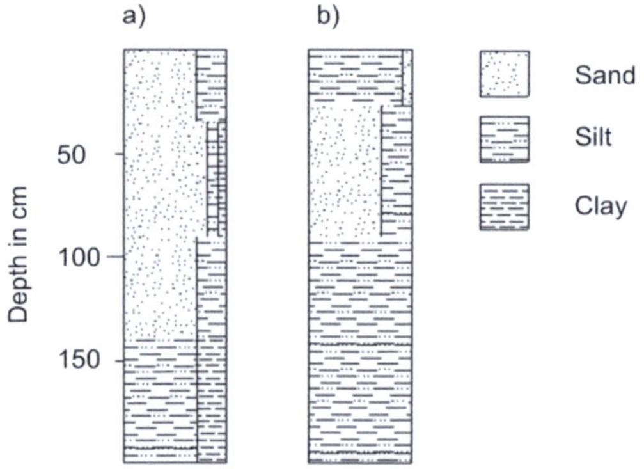

Source: own illustration

indicate that the top layers are colluvial and can be described as loess mixed with saprolite and other aeolian sediments of the Main Valley. These layers are debased. Underneath a depth of 140 cm, the sand disappears, and only clayey silt is left. Here, the carbonate content is higher. Within A1_BH_b (Figure 137b), almost the same features are visible. However, the sand-rich layer is thinner, and the clay content is even lower. Within the uppermost 140 cm in -a and within the uppermost 90 cm

Figure 138: Outcrops – A1_OC_a

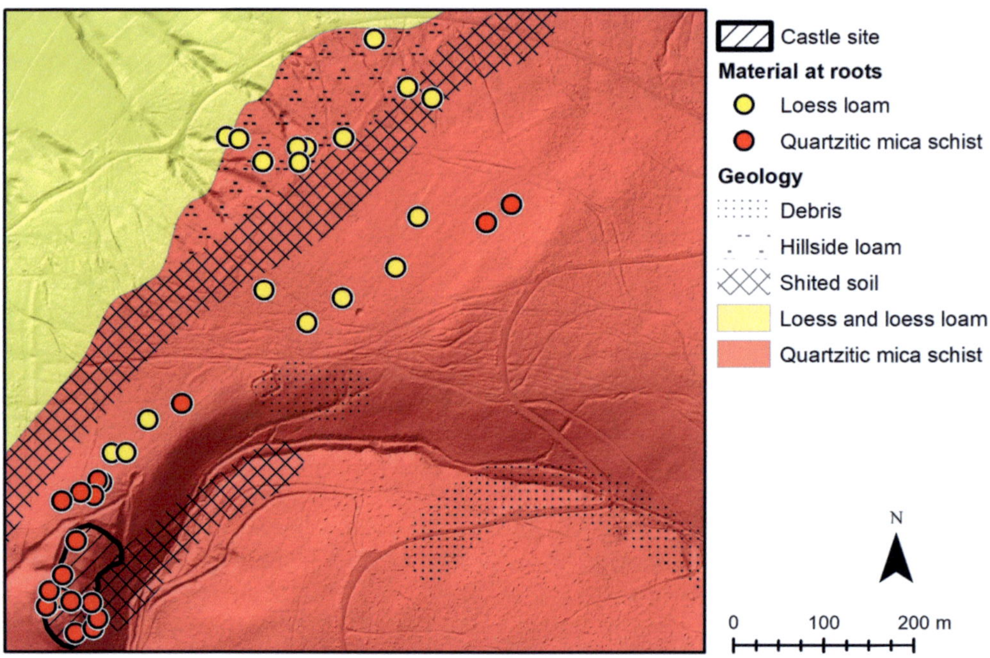

Sources: Okrusch et al., 1967; Bavarian Agency for Digitalisation, High-Speed Internet and Surveying, 2014
Map creation: J. Trappe

in -b, features of initial soil development, such as clay cutaneous, can be identified. However, these features seem to be mostly relocated.

Outcrop

After a massive storm in August 2019, many trees were uprooted. For A1_OC_a (Figure 138), the substrate at the roots of those trees was documented. Next to the described main substrate, other included materials, anthropogenic debris and the type of tree are described (not shown).

It is visible that the substrate consists of the weathering products of quartzitic mica schist or has a recognizable loess content (e.g. Figure 138). In combination with the geological map, it becomes visible that the content of loess loam does not fully correlate with the lithology types that include loess loam.

4.4.3 A1 – Loess coverage: interpretation

The interpretation of the data was performed according to workflow Step 5 (S5) – *Interpretation of the data.*

Synthesis

In the ERT results, an r-layer with higher resistivity values (about 130 Ωm in ERT_d and 190 Ωm in ERT_c) is partly located on top of an r-layer with lower resistivity values (about 100 Ωm). The boreholes (Figure 137) display loess loam in these positions. The loess' top layers often had a higher sand content, and colluvial material is mixed. This can cause higher resistivity values (Reynolds, 2011). Underneath this layer and in the uphill parts of A1_ERT_b and -c, a low-resistive layer is located (with about 50 Ωm), which are associated most likely with the weathered mica schist. Bottomward, the values increase to less-weathered mica schist (Reynolds, 2011). In the bottom layer, some higher resistivity values are located at the edges. They would fit into the suggestion of the less-weathered mica schist, but most likely, they can be associated with inversion artifacts.

Interpretation

What is the extent of loess cover in the study area?

With a combination of the results of the geophysical measurements, the boreholes (Figure 137) and outcrops (Figure 138) with the geological information of the area, it becomes visible that the spatial distribution of the loess loam deposits and other lithological units is more heterogeneous than expected.

The conceptual model of A1 (Figure 139) illustrates the dropping out of the loess loam on top of the saprolite uphill. Depending on its position, the mica schist's weathering is of different intensity. In some areas, it decreases gradually with a continuous increase of the resistivity values, whereas the saprolite layer seems to be very thin in other parts. With especially A1_ERT_b (Figure 131) and -c (Figure 133), the paleo-surface, before the loess accumulation, varies strongly from the present-day one. The sand enrichment occurs approximately due to the position of the slope exposed toward and near the Main Valley. It can be estimated, this substrate mainly consists of aeolian sands or is mixed with the weathering products of the schist. However, the colluvial enriched material and the intense gully erosion indicate an agricultural usage of the area. Such agricultural usage was most likely combined with deforestation of the area, including higher erosion rates.

Figure 139: Conceptual model – A1

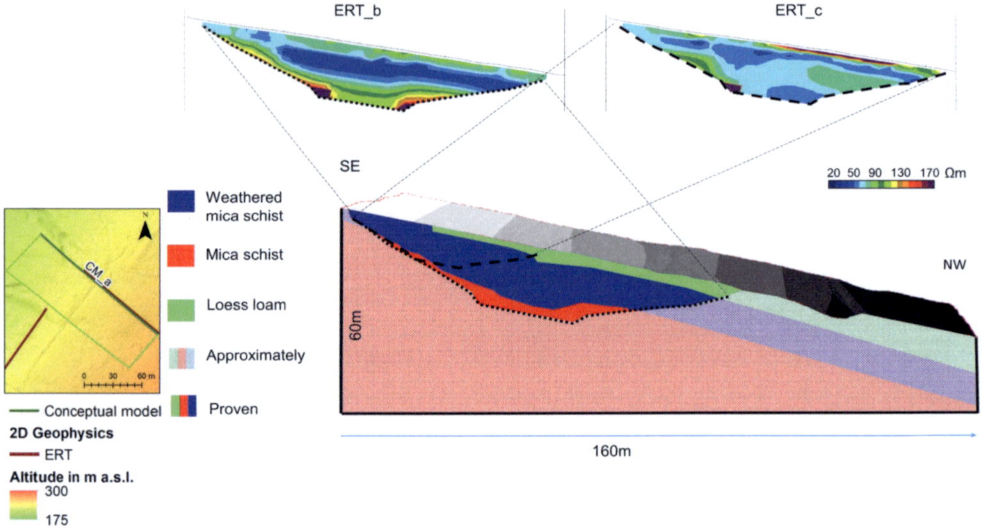

Source: own illustration, based on Bavarian Agency for Digitalisation, High-Speed Internet and Surveying, 2014

4.4.4 A4.1 – Building structures on-site: results

Where are buried building structures on the castle site?

The topography of *Rannenburg Castle* is very rough. Therefore, the castle site was divided into five different subsites (cf. Figure 140). In contrast to the corresponding subsite, the rising letters of the measurements are not necessarily continuous, but they depend on the subsite. However, subsite "a" is located in a flattened area west of the castle hill itself. The subsites "b," "c" and "d" are located at single spots on the castle hill subdivided by dug ditches, "b" includes the first neck ditch as well. Subsite "e" is located outside the castle site itself, surrounded by deep incisions of ditches.

4.4.4.1 Geophysical data

The geophysical data of A4.1 are associated with workflow Step 3 (S3) – *Geophysical methods*.

ERT

First, the inverted resistivity distribution of the profiles is described. Then, their relative distribution is discussed according to the guidelines for ERT description.

Figure 140: Location of measurements – A4.1

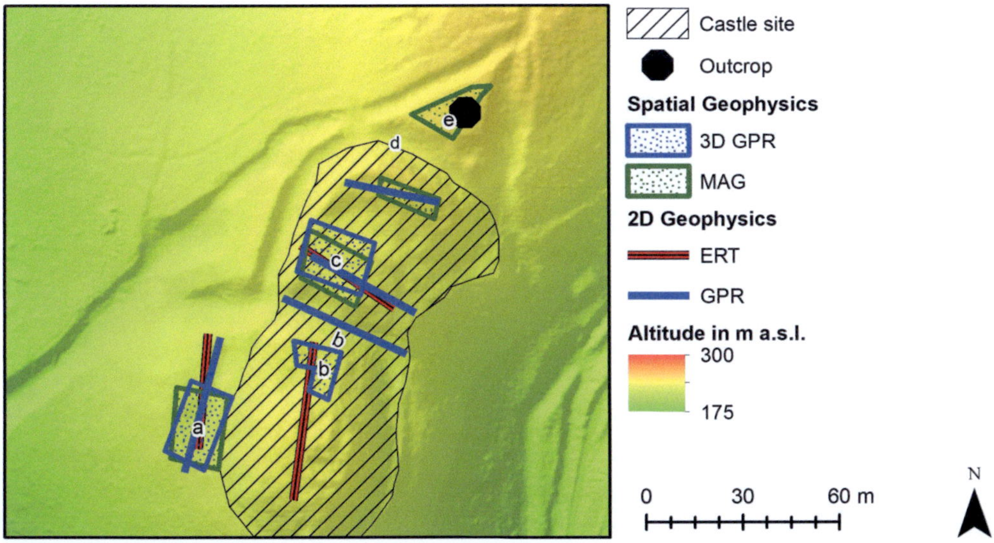

Source: Bavarian Agency for Digitalisation, High-Speed Internet and Surveying, 2014
Map creation: J. Trappe

A4.1_ERT_a resistivity distribution

A4.1_ERT_a is located on a small, flattened area, west of the castle site itself, surrounded by steep slopes.

Figure 141: Electrical resistivity tomography – A4.1_ERT_a

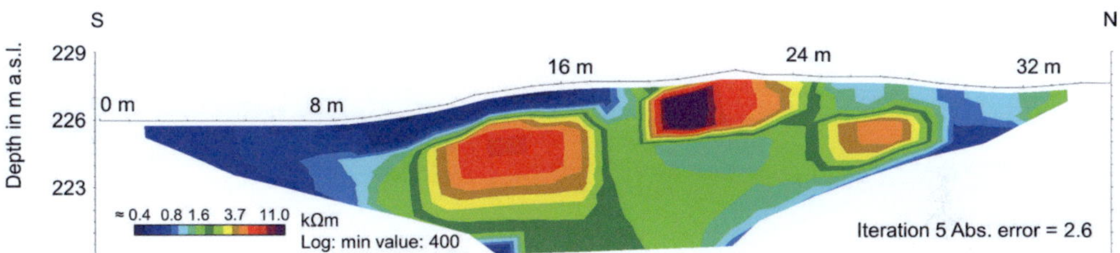

Source: own illustration

Within A4.1_ERT_a (Figure 141), three high-resistive lenses with resistivity values between 3.0 and 12.0 kΩm are located. Those have an altitude of between 2 and 3 m and a length of between 4 and 6 m. The surrounding matrix has resistivity values of between 1.0 and 2.0 kΩm. On top of the southern lens, in the first 10 m of the profile and at the northern edge, the resistivity values are far lower, around a few hundred Ωm.

A4.1_ERT_a description after the guidelines

With the addition of the boundary analysis' information (Figure 142) to the resistivity distribution, three r-anomalies located within a single r-layer become visible (A; Table 31).

Figure 142: Boundary analysis – A4.1_ERT_a: a) sharpness of r-layer boundary including the position of r-layers and r-anomalies; b) orientation of the boundary gradient

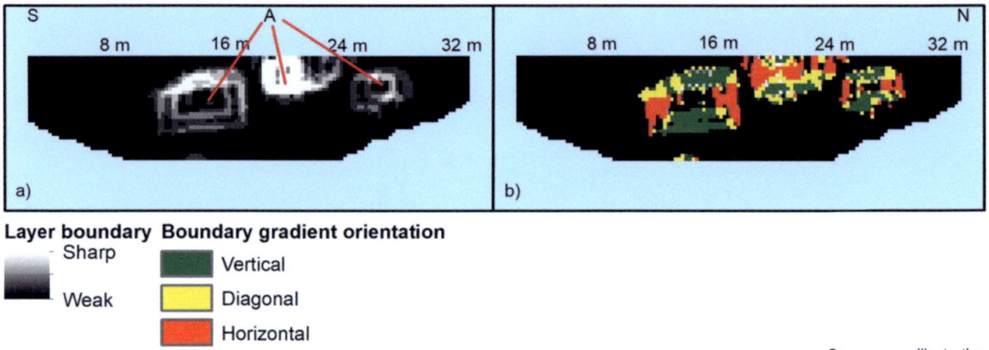

Source: own illustration

The homogeneity within this r-layer is high besides the r-anomalies and only a few other boundaries are visible. It displays that the northern and southern r-anomaly boundaries are medium-to-sharp, while the central r-anomaly has sharp boundaries surrounding it.

Table 31: ERT guidelines – A4.1_ERT_a

R-layer	*Homogeneity*	*R-layer thickness*	*Layering*	*Horizontal changes*	*Boundary of r-layers*	*R-layer orientation*
R-layer 1	4	-	-	-	-	-
R-anomaly	*Number of r-anomalies*	*R-anomaly location*	*R-anomaly form*	*R-anomaly size*	*R-anomaly boundary*	*Boundary orientation of r-anomalies*
R-anomaly Type A	3	1–2	1	1	4–5	1

Source: own research

A4.1_ERT_b resistivity distribution

A4.1_ERT_b is located at the southern intersection of the castle hill.

Figure 143: Electrical resistivity tomography – A4.1_ERT_b

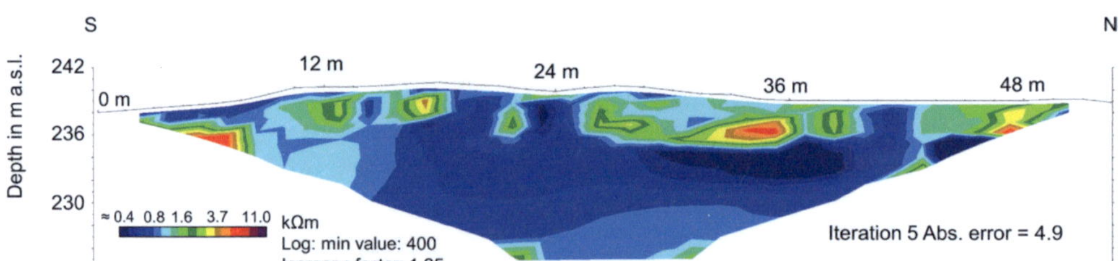

Source: own illustration

A4.1_ERT_b (Figure 143) was measured using the Dipole-Dipole configuration. Therefore, several small-sized lenses could be more easily differentiated from the surrounding matrix. Many of those are located within the first 3–4 m below the surface. These lenses have different sizes and forms. The resistivity values within them vary between less than 2.0 kΩm and several kΩm. Their location ends at 3 or 4 m below the surface. The matrix surrounding them has resistivity values around a few hundred Ωm.

A4.1_ERT_b description after the guidelines

With the addition of the information of the boundary analysis (Figure 144) and the resistivity distribution, a single r-layer (1), including a high number of r-anomalies (A), appears (Table 32).

Figure 144: Boundary analysis – A4.1_ERT_b: a) sharpness of r-layer boundary; b) orientation of the boundary gradient

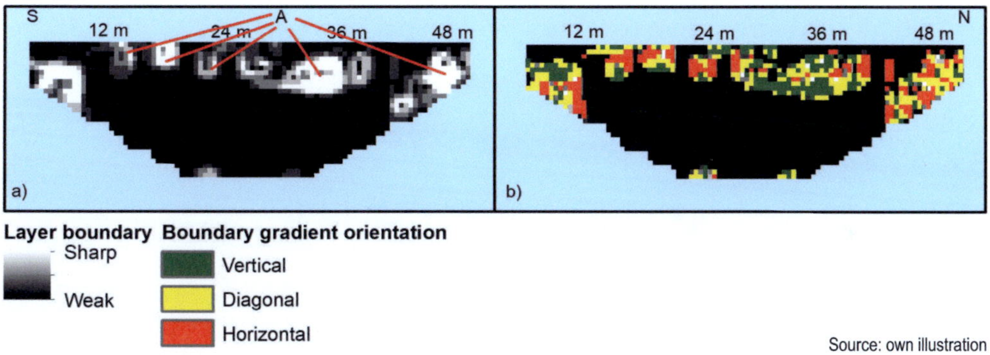

Source: own illustration

The homogeneity within the r-layer is high; only a few boundaries besides the r-anomalies are visible. The boundary surrounding the small-sized r-anomalies is mostly sharp, but a few of these boundaries are also weaker.

Table 32: ERT guidelines – A4.1_ERT_b

R-layer	*Homogeneity*	*R-layer thickness*	*Layering*	*Horizontal changes*	*Boundary of r-layers*	*R-layer orientation*
R-layer 1	4	-	-	-	-	-
R-anomaly	*Number of r-anomalies*	*R-anomaly location*	*R-anomaly form*	*R-anomaly size*	*R-anomaly boundary*	*Boundary orientation of r-anomalies*
R-anomaly Type A	3	1	1	1	5	1

Source: own research

A4.1_ERT_c resistivity distribution

A4.1_ERT_c is located at the central castle hill.

Within A4.1_ERT_c (Figure 145), two areas of different resistivity distribution are visible. The top has higher values than the bottom part does. Here, the resistivity values lie in the eastern half and between meters 29 and 33 of the profile over 3.0 kΩm,

Figure 145: Electrical resistivity tomography – A4.1_ERT_c

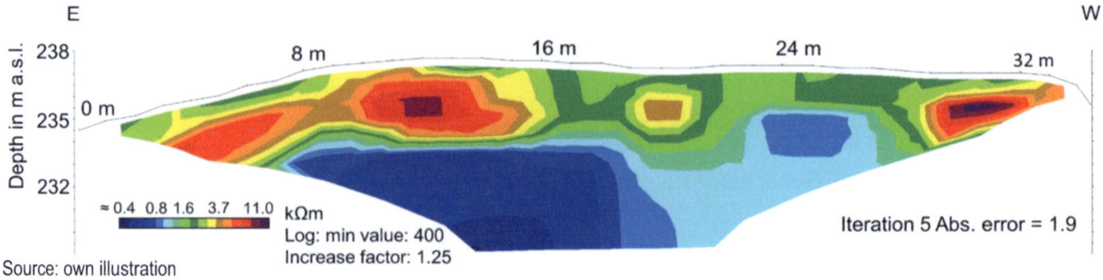

Source: own illustration

whereas in between lower resistivity values between 1.5 and 2.0 kΩm occur. Within this area, a lens with 3.0–11.0 kΩm exists. The size of the top area varies around 5 m. The bottom area is characterized by resistivity values between about 0.5 and 1.3 kΩm.

A4.1_ERT_c description after the guidelines

With the combination of the resistivity distribution with the boundary analysis (Figure 146), two r-layers occur (Table 33).

Figure 146: Boundary analysis – A4.1_ERT_c: a) sharpness of r-layer boundary including the position of r-layers and r-anomalies; b) orientation of the boundary gradient

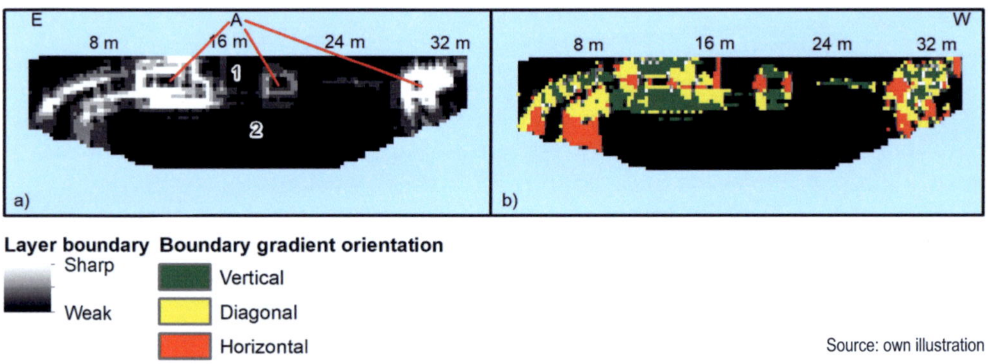

Source: own illustration

Table 33: ERT guidelines – A4.1_ERT_c

R-layer	*Homogeneity*	*R-layer thickness*	*Layering*	*Horizontal changes*	*Boundary of r-layers*	*R-layer orientation*
R-layer 1	2	1	5	0	0–5	7
R-layer 2	5	1	-	-	-	-
R-anomaly	*Number of r-anomalies*	*R-anomaly location*	*R-anomaly form*	*R-anomaly size*	*R-anomaly boundary*	*Boundary orientation of r-anomalies*
R-anomaly Type A	3	1	1	1	5	1

Source: own research

Within the first (1) r-layer, several r-anomalies (A) are embedded, which are somewhat difficult to distinguish from the surrounding matrix. The boundary between the top and the bottom (2) r-layer is partly sharp and partly weak. The top r-layer is classified by a low homogeneity, whereas the bottom r-layer is very homogeneous. The boundary's orientation mostly follows the topography; only at both edges does it seems steeper than the surface. The r-anomalies are surrounded by sharp boundaries.

GPR

A4.1_GPR_a is located on the flattened area, west of the castle hill. A4.1_GPR_b crosses the neck ditch intersecting the southern hill from the ones located further north. A4.1_GPR_c covers the less-exposed eastern slope and the central castle hill, and A4.1_GPR_d is located at the most northern part of the castle hills.

A4.1_GPR_a

Within A4.1_GPR_a (Figure 147), according to Neal (2004), the bottom and the top boundaries have a concordant reflection geometry.

Figure 147: Ground-penetrating radar measurement – A4.1_GPR_a: a) data; b) data with marked changes in signal intensity and structure

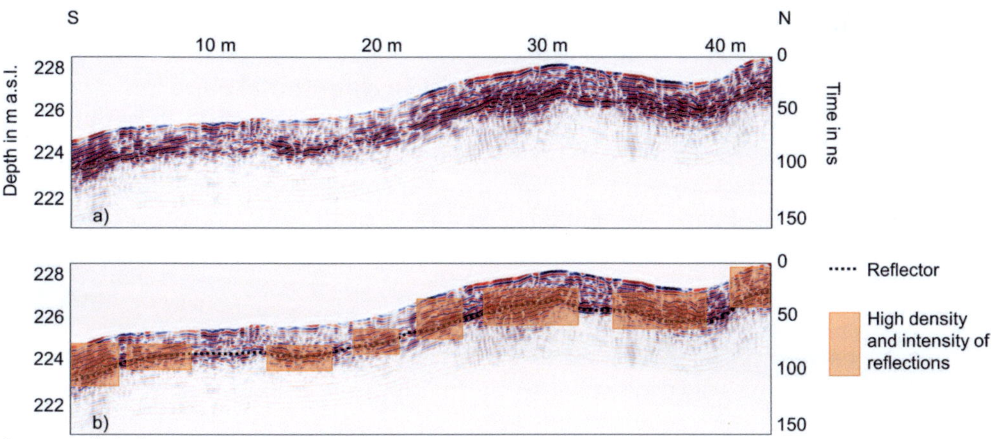

Source: own illustration

A planar/even surface-parallel layer is visible at a depth of about 150–200 cm. This layer is more or less continuous but also has some discontinuous areas. The intensity of the reflectors in the areas above varies. Some areas of different forms and sizes have a clear planar/even and continuous continuity, whereas other areas are absolutely discontinuous and seem diffuse. In other parts, for example, the uppermost 1.5 m between 10 and 20 m, almost no strong reflection is visible. In the northern 20 m and the southern 5 m of the profile, the reflections are more distinctive, still varying in continuity. The blocks of parallel, planar reflections can occur through the surface-parallel layer. Mostly a horizontal or at least surface-parallel dipping is visible, and the relationship between the reflections is classified as parallel.

A4.1_GPR_a

Analog to A4.1_GPR_a, within A4.1_GPR_b (Figure 148), both boundaries have a concordant reflection geometry, and a single, surface-parallel reflector is visible.

Figure 148: Ground-penetrating radar measurement – A4.1_GPR_b: a) data; b) data with marked changes in signal intensity and structure

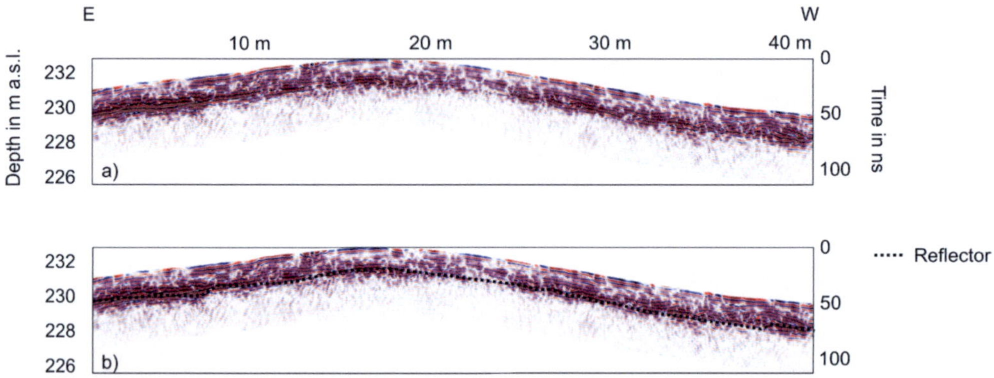

Source: own illustration

This reflector occurs at a depth between 0.8 and 1.5 m. The reflector is continuous in the eastern 10 m and moderately continuous between 20 m and the end of the profile. At this position, the mostly planar/even shape turns more into a curved, sigmoidal one. In between these two areas, the layer is much more discontinuous. On top of the reflector, there are many diffuse and mostly discontinuous reflectors located. In some parts, they seem to create a block, analog to A4.1_GPR_a. However, these blocks are less clearly developed. The dipping can be classified as being surface parallel.

A4.1_GPR_c

Within the eastern parts of A4.1_GPR_c (Figure 149), the upper and lower boundaries are classified as concordant, whereby the upper boundary turns to erosional truncation in the western part.

A continuous layer is visible in the eastern part, comparable to the one in A4.1_GPR_a and -b. Opposite of the comparable layers in A4.1_GPR_a and -b, the layer ends more or less suddenly and crops out to the surface at 20 m. Additionally, a couple of blocks of continuous reflectors exist in a discontinuous, chaotic matrix of different intense reflectors. Starting at 20 and up to 30 m, several reflectors crop out to the surface at a steep angle. Further downslope, the angle is smoothed more to the angle of the surface. A few of the described blocks occur in this area as well but are less well developed. In the entire profile, the blocks do not end necessarily at the position of the strong reflector. The dip is varying, whereby the outcropping reflectors are mostly horizontally located or even dip in the direction of the surface. The other reflectors dip mostly parallel to the surface.

Figure 149: Ground-penetrating radar measurement – A4.1_GPR_c: a) data; b) data with marked changes in signal intensity and structure

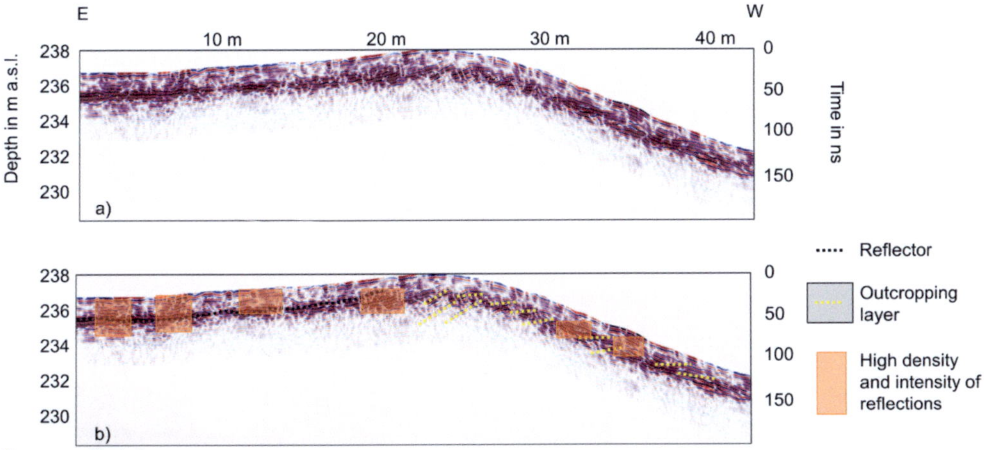

Source: own illustration

A4.1_GPR_d

Within A4.1_GPR_d (Figure 150), both boundaries have mostly a concordant reflection geometry.

Figure 150: Ground-penetrating radar measurement – A4.1_GPR_d: a) data; b) data with marked changes in signal intensity and structure

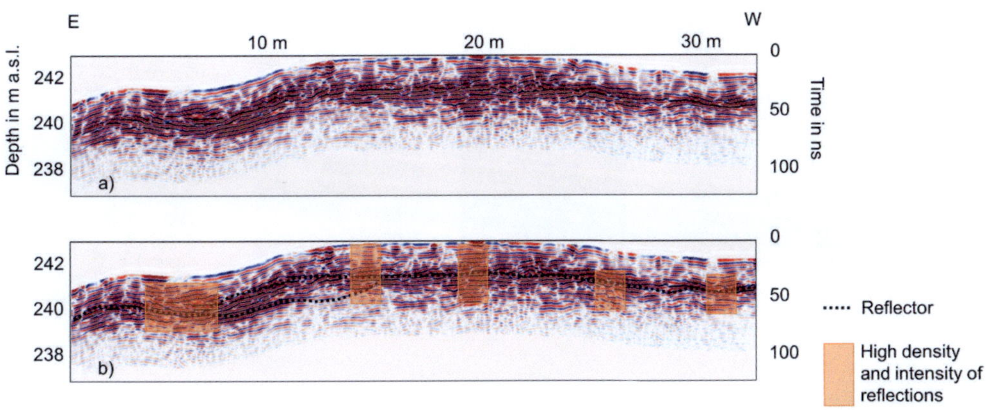

Source: own illustration

In contrast to the other GPR profiles of A4.1, in A4.1_GPR_d, the intense reflector is still visible but occurs as a massive bunch of reflectors. The auxiliary line in the interpretation is located at the intensest reflection within the bunch. The reflector is located closer to the surface as in the other GPR profiles, with a depth of about 1.5 m. The profile is the only one in which the blocks are planar/even. In some cases, curved, sigmoidal and parallel blocks occur up to the surface and not only within the subsurface. The reflection continuity is the most continuous of all the profiles.

3D GPR

A4.1_3DGPR_a is located on the flattened area. In addition, A4.1_3DGPR_b is located on the southernmost castle hill, and A4.1_3DGPR_c is located on top of the central castle hill.

A4.1_3DGPR_a

Within the first layer of A4.1_3DGPR_a (Figure 151), several high reflection intensities are visible.

Figure 151: 3D ground-penetrating radar measurement – A4.1_3DGPR_a

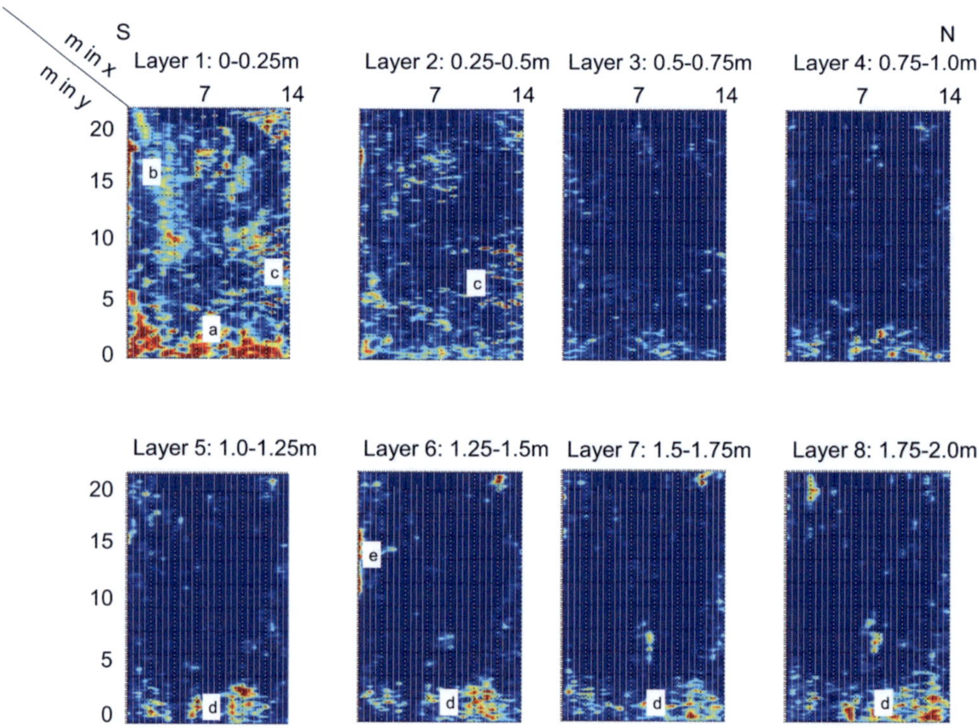

Source: own illustration

Especially in the eastern part of the grid, a vast polygonal area with strong reflections is visible (a) in the uppermost layer (0-0.25 m of depth). In this layer, another polygonal structure is visible between 10-17 m along the y-axis and 3-8 m along the x-axis (b). Starting in the first layer, another almost rectangle reflector (c) is located on 4-10 m along the y-axis and 10-14 m along the x-axis. This layer is even more visible in layer 2 (0.25-0.50 m). Other reflectors almost do not exist up to layer 4 (0.75-1.0 m). But further punctual reflectors are located within the bottom four layers (1.0-2.0 m), too (e.g. in layer 6 at 6-8 m along the y-axis and 7.5 m along the x-axis (e) or at 1-4 m along the y-axis and the entire x-axis in layers 5- layer 8(d)).

A4.1_3DGPR_b

Within the top layer of A4.1_3DGPR_b (Figure 152), several strong reflectors are visible. Some of these reflections occur to a depth of 0.75 m. However, in deeper layers, still different reflectors of different forms and sizes occur without being orientated specifically. These strong reflectors appear mainly in the eastern part: e.g., in layer 4 (0.75-1.0 m of depth) at about 3 m along the y-axis and 5 m along the x-axis or in layer 7 to layer 9 (1.5-2.25 m of depth) at 5 m along the y-axis and 2.5 m along the x-axis.

Figure 152: 3D ground-penetrating radar measurement – A4.1_3DGPR_b

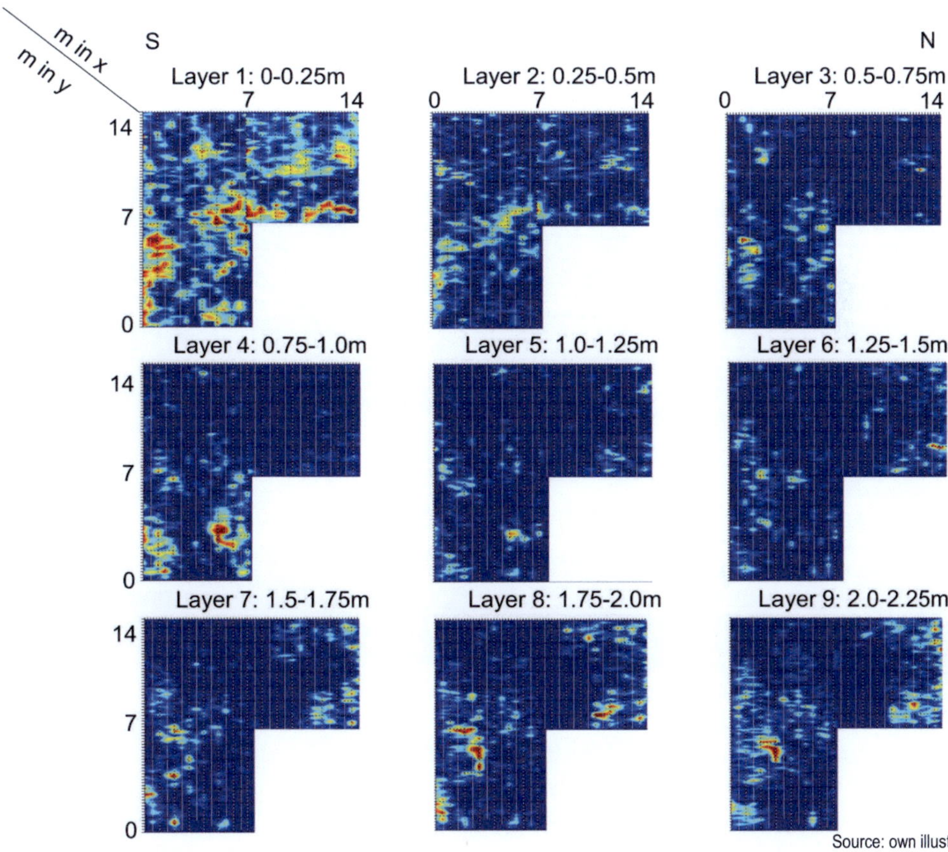

Source: own illustration

A4.1_3DGPR_c

A4.1_3DGPR_c (Figure 153) displays different reflectors within the top layer (0-0.25 m of depth). Especially in the eastern part at 0-2 m along the y-axis and x-axis or in the intermediate part between 5-12 m along the y-axis and 3-9 m along the x-axis. In deeper areas, only a few reflections occur diffusely anymore. Those are most present in layer 5 to layer 8 (1.0-2.0 m of depth) in the eastern part.

Figure 153: 3D ground-penetrating radar measurement – A4.1_3DGPR_c

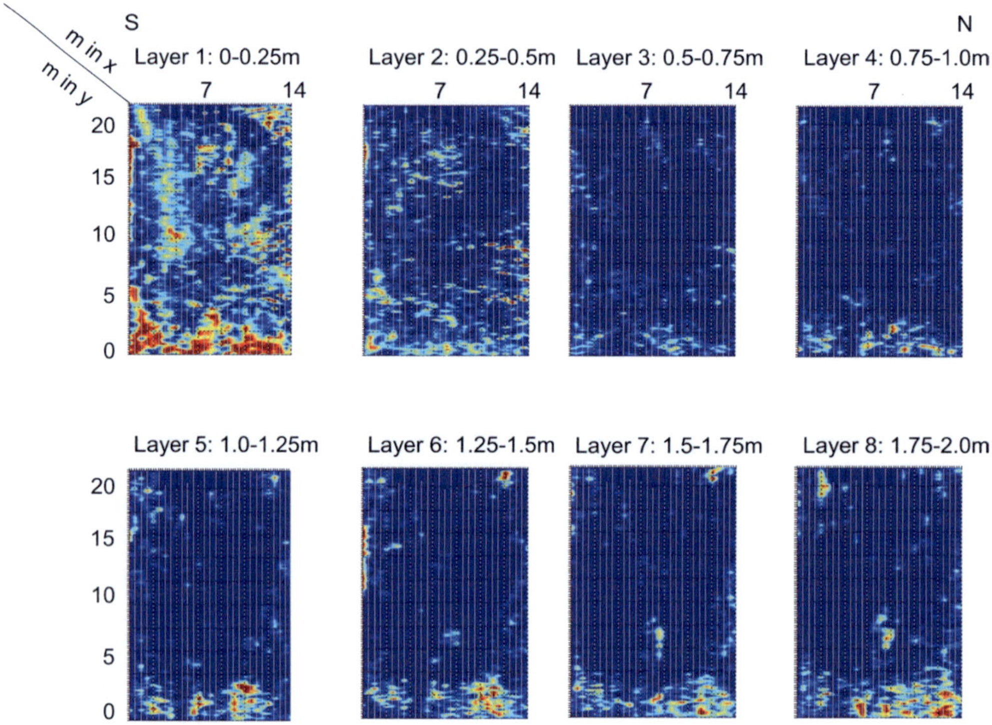

Source: own illustration

Magnetic survey

The magnetic surveys of A4.1 are visible in Figure 154. While A4.1_MAG_a is located at the flattened area, A4.1_MAG_c and -d extend on top of the central and the northern hill. A4.1_MAG_e is located on a triangular, more or less flat area, north of the castle. However, a couple of linear depressions surround this area, indicating that it belongs to the castle.

At *Rannenburg Castle* near Alzenau, several trees and exposed terrain made magnetic surveying complicated and fragmented. Those obstacles are visible in the magnetic data as black polygons. Small dipole signatures indicate iron within the ground. In all four profiles, a few of these signatures are visible. Within A4.1_MAG_a and especially A4.1_MAG_e, linear structures of lower nT values are visible. Different diffusely located high and low nT structures are visible without forming specific objects in all the surveys.

4.4.4.1 Invasive methods and outcrops

As mentioned above, *Rannenburg Castle* was not excavated in the context of this work, and after the prohibition of the excavation, no further invasive work could be performed. Therefore, almost no ground-truth can be delivered.

Figure 154: Magnetic surveys – A4.1_MAG

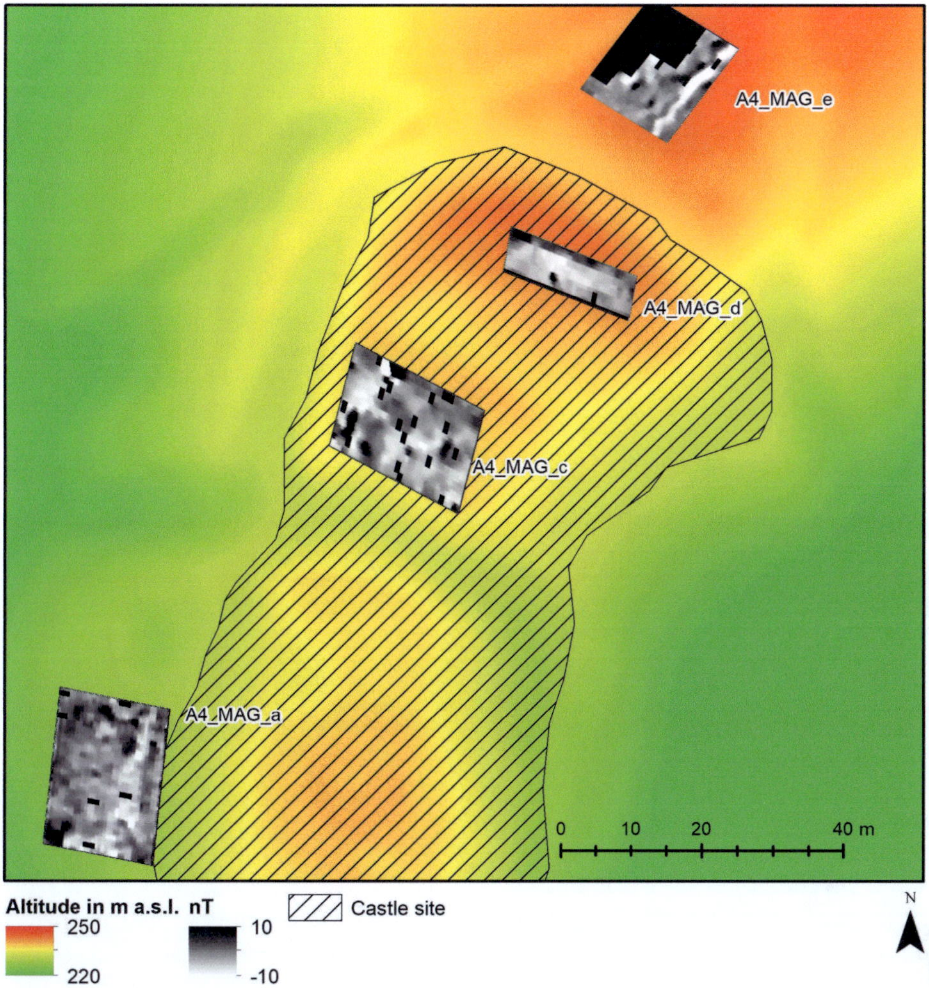

Source: Bavarian Agency for Digitalisation, High-Speed Internet and Surveying, 2014
Map creation: J. Trappe

Outcrop

Nevertheless, outcrop A4.1_OC_e (Figure 155), located in the eastern part of the subsite "e," illustrates the bedrock that crops out at the top. At the small pathway, bare, weathered mica schist is visible.

4.4.5 A4.1 – Building structures on-site: interpretation

The interpretation of the data was performed according to workflow Step 5 (S5) – *Interpretation of the data.*

Figure 155: Field picture – Outcrop A4.1_OC_e

Source: own picutre, 2020

Synthesis

The different geophysical data displays a heterogeneous image of the castle site at *Rannenburg Castle* in Alzenau. As mentioned above, the ground-truth was not able to be determined. Therefore, the data are still ambiguous and different interpretations are possible. However, the ERT data of A1 displays the saprolite of the mica schist as a very low-resistive r-layer. Even though the average resistivity within the profiles of A4.1 is far higher, the low-resistive part is associated with the quarzitic mica schist, as the geological map (Okrusch *et al.*, 1967) indicates it as bedrock at the study site and the outcrop shows it located directly at the top. Several high-resistive r-anomalies are embedded within this low-resistive matrix, surrounded with a sharp boundary, without a distinct orientation. Therefore, caused by their position, these r-anomalies can be interpreted as manmade wall structures. Whether these structures were built in-situ, have been moved or have collapsed during the destruction of the castle cannot be said at this time. Similar structures can be found within the GPR data. Blocks of planar and almost parallel layered reflectors are visible. On the one hand, this can be caused by a high frequency of layers with different dielectric constants, such as in cemented walls; on the other hand, this can be caused by multiple refractions of the radar wavetrain due to anomalies lying above another strong reflector. This is possible in the described setting, too, and do not disagree with such described wall structures. In the 3D GPR, these structures are visible as single blocks with a strong amplitude.

Interpretation

Where are buried building structures on the castle site?

With the combination of this information, at the flattened area west of the hill, subsite "a," the occupation layer starts in general at a depth of about 1 to 1.5 m. Here, both GPR datasets (Figures 147 and 151) show a more or less continuous layer. Within the ERT (Figure 141) in the first 18 meters in the direction of the x-axis, this layer is located on top of one of these r-anomalies. At this position, these might belong to some buried basement fragments, or parts of the collapsed walls, once located uphill in this area. Several more high-resistive r-anomalies or areas of high intensity of reflections are located within the location. Only in the northern part of the 3D GPR_a (Figure 151), they form a linear structure, which is associated with in-situ walls. All other structures are randomly distributed in the area and are likely collapsed and fallen wall structures from the castle upslope. The magnetic survey (cf. Figure 154) at this position does not contradict these results. However, the most visible anomaly is a linear structure in the eastern part of the survey. Combined with the results of the 3D GPR (Figure 151), the outcrop (Figure 155) and A4.1_MAG_e (Figure 154), this seems to consist of bedrock that crops out at this position. Maybe due to the flattened surface of the area or the steeping of the surrounding sites of the castle hill itself, here it was dug to the bedrock.

Within the neck ditch, only A4.1_GPR_b (Figure 148) is located. The bottom of the dug ditch can be expected at a depth of about 1.5–2 m. Above this layer, several randomly located anomalies are distributed, which are associated with collapsed wall structures and debris as infilling of the ditch.

At the southern hill, subsite "b," A4.1_ERT_b (Figure 143) and A4.1_3DGPR_b (Figure 152) indicate an average depth of the occupation layer of 2–3 m. Starting in this depth, the number of r-anomalies decreases in the ERT, and within the GPR data, diffuse reflections start to occur. Due to the strong topography within the area, it can be expected that these reflections can be associated with it. On top of the occupation layer, many randomly located anomalies are located as being associated with collapsed walls, basements or the foundation of them. Only at depths between 0.75 and 1 m a parallel structure is visible within the western part of the GPR, which can be associated with in-situ wall structures.

At the second hill, subsite "c," A4.1_ERT_c (Figure 145) and A4.1_GPR_c (Figure 149) indicate a depth of the occupation layer of about 2–3 m, at least in the western part of this hill. At the same time, the ERT estimates a greater depth. However, on top of this sharp border, several structures associated with building structures are located. At about meter 22, in both the ERT and GPR, the bottom layer crops out closer to the surface. Farther westward, the GPR displays several layers cropping out to the steep surface. In addition, the ERT ends at the position with higher resistivity values at the uppermost layer. The 3D GPR (Figure 153) displays the spatial distribution of the strong reflectors grouping at a depth of about 0.25 m, without building any polygonal structures, which could be estimated with in-situ foundations or wall structures. A4.1_MAG_c (Figure 155) has a heterogeneous distribution of the magnetic field, too. In the western part, it can be estimated that the diggable saprolite of the mica schist was likely removed to construct a stable and steep slope. The material could be put on top of the hill to flatten it. However, without ground-truth, this is still a reasonable but ambiguous estimation.

At the northern hill, there are fewer reflections and magnetic anomalies in the uppermost areas. However, there were still a few, which could be associated with building structures in A4.1_GPR_d (Figure 150). The occupation layer is located at a lower depth, about 1 m deeper at this hill than at the other hills, as A4.1_GPR_d suggests. The bunches of reflections in this profile can most likely be associated with the bedrock of different weathering statuses; however, there is still the possibility that a massive wall was located here. At least the reflections seem to be comparable to those of wall structures in Mömbris.

North of the castle hill, a small triangle surrounded by a deep incision of ditches is located. Here, A4.1_MAG_e (Figure 154) describes a few diffuse located anomalies and a linear structure with lower nT values. A4.1_OC_e (Figure 155) displays the bedrock directly at the surface at the location of the linear anomaly. In agreement with the argumentation for a flattened area, some material was approximately raised in the western parts to create a leveled area.

4.4.6 Quantification of the medieval, anthropogenic influence in Alzenau

This chapter is associated with workflow Step 6 (S6) – *Quantification of the anthropogenic influence.*

First, all results and interpretations at *Rannenburg Castle* near Alzenau are briefly summarized, and the medieval, anthropogenic influence on the relief development is estimated according to Section 3.5.

A1 – Loess coverage

Detected with ERT measurements and boreholes, partly huge loess deposits could be found at the study site in Alzenau. Colluvial material within the boreholes indicates a relocation of the material, suggesting agricultural or at least forestry use of this area. Therefore, the usage northeast of the castle can be classified in Category 2 (cf. Figure 156). However, the intense gully erosion, combined with potential deforestation due to agricultural use, allows for the assumption that relief development is passively, massively affected (Category 3) by anthropogenic usage (Figure 156).

A4.1 – Building structures on-site

Different geophysical data indicates a massive impact (Category 4) of the anthropogenic building activity on the relief development on-site (cf. Figure 156). Additionally, intense earthworks were restructuring the relief at the site. Beside the dug neck ditches, the removed material from those was most likely placed on top of the different sections of the castle hill to level it.

Figure 156: Intensity and type of medieval, anthropogenic influence – Alzenau

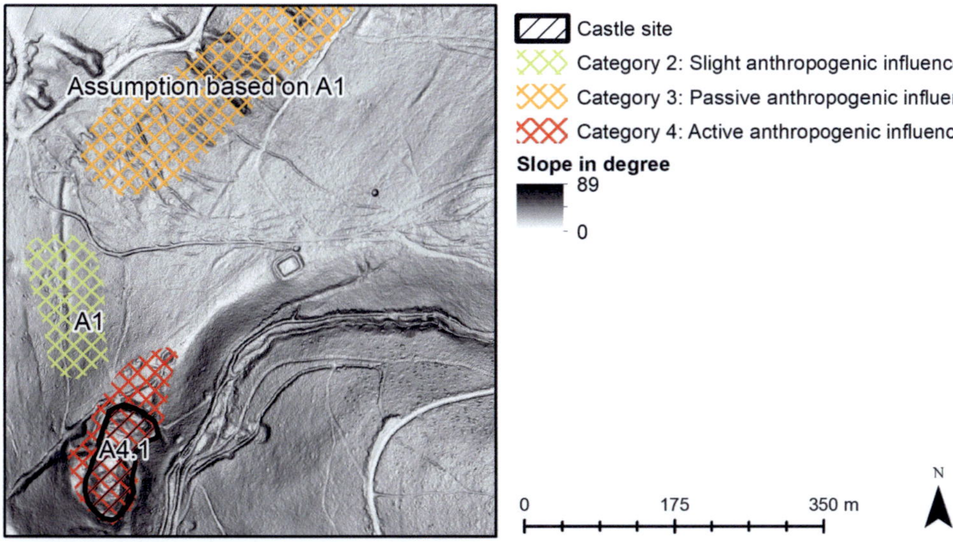

Source: Bavarian Agency for Digitalisation, High-Speed Internet and Surveying, 2014
Map creation: J. Trappe

4.5 Eisenberg

First, site-specific questions were derived with the help of the GIS-based terrain analyses at Eisenberg. The results for each question are provided and interpreted. Then, an estimation of quantification of the anthropogenic influence is made. A technical overview of the used application of the geophysical methods is given in Appendix 14 to Appendix 16.

4.5.1 Morphometric analyses for the creation of site-specific questions

The morphometric analyses for creating site-specific questions are associated with workflow Step 2 (S2) – *GIS-based terrain analyses.*

The area of interest is mostly classified as a culmination area, but single spots are also identified as top, shoulder by the TPI landform classification (Figure 157a). At lower elevations, the topographic position is classified as lowlands, larger depression and lowlands, small depression inside the creek valleys. Surrounding the Eisenkopf, hollow ways are deeply caught through the terrain. At all slopes and even on top of lower peaks, wet meadows are located. Most prominent are the Neudorfwiesen, west of the Eisenkopf. At these meadows, for example, west of the hill, more often, TWI values (Figure 157b) are higher than in the surrounding area, and linear structures are visible in its distribution.

Figure 157: Morphometric analyses – Eisenberg: a) landform classification and slope; b) topographic wetness index

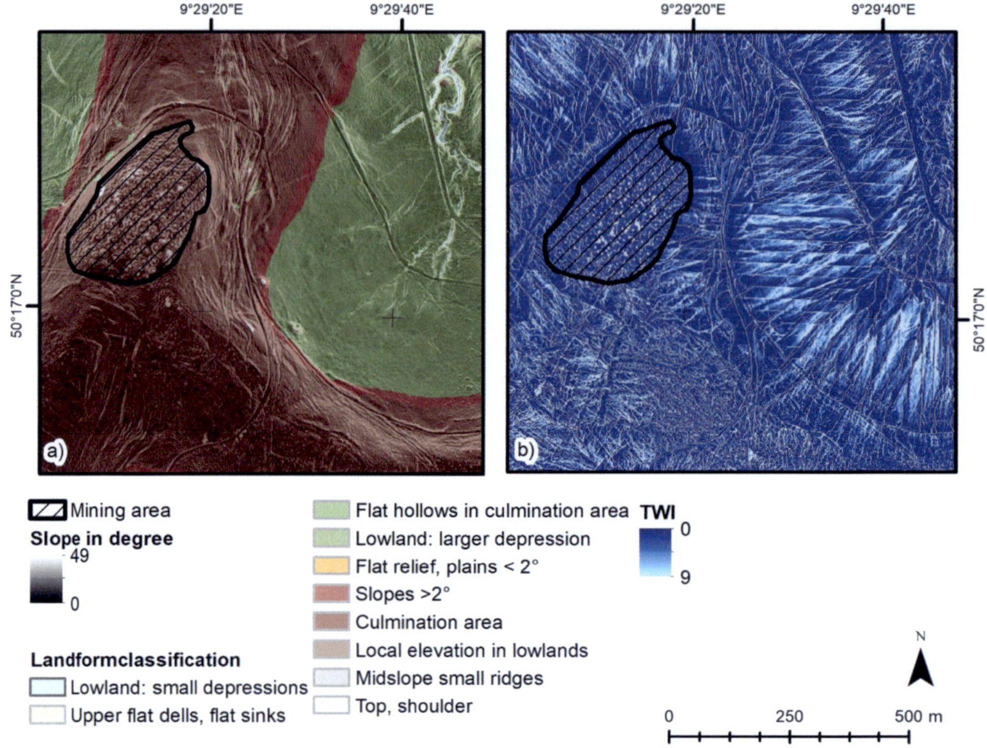

Sources: own illustration, based on Hessian Administration for Land Management and Geoinformation, 2007
Map creation: J. Trappe

With respect to the specific position of the mining site surrounded by wet meadows, including several collapsed shafts, the following questions are derived (an overview is in Table 34).

Question E2 – surficial, anthropogenic features:

What is the internal structure of the collapsed shafts? Where is the ore body?

The depth, form and internal structure of the collapsed shafts are still unknown and need further investigation for both the archaeological interpretation and understanding of its influence on landscape development.

Question E3 – buried, natural features:

What is the hydrological situation in the meadows around Eisenberg? How much are they anthropogenically influenced?

A movement of the heavy ore over a wider distance is unlikely. Therefore, it must be expected that buildings belonging to the metal industry are located in the surroundings. As access to water was crucial for medieval smelting locations, the wet meadows are further investigated for both archaeological and hydrological knowledge.

Table 34: Site-specific questions at the Eisenberg

Category	*Questions*
E2	What is the internal structure of the collapsed shafts? Where is the ore body?
E3	What is the hydrological situation in the meadows around Eisenberg? How much are they anthropogenically influenced?

Source: Own table

4.5.2 E2 – Collapsed shafts: results

What is the internal structure of the collapsed shafts? Where is the ore body?

Cross-profiles of ERT and GPR were placed at the edge of the mining area, characterized by collapsed shafts. Additionally, two pits were dug with an excavator – one at a collapsed shaft, another slightly outside the mining areas (cf. Figure 158). Further drills were hindered by the massive basalt layers almost directly below the surface.

Figure 158: Location of measurements – E2

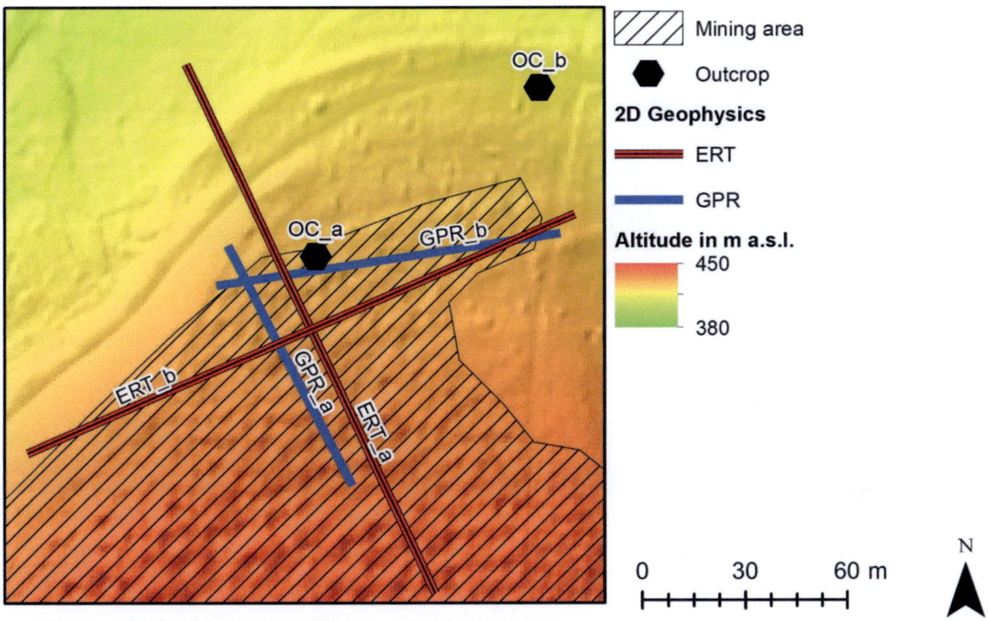

Source: Hessian Administration for Land Management and Geoinformation, 2007
Map creation: J. Trappe

4.5.2.1 Geophysical data

The geophysical data of E2 are associated with workflow Step 3 (S3) – *Geophysical methods*.

ERT

First, the inverted resistivity distribution of the profiles is described. Then, their relative distribution is discussed according to the guidelines for ERT description.

E2_ERT_a resistivity distribution

E2_ERT_a stretches in a south-north direction across the *Eisenberg mining area,* including the collapsed shafts.

Figure 159: Electrical resistivity tomography – E2_ERT_a

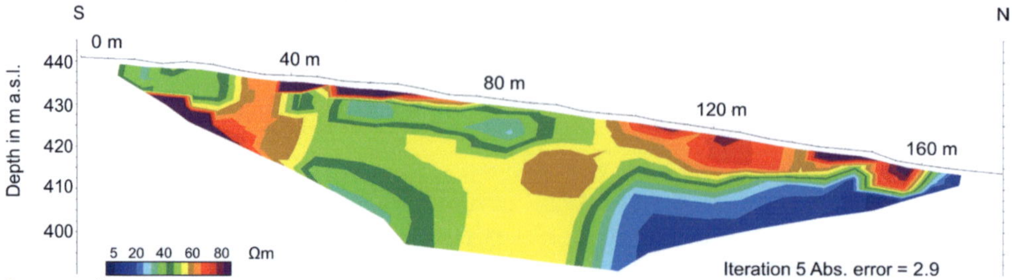

Source: own illustration

In the southern part of E2_ERT_a (Figure 159), areas with higher resistivity values of up to 150 Ωm alternate with areas of lower resistivity values of between 30 to 50 Ωm. Areas with higher resistivity values occur at the bottom of the profile, between meters 10 and 40 at a depth of 5–20 m and in the uppermost part between meters 40 and 75. Lower resistivity values occur in lenses surrounded by a matrix of slightly higher resistivity values. At a depth of 5–10 m, in the central part between meter 80 and meter 100, an area with constant resistivity values of about 50–60 Ωm exists. In the northern part, two areas of different resistivity values are visible. Resistivity values directly below the surface are between 70 and 90 Ωm and decrease to 5–10 Ωm at greater depths. The upper, horizontal boundary of this structure with lower resistivity values intersects with the inclining surface at about meter 160.

E2_ERT_a description after the guidelines

Adding to the chaotic appearing resistivity distribution the information of the boundary analysis (Figure 160), three layers are classified (Table 35).

Figure 160: Boundary analysis – E2_ERT_a: a) sharpness of r-layer boundary including the position of r-layers and r-anomalies; b) orientation of the boundary gradient

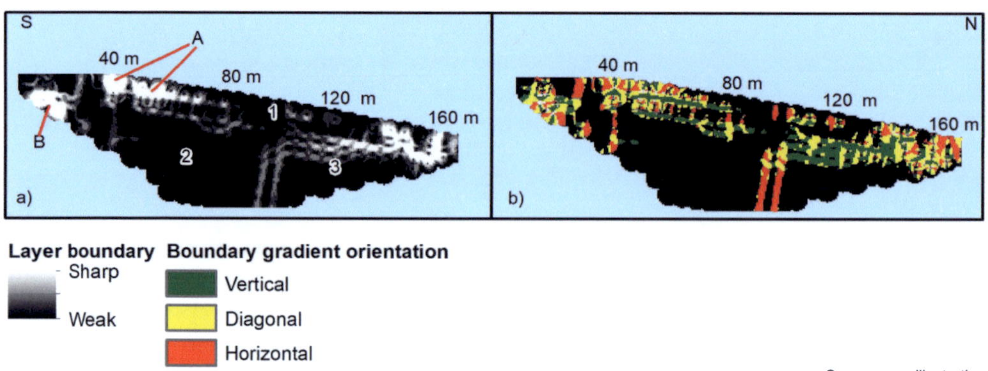

Source: own illustration

At the top, an r-layer (1) with higher resistivity values, including several gaps, is delimited with a sharp boundary to the bottom r-layer. The thickness of this r-layer varies but stays mostly small. Underneath this r-layer and at the top layer's gaps, directly underneath the surface, a second r-layer (2) is located. This r-layer has a vertical, medium-sharp boundary to the northern, third r-layer that has lower resistivity values. The first two r-layers show multiple inhomogeneities. In contrast, the third (3) r-layer is homogeneous. Within the top two r-layers, several r-anomalies (A) are embedded. The r-anomalies are visible through even higher resistivity values than that of the surroundings and are separated with sharp boundaries to the second and medium-sharp boundaries to the first r-layer. These boundaries surround the r-anomalies nearly completely; only the top is left out. In deeper areas, some blocky r-anomalies (B) are located. Weak-to-sharp boundaries surround them.

Table 35: ERT guidelines – E2_ERT_a

R-layer	*Homogeneity*	*R-layer thickness*	*Layering*	*Horizontal changes*	*Boundary of r-layers*	*R-layer orientation*
R-layer 1	3	1	3	2	5	7
R-layer 2	2	3	4	2	3	6
R-layer 3	5	2	-	-	-	-
R-anomaly	*Number of r-anomalies*	*R-anomaly location*	*R-anomaly form*	*R-anomaly size*	*R-anomaly boundary*	*Boundary orientation of r-anomalies*
R-anomaly Type A	2	1	2	2	5	2
R-anomaly Type B	1	2	1	2	1–5	1

Source: own research

E2_ERT_b resistivity distribution

E2_ERT_b covers the northern part of the mining area of Eisenberg, orthogonally to E2_ERT_a.

In the western part of E2_ERT_b (Figure 161), up to meter 80 of the profile, relatively higher resistivity values occur in an uppermost part between 5 and 10 m thickness. At meter 50 and meter 110, lenses of this value range are embedded

Figure 161: Electrical resistivity tomography – E2_ERT_b

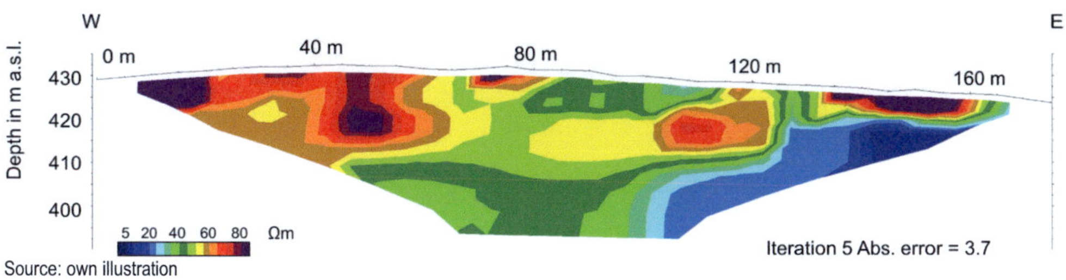

Source: own illustration

in deeper areas up to 20 m of depth. In the western part, until a depth of about 20 m, the lenses are surrounded by a matrix with resistivity values between 50 and 60 Ωm. At a depth of 20 m, the resistivity values decrease to around 40 Ωm and reach the surface between meter 80 and meter 115. However, underneath a depth of 5–10 m, in the eastern part, resistivity values decrease to values below 20 Ωm.

E2_ERT_b description after the guidelines

With the combination of the resistivity distribution information with those of the boundary analysis (Figure 162), three r-layers appear, similar to E2_ERT_a (Table 36).

Figure 162: Boundary analysis – E2_ERT_b: a) sharpness of r-layer boundary including the position of r-layers and r-anomalies; b) orientation of the boundary gradient

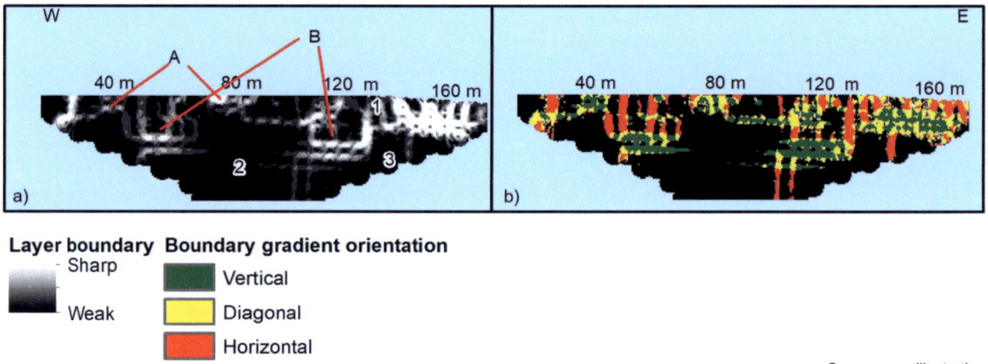

Source: own illustration

An uppermost (1) r-layer, including a couple of gaps surrounded by sharp boundaries, followed by a second (2) r-layer, in the western part, is split by a medium-weak horizontal boundary from the third (3) r-layer in the eastern part. The layering be-

Table 36: ERT guidelines – E2_ERT_b

R-layer	*Homogeneity*	*R-layer thickness*	*Layering*	*Horizontal changes*	*Boundary of r-layers*	*R-layer orientation*
R-layer 1	3	1	3	2	4	7
R-layer 2	3	3	3	2	2	3
R-layer 3	5	2				
R-anomaly	*Number of r-anomalies*	*R-anomaly location*	*R-anomaly form*	*R-anomaly size*	*R-anomaly boundary*	*Boundary orientation of r-anomalies*
R-anomaly Type A	2	1	2	1	5	2
R-anomaly Type B	2	2	1	2–3	4	3

Source: own research

tween these r-layers is visible but not that well developed. The first two r-layers have a medium homogeneity and various boundaries. However, these boundaries partly belong to r-anomalies (A) located within the top and the second r-layer. The uppermost r-anomalies are located directly underneath the surface and are characterized by a sharp boundary with the second r-layer and weaker boundaries with the top r-layer. The deeper r-anomalies (B) are drop-like, with a bottom, blocky part but still with some kind of limb at the top. These are located within the second r-layer and are surrounded by medium-weak boundaries; only the top is left out.

GPR

The GPR data is described in line with Neal (2004).

E2_GPR_a and _b

E2_GPR_a extends downslope on the northern part of the hill.

Figure 163: Ground-penetrating radar measurement – E2_GPR_a: a) data; b) data with marked changes in signal intensity and structure

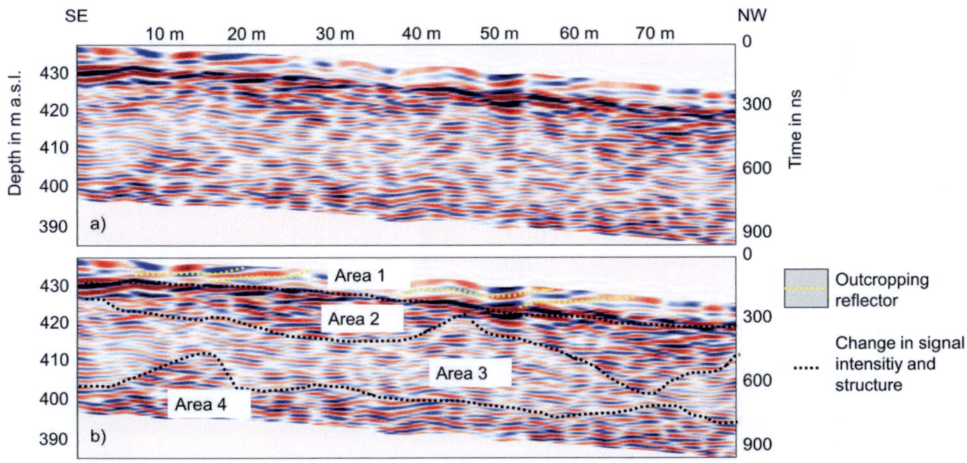

Source: own illustration

Within E2_GPR_a (Figure 163), the upper and lower boundary of the profile are classified as concordant and slightly erosional truncation, with a couple of reflections cropping out. Four different areas can be distinguished. Reflections of the uppermost area (1) are mostly planar/even, horizontally dipping and with a subparallel relationship between the reflections and a continuous reflection continuity. The area has a size of up to 5 m. A clearly visible discontinuous boundary delimits this area from the following area (2). Discontinuous and partly chaotic reflectors dominate this area. They are partly classified as curved, sigmoidal, partly as subparallel. The thickness of this area varies strongly and has a maximum extent of up to 15–20 m. The reflections of the third area (3) are less strong, but they are more continuous, less chaotic and subparallel. However, reflection continuity is still moderately continuous. The bottom area (4) starts below a depth of 405–410 m a.s.l. and displays stronger reflections again. The single reflections are still moderately continuous, but also several-meters tall ones are visible. E2_GPR_b covers the northern part orthogonally to E2_GPR_a.

Figure 164: Ground-penetrating radar measurement – E2_GPR_b: a) data; b) data with marked changes in signal intensity and structure

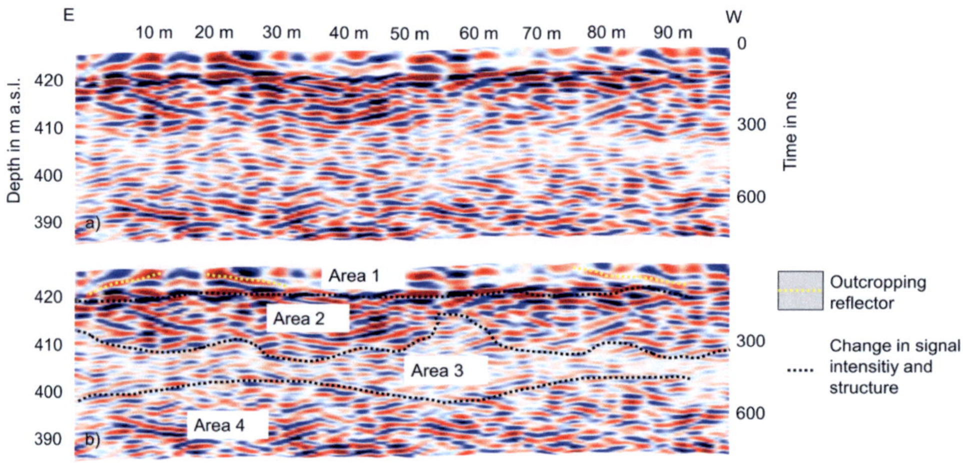

Source: own illustration

The same four areas described for E2_GPR_a are also visible in E2_GPR_b (Figure 164). They are even more recognizable than those in E2_GPR_a. The size of the uppermost area is comparable to that in E2_GPR_a. The bottom area is located at almost the same elevation. Therefore area 3, including the weak reflections, is thinner.

4.5.2.2 Invasive methods and outcrops

The results within this subsection invasive methods and outcrops are associated with workflow Step 4 (S4) – *Ground-truth.*

Outcrop

E2_OC_a (Figure 165a) is an excavated, collapsed shaft, and E2_OC_b (Figure 165b) is an outcrop north of the expected position of the ore body. Both pits were performed using an excavator.

The pits E2_OC_a and E2_OC_b were both made with an excavator. The location of E2_OC_a is directly at a collapsed shaft. Due to the shaft's funnel-shaped outline, the upper 3.5 m of the shaft are filled with debris. The debris is enriched with fragmented basalt rocks and even blocks. Underneath this layer, a clay layer is developed. Within this layer, sand lenses and also fragmented basalt (cf. Figure 166a, b) pieces are mixed in the matrix.

Figure 165: Outcrops – E2: a) E2_OC_a; b) E2_OC_b

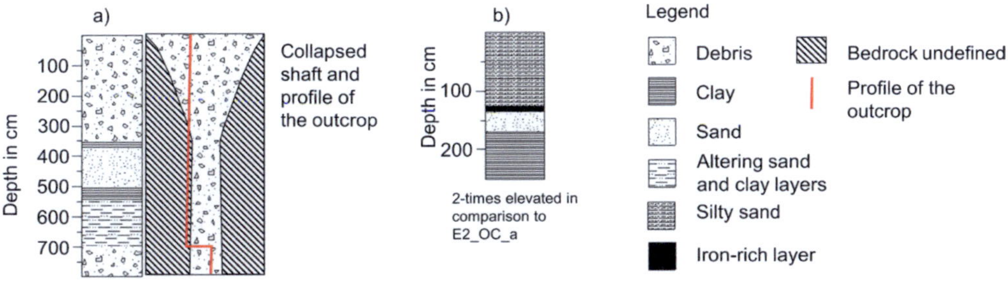

Source: own illustration

Figure 166: Field pictures – Outcrops E2: a) E2_OC_a; b) E2_OC_a – illustrating the Basalt; c) E2_OC_b

Photographer: A. Schreck, 2020

Up to a depth of 5 m, sand is the dominant grain-size. Clay-richer areas are partly located within this sandy layer. Here, basalt fragments up to peddle size are still mixed in the surrounding matrix. The sand-rich layer indicates oxidation features. To a depth of 5.5 m, a massive, strongly reduced clay layer is located. Underneath this layer, the excavator was not able to dig further. Therefore, within the outcrop, a Pürckhauer sounding was performed. Until this depth, the description was performed at the walls of the dug hole. Underneath 7 m depth, due to the Pürckhauer sounding, it was in the middle of the hole. The sediments within the core still consist of loose sand and basalt fragments. It was loose and stacked, and therefore it was described as debris, most likely of the shaft itself.

The photographs (Figure 166a, b) were collected after the outcrop of the collapsed shaft was closed again. The described debris was removed as far as the bedrock occurred. It is visible that the bedrock consists of basalt in the upper parts.

E2_OC_b (Figure 165b and 166c) was located further downhill out of the area of the collapsed shafts. Opposite of the description of E2_OC_a, a more detailed description could be performed. In the first 130 cm, silty, partly loamy sand layers are located. This package is separated from the underlying layers by a few centimeter-sized layer of iron oxides. This layer is cemented and dark brown to black in color. Underneath this layer, an almost pure fine sand layer is located. The bottom layer consists of partly loamy clay. At a depth of about 100–170 cm, the substrate is oxidized and reduced at root paths, whereas the clay is reduced and only at small areas oxidized.

4.5.3 E2 – Collapsed shafts: interpretation

The interpretation of the data was performed according to workflow Step 5 (S5) – *Interpretation of the data.*

Synthesis

With the combination of all the different datasets, at least three lithological units can be distinguished, and the location and depth of the shafts can be estimated. At the top, relatively high resistivity values and medium-high values alternate. The medium-high resistivity values can be associated with loess loam or loess loam-rich deposits, which could be observed in E2_OC_b (Figure 165b) and are described in the geological map (Diederich & Ehrenberg, 1998). The observed resistivity values of around 30–50 Ωm align with the literature values (Reynolds, 2011). Most likely caused by local differences in material, water chemistry and saturation, the values are recognizable lower than those of the loess-rich deposits at other areas of investigation in this study.

Partly underneath loess loam-rich deposits, the basalt coverage is located. Whereas the strong reflector within the GPR data indicates a more or less continuous layer, the higher resistivity values associated with the basalt, are less continuous. Furthermore, the basalt has a vesicular texture. Due to different water content and a different mineral enrichment within these pores, a different conductivity can be caused. Literature values show a wide range of values for this rock type (Reynolds, 2011), but their relative position and the in-situ data are in line with this argumentation. Possible measurement errors can occur by the existence of the ore as a conductive material. The top two layers are difficult to distinguish with GPR, but at least parts of the basaltic bedrock are resolved within Area 2. However, these seem not to be such dominant in the data. Possibly wide parts of the ore are mined, and an orebody underneath the basalt is expected (both discussed in Chapter 5).

Underneath the basalt, Miocene sediments occur. These sediments crop out at E2_OC_b (Figure 165b). Their materials can also be found as debris in the collapsed shafts (Figure 165a). However, these sediments are described as sand and clay. Therefore, their resistivity values differ. The GPR results display no strong reflections in Area c. Most likely, the size of the clay-rich layers which are described within the Miocene sediments are too small, as indicated by E2_OC_b, or groundwater superimposes a differentiation within the lithological units.

Below the Miocene sediments, the Buntsandstein is described in the geological map (Diederich & Ehrenberg, 1998). Evidence for this unit is barely visible in the data. However, the southernmost part of E2_ERT_a (Figure 159) suggests the first outcrop, and the approximate change of the GPRs in Area 3 in comparison to Area 4 indicates its existence. Nevertheless, these results are not proven, but the description underneath the Miocene sediments would fit these results.

For safety reasons, a maximum depth of 8 m could only be reached in E2_OC_b, therefore the collapsed shafts are located to depths of at least 8 m. It can be expected that the orebody underneath the basalt was mined more or less area-wide. The ERT describes a couple of r-anomalies with drop-like or blocky forms. These anomalies possibly represent voids filled with overburden of the subsequently dug shafts (e.g. E2_OC_a). Depending on the type of material, the water content and air content, such voids can be identified in the ERT or not. However, such structures can be found in the GPR as well.

Interpretation

What is the internal structure of the collapsed shafts? Where is the ore body?

Resulting from this, the interlocking of geomorphology and geology with the anthropogenic use of the area in the Middle Ages becomes visible. In some areas of the Eisenberg, loess loam deposits occur. Due to the used spacing of the geophysical data, these deposits are not well recognized within the data.

As the conceptual model E2 shows (Figure 167), the basalt coverage size, located mostly at the uppermost part, can be estimated at 3–8 m, depending on the position on the hill. Within both geophysical methods, those sizes of the layer can be approximated. Whereas the GPR displays a more or less constant size of about 3–5 m downhill, the top of E2_GPR_a (Figure 163) and the ERT data indicate this layer's further extent uphill. Additionally, the outcrop E2_OC_a, at the foot of the hill, describes a layer of the loose sediments at a depth of 3.5 m, and the basalt crops out at the top as well (Figure 166a, b).

Underlying the basalt and also further downhill, Miocene clay and sand are located. In general, those deposits are only located in favorable locations (Ehrenberg & Hickethier, 1982); at the Eisenberg, they seem to be located area-wide, at least in the northern part. On the one hand, the E2_OC_a (Figure 165a) displays these sediments at least until a depth of 8 m, and the data of both geophysical methods underline this

Figure 167: Conceptual model – E2

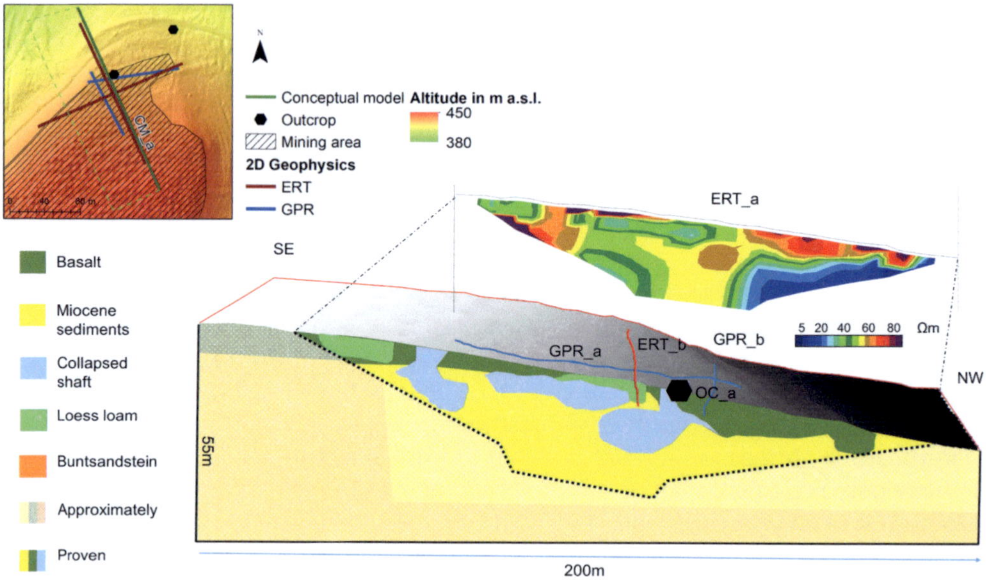

Sources: own illustration, based on Hessian Administration for Land Management and Geoinformation, 2007

estimation. On the other hand, the transect of the geological map (Figure 168) estimates such a layer to be underneath a neighboring hill only a few hundred meters away. The fine structure within these loose sediments, visible within the outcrops, cannot be resolved in the geophysical data. While the ERT data allows for, at least, a differentiation in a northeastern part with resistivity values around 10 Ωm and a southwestern part with resistivity values around 50 Ωm, but within the GPR data, there is no indication of a differentiation.

Figure 168: Geological profile

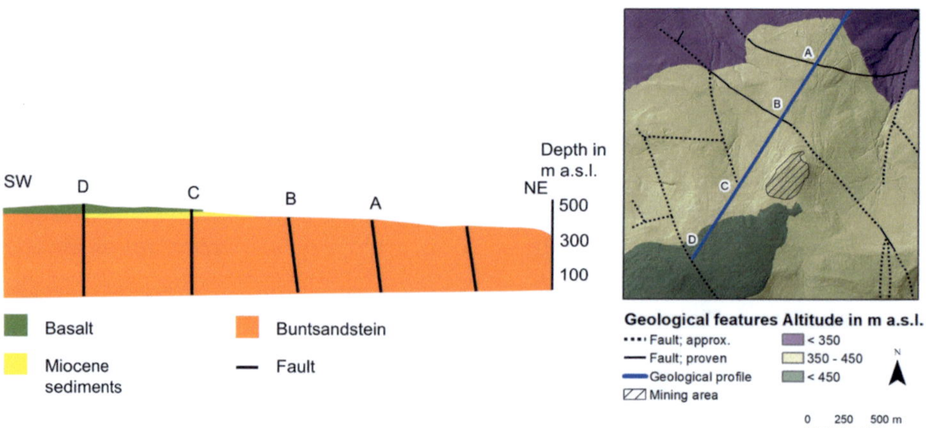

Source: own illustration, based on Diederich & Ehrenberg, 1998

Until these loose sediments, the shafts were dug. Derived from the geophysical data, these shafts could be dug to a depth of about 15–20 m. However, at least 8 m of depth can be proven with the maximum depth of E2_OC_a. They are visible in the ERT by high-resistive r-anomalies, surrounded by variable orientated and pronounced boundaries and by areas of discontinuous reflectors within the GPR. The shafts are collapsed; therefore, the fallen material delivers discontinuous reflections, including a high pore volume filled with air or water that delivers anomalies of different resistivity values.

At the bottom of the profiles, Buntsandstein can be expected. Whereas the GPR indicates a more or less continuous border at an elevation of 400 m a.s.l., the only hint within the ERT data is located at the southern part of E2_ERT_a. Here, a high-resistive quasi-layer is located at an elevation of about 430 m a.s.l. Combining this information with the geological profile western of the area of interest leads to the conclusion that faulting of the Buntsandstein can be the cause, even though it is not described on the map.

According to the question, it is visible that the shafts were dug through the basalt. Here the material was removed in wide areas in the surroundings of the shaft. Because of this, the ore body can be expected underneath the basalt and in or on top of the Miocene sediments.

Figure 169: Location of measurements – E3

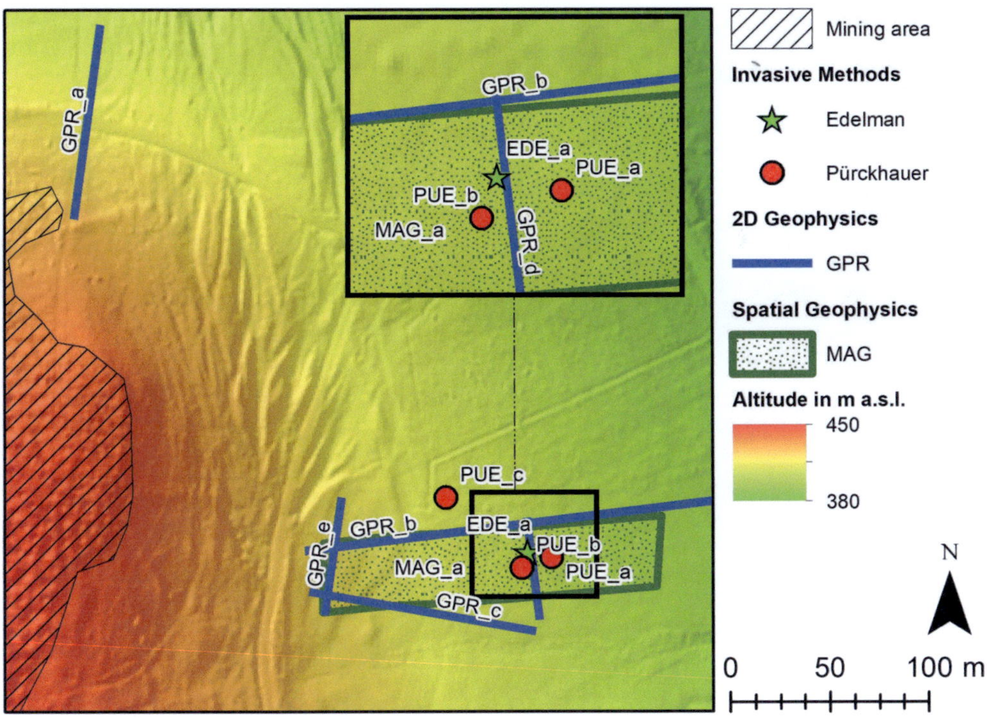

Source: Hessian Administration for Land Management and Geoinformation, 2007
Map creation: J. Trappe

4.5.4 E3 – Hydrological situation: results

What is the hydrological situation in the meadows around Eisenberg? How much are they anthropogenically influenced?

For answering the question about the hydrological situation, GPR profiles were placed at the Neudorfwiesen and also at a small meadow north of the mining area (cf. Figure 169). Additionally, a huge magnetic survey was conducted, and invasive methods were utilized at the Neudorfwiesen, too.

4.5.4.1 Geophysical data

The geophysical data of E3 is associated with workflow Step 3 (S3) – *Geophysical methods*.

GPR

E3_GPR_a is located within Eisenberg's northern part, covering the mining area's downslope parts and the northernmost meadow. E3_GPR_b and -c are located downslope in an east-west direction at the meadows of the Neudorfwiesen. In contrast, E3_GPR_d and E3_GPR_e are parallel to each other, slope-parallel orientated and orthogonally to E3_GPR_b and E3_GPR_c.

E3_GPR_a

Within E3_GPR_a (Figure 170), the upper and the lower boundary are classified as concordant to erosional truncation.

Figure 170: Ground-penetrating radar measurement – E3_GPR_a: a) data; b) data with marked changes in signal intensity and structure

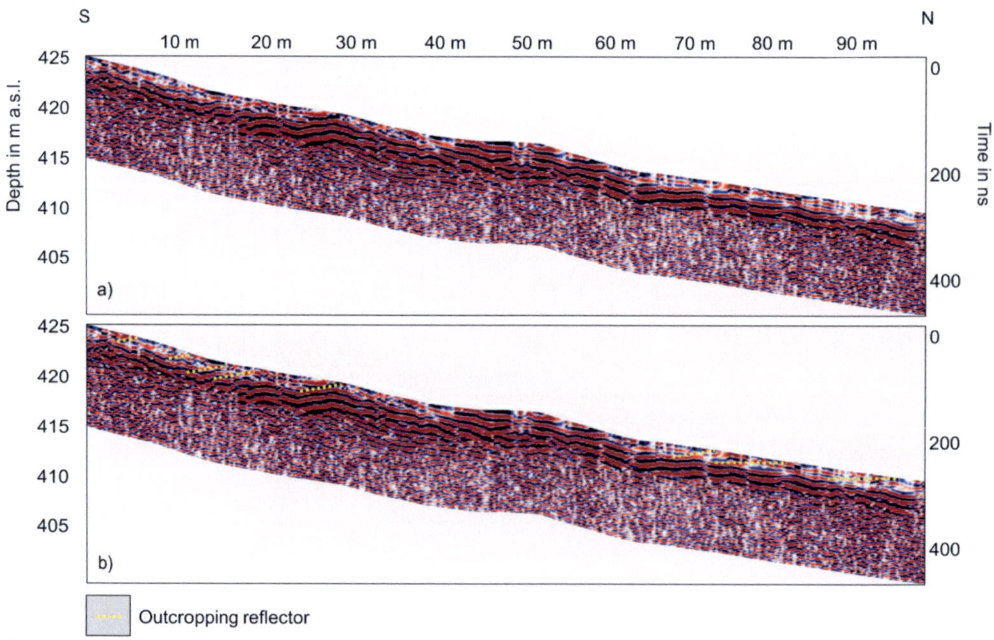

Source: own illustration

Several moderately continuous, subparallel or parallel reflectors are visible within the first few meters. These are horizontally orientated and, therefore, crop out to the surface of the smooth slope. Especially within the uppermost meter, the reflections become discontinuous. Underneath a depth of about 5 m, only in some parts moderately continuous reflections can be found. Deeper than about 6 m, reflectors can be seen only sporadically within the matrix.

E3_GPR_b and -c

Due to its position within the wet meadows of the Neudorfwiesen, E3_GPR_b (Figure 171) and -c (Figure 172) include weaker reflections in total than E3_GPR_a does.

However, similar outcropping reflectors are visible, including the moderately continuous, parallel reflectors. The reflections are better developed in the upper-

Figure 171: Ground-penetrating radar measurement – E3_GPR_b: a) data; b) data with marked changes in signal intensity and structure

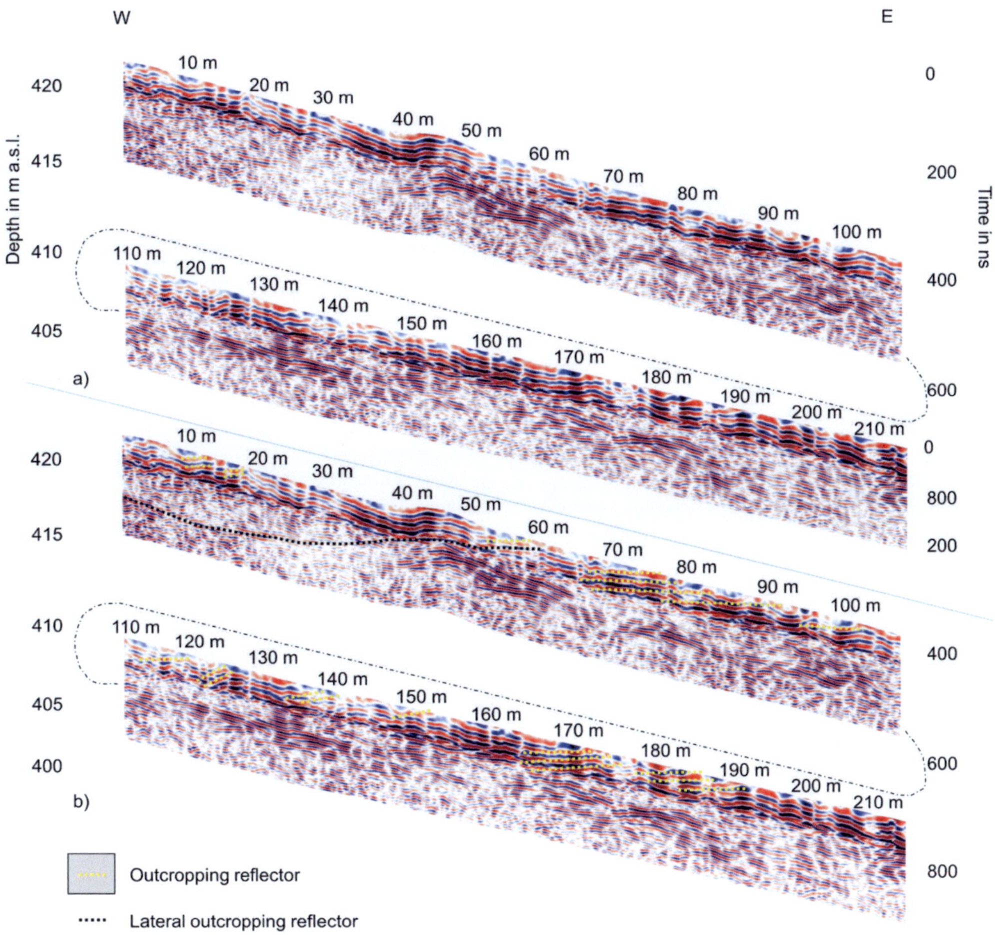

Source: own illustration

most area in E3_GPR_a, but the penetration depth is lower due to the higher water content and conductivity at the meadows. Within E3_GPR_b, there are parts of a reflector that are visible, which is moderately continuous and crops out underneath an area of a higher slope at about meter 60.

Figure 172: Ground-penetrating radar measurement – E3_GPR_c: a) data; b) data with marked changes in signal intensity and structure

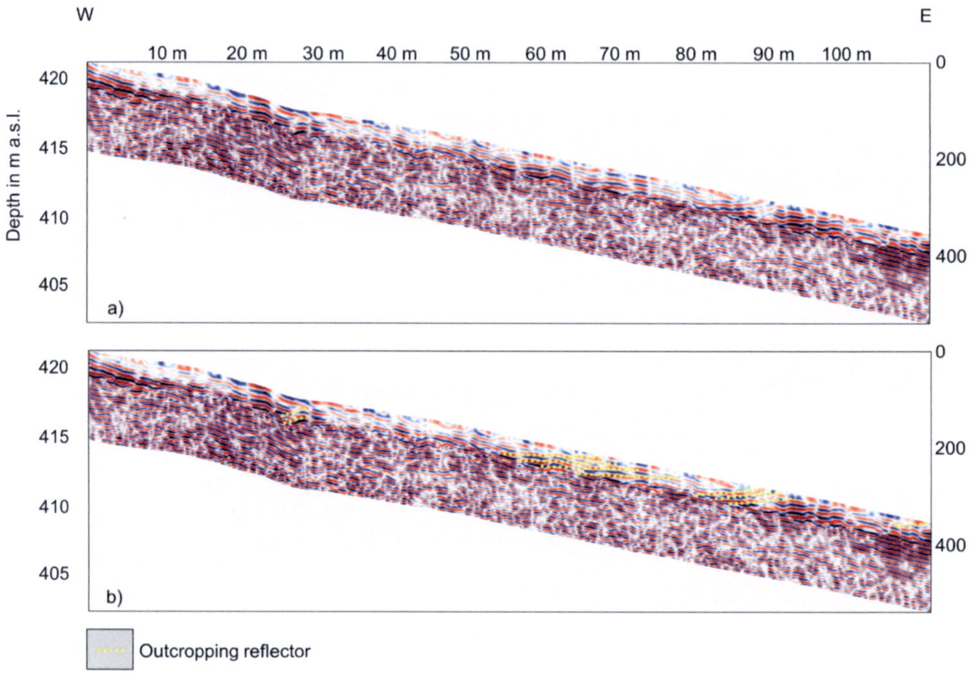

Source: own illustration

E3_GPR_ d and -e

Within E3_GPR_d (Figure 173), a concordant structure is weakly developed but still visible. However, in E3_GPR_e (Figure 174), almost continuous reflectors crop out to the surface.

The radargrams' upper boundary of E3_GPR_d and -e can mostly be classified as erosional truncation by outcropping reflectors and the bottom boundary is not resolved.

Magnetic survey

E3_MAG_a covers a huge area at the Neudorfwiesen, intersecting with several profiles of the GPR.

Within E3_MAG_a (Figure 175), different magnetic anomalies are visible. On the one hand, many small dipole anomalies are visible, forming a curved, linear structure at 35 m along the x-axis and 35 m along the y-axis (a), ending at about 20 m along the x-axis and 0 m along the y-axis. Additionally, there are many more areas

Figure 173: Ground-penetrating radar measurement – E3_GPR_d: a) data; b) data and interpretation

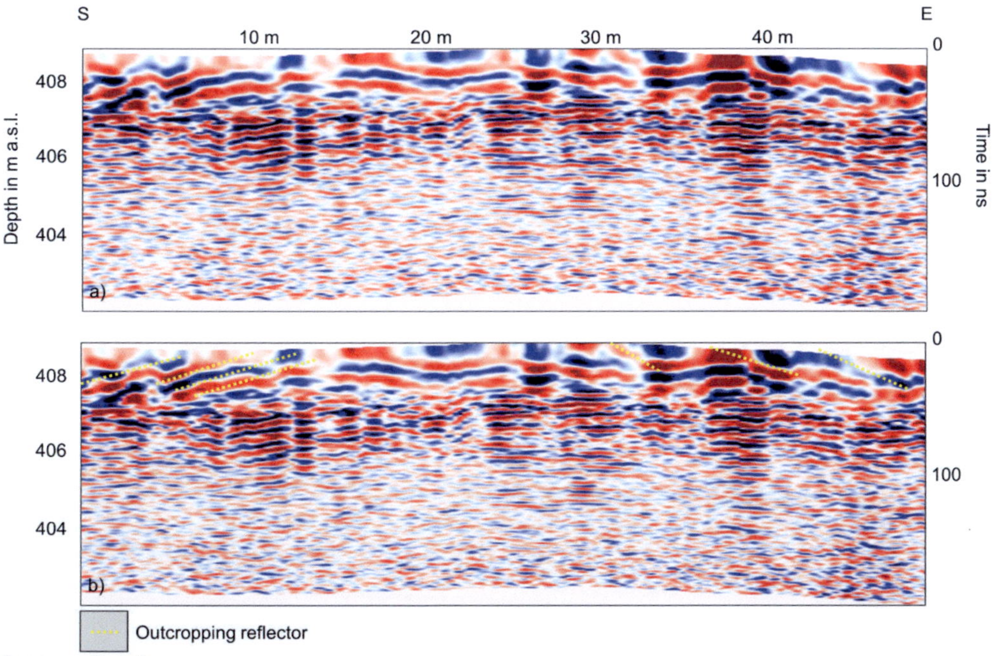

Source: own illustration

Figure 174: Ground-penetrating radar measurement – E3_GPR_e: a) data; b) data and interpretation

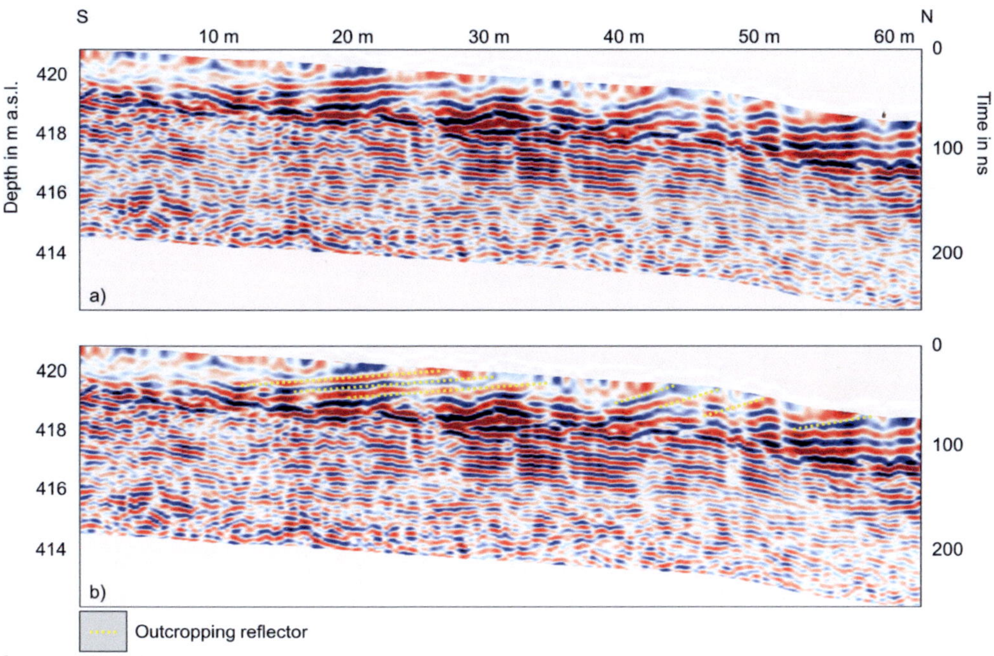

Source: own illustration

Figure 175: Magnetic survey – E3_MAG_a

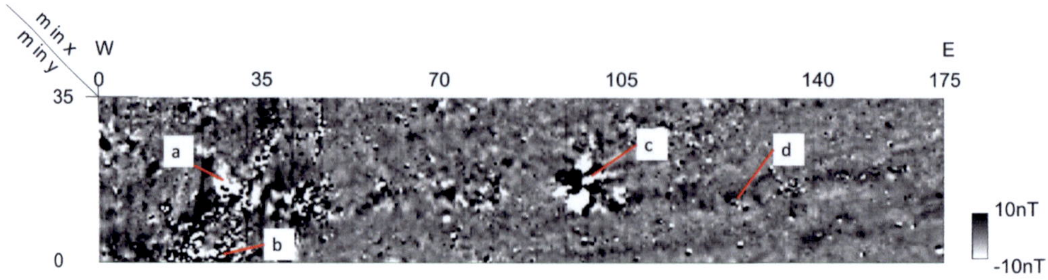

Source: own illustration

where these types of anomalies cluster, for example, in the southern half, at about 30–45 m along the x-axis (b). However, punctual dipole anomalies occur all over the magnetogram, whereby many more occur in the western part of the survey. At about 100 m along the x-axis and 10–20 m along the y-axis, the most significant dipole anomaly exists (c). Opposite to the other dipole anomalies where small high and low nT values occur, huge high and low nT areas exist. Additionally, less intense shades exist, for example, between 100 and 175 m along the x-axis at about 5–20 m along the y-axis (d).

4.5.4.2 Invasive methods and outcrops

The results within this subsection invasive methods and outcrops are associated with workflow Step 4 (S4) – *Ground-truth.*

Pürckhauer and Edelman

Only E3_PUE_c is north of the meadow within the surrounding forest. All other data is located in the wet areas of the meadows, delivering the ground-truth for the magnetic anomaly of E3_MAG_a and the GPR.

Figure 176: Edelman and Pürckhauer soundings – E3: a) E3_EDE_a; b) E3_PUE_a; c) E3_PUE_b; d) E3_PUE_c

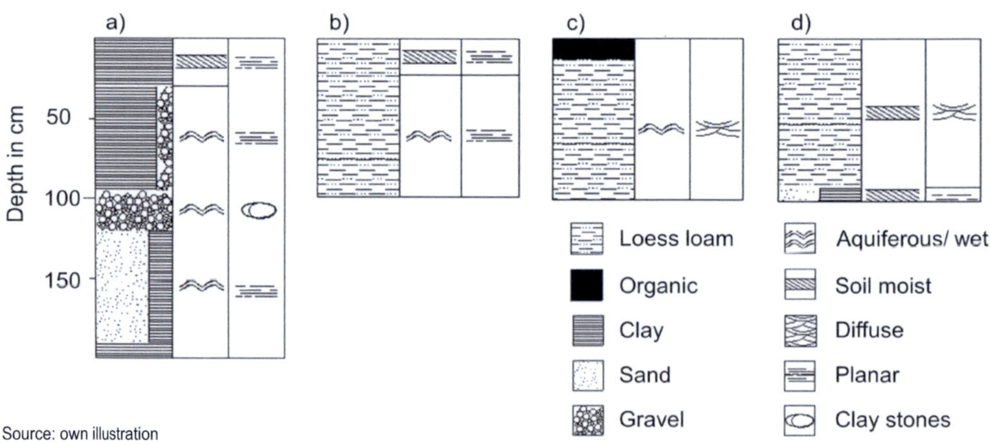

Source: own illustration

All invasive profiles of E3 are located within or in the surroundings of the wet meadows of the Neudorfwiesen. The material in E3_PUE_a (Figure 176b) consists of loess loam. The uppermost layer consists of silty sand, and the material is moist. Underneath a depth of 23 cm, the silt and clay content increases, whereas reduction features are intensely visible. The reduction features are less intensely visible in the bottom 50 cm of the profile than they are in the layer located above, but they are still recognizable. Within the bottommost 23 cm, the reduction changes from pure bleaching to a gleyic reduction. Therefore, caused by the changes in reduction bottomward, different flows of water seem to affect this profile. Because of that, the soil can be described as stagnic gleysol. The amount of water is high, indicating that the layer is aquiferous. The planar lamination is not well-developed; nevertheless, there are still differences to the colluvial soils in the surroundings.

Within E3_PUE_b (Figure 176c), the uppermost layer has a high organic content and is classified as half-bog. Underneath this layer, up to a depth of 40 cm, a silty layer occurs. This layer is intensely reduced, and many charcoal fragments and coarser grains are located in it. The lamination is diffuse and must be necessarily classified as colluvial. Underneath this layer, the reduction features decrease while the clay content and oxidation increase. Within this layer, different fragments of grus of Buntsandstein and Basalt are located. Due to its different reduction features that are unusual for a single aquiferous layer, the profile is classified as colluvial, stagnic gleysol. However, the material mainly consists of loess loam.

Opposite of the other profiles, within E3_PUE_c (Figure 176d), (stagno)gleyic features play a less important role. Underneath the topsoil until a depth of 47 cm, the colluvial material starts to gain cambic features. The sand content increases to a depth of 90 cm. Within this layer are basalt and Buntsandstein fragments mixed into the matrix with a maximum size of that of gravel. Here, no soil-forming processes are recognizable. In the bottommost part of the profile, sand and clay layers alter in a well-developed lamination.

E3_EDE_a (Figure 176a) is located within the magnetic anomaly of E3_MAG_a. Within the uppermost layers, there is a high clay content. Whereas the clay is mixed with silt in the uppermost 30 cm, in deeper parts, there is a higher sand content. Between 30 and 95 cm, the material is intensely reduced and slightly oxidized. Starting at a depth of 95 cm, a massive clay grus layer occurs. The single clay fragments have massive iron concretions surrounding them. Within this layer, there is high water content, and strong oxidation combined with a lower content of reduction features are visible. Underneath this layer, clayey sand is located. To a depth of about 190 cm, oxidation and reduction features alter in combination with different contents of the building substrate. Underneath, a pure clay layer is located. This layer is intensely reduced.

4.5.5 E3 – Hydrological situation: interpretation

The interpretation of the data was performed according to the workflow Step 5 (S5) – *Interpretation of the data.*

Synthesis

In the surroundings of the Eisenberg, the results of the radar illustrate several reflectors that crop out at the top of the profiles. Within E3_GPR_b (Figure 171), there is a discontinuous but also planar and horizontal layer visible. Because of the field observation and the wet and stagnic conditions downslope, this should be a layer of untreated groundwater. On the one hand, the other outcropping reflections consist of loess loam deposits or loess loam-rich, colluvial deposits. On the other hand, these outcropping layers consist of Miocene sediments. These sediments are documented in the geological map and are best described in E2_OC_b (Figure 165b).

Interpretation

What is the hydrological situation in the meadows around Eisenberg? How much are they anthropogenically influenced?

The described Miocene sediments consist of several small layers, partly in-situ and weathered, partly removed to their recent position. These layers cause stagnic conditions on top of clay-rich layers, visible in E3_PUE_a, -b and E3_EDE_a (Figure 176). Underneath the stagnic conditions, there are different other aquiferous layers, cropping out to the surface and feeding the wet meadows with water, partly creating half-boggy or even boggy conditions. Additionally, many creeks originate from this area. Of course, the uppermost parts of the area surrounding the Eisenberg are influenced and used by humans, with, for example, colluvial layers or captured springs. However, these features are not area-wide; the magnetic anomalies seem to be natural. At the locations of the most significant magnetic anomalies, layers of iron concretions occur. In the surroundings of the iron-rich layers, not much anthropogenic use is visible. There was no indication of a building structure or something comparable upward from the concretions. Therefore, it seems most likely that the most significant magnetic anomaly visible in E3_MAG_a (Figure 175) is of natural origin, which shows an accumulation of grus surrounded by an iron crust.

According to the question, it becomes visible that humans partly used the meadows surrounding the Eisenberg. The colluvial layers indicate mainly agricultural use, and only in the western area is metal processing expectable due to several small and dipole magnetic anomalies. More important for the development of the wet meadows are dense and aquiferous layers, mainly of the Miocene sediments. These layers crop out to the surface, and slope water and untreated groundwater located on top or within these layers flows into the topsoil or even onto it. Clay-rich, dense layers prevent the water from reinfiltrating into deeper layers and let it stay nearby or at the surface, creating these soils profoundly affected by water. Even though the high number of magnetic anomalies indicates an intense anthropogenic effect, only at the uphill parts were smelting and metal processing plants located (Hasenstein *et al.* 2018). Downslope, within the uppermost layers, some material is partly mixed in, whereby deeper areas are free of artifacts and include a natural enrichment of iron.

4.5.6 Quantification of the medieval, anthropogenic influence at the Eisenberg

This subsection is associated with workflow Step 6 (S6) – Quantification of the anthropogenic influence.

First, all results and interpretations at the Eisenberg at Steinau an der Straße are briefly summarized, and the medieval, anthropogenic influence on the relief development is estimated according to Section 3.5.

E2 – Collapsed shafts

In combination with dug outcrops, ERT and GPR measurements indicate a massive, active effect (Category 4) on the topography at the Eisenberg (cf. Figure 177). In combination with the geological processes creating the ore, the paleo-relief plays a significant role in the use of the area as a mining area.

E3 – Hydrological situation

The hydrological situation and its importance for medieval use were prospected with the help of 2D GPR measurements, magnetic surveying and different invasive methods. Even though the magnetic survey shows huge magnetic anomalies, those appear as natural iron concretions; only the western part of the Neudorfwiesen was used for smelting (Hasenstein *et al.*, 2018). Because of that, a slight anthropogenic impact on the relief during the medieval period is estimated to be in the western part (Category 2), whereas no evidence for medieval usage (Category 1) could be found eastward (cf. Figure 177).

Figure 177: Intensity and type of medieval, anthropogenic influence – Eisenberg

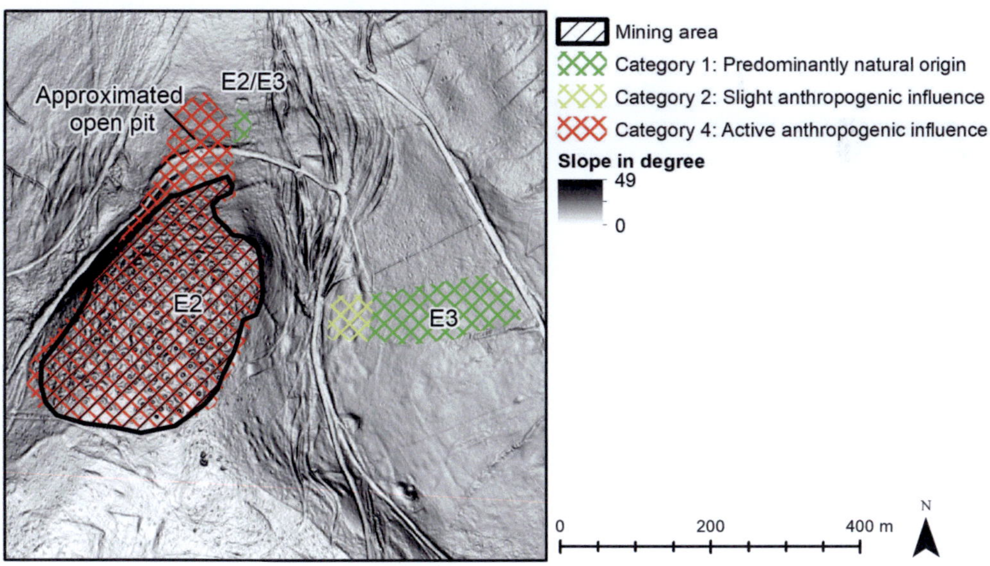

Source: Hessian Administration for Land Management and Geoinformation, 2007
Map creation: J. Trappe

5 Discussion and outlook

The discussion is structured according to the objectives: first, the site-specific, regional bunch of objectives is discussed. Second, the data transferability is discussed, followed by a methodical discussion and outlook. The questions raised via the objectives are answered, and the results, interpretations and applied methods are critically reflected.

5.1 Site-specific

First, in the site-specific discussion, some general statements are made according to the analyses of the individual study sites. The first two objectives are related to each other and, therefore, are answered together. Subsequently, the questions for the single sites are discussed.

- *O1.1.1 Is it possible to decide which parts of the present relief are predominantly of natural origin and which have been intensely anthropogenically influenced at the individual study sites?*

- *O1.1.2 What specific interactions of relief development and medieval building activity exist at the single study sites? How can these be detected, measured and quantified to answer the site-specific questions according to the following categories?*

Fifteen site-specific questions were derived distributed across the five study sites. These questions helped decipher which parts of the study sites are predominantly of natural origins and which are affected by medieval building activity. These questions are related to different specific interactions between relief development and building activity from different perspectives. Therefore, some are focused more on archaeological features, for example, to find buried building structures. Others focused on relief development exemplary at unique geomorphological positions of the castle. Most of the questions can be answered by using the multi-method approach.

At four of these study sites, before the measurements, during or afterward the measurements, archaeological excavations were conducted. Some of this study's results were used for more targeted excavations, but both the excavation and this research can be put in the context of each other.

At most of the study sites, the interaction was not measured directly. In most of these cases, predominantly natural relief features and anthropogenic building structures were detected using geophysical and sedimentological investigations. However, their interaction could only be derived from causal relationships between these features.

For all of the study sites, the origin of the relief was described qualitatively; however, for the anthropogenic influence, an ordinal scale was derived to quantify the impact. Further quantification of this impact was not the aim of this work and could not be performed using the multi-method approach.

In contrast to studies that focus on a precise and measurable quantification (e.g. Rózsa *et al.*, 2020), this study only performed basic ordinal-scaled quantification and the actual influence was not measured. The anthropogenic influence of the past during a specific period was estimated based on the inventory of anthropogenic features within the subsurface and the detectable earthworks. For a more detailed estimation, classification and quantification, a combination with different geochemical proxy analyses and further dating methods would be helpful (e.g. Mayoral *et al.*, 2020). However, such analyses are not the scope of this work.

However, the general deviation into the active and passive anthropogenic effect on a landscape (Herget, 2006; Schütt, 2006b) could be assessed. Nevertheless, primarily passive impacts, such as soil erosion, often cannot be estimated due to the lack of archives (Anselmetti *et al.*, 2007). Instead, they can only be described whenever definitely seen or clearly approximated. Therefore, such impacts are most likely significantly underestimated.

In the following, the question and answer of each site-specific question of each study site are summarized. Then, the mutual impact at the location of the site-specific questions is discussed according to Objective O1.1.3 – *What general interactions can be derived from the site-specific questions to reveal the mutual influence of the paleo-surface and medieval building activity at each study site.*

5.1.1 Waldaschaff

At *Wahlmich Castle* near Waldaschaff, five site-specific questions were derived. The rough relief south of the castle hill was proven as being predominantly of natural origin (W1). Directly at the castle site, deep neck ditches had been dug previously through the diggable weathering products of the rocks, creating this relief (W2). Several faults are located in the area of investigation (W3), which, most likely, influenced the creation of the rough relief focused on in W1. At the castle site itself, wall structures were detected (W4.1) and proven by the excavation. However, the outer bailey could not be found (W4.2). Its position is most likely farther along the valley, as suspected (Rosmanitz & Bachmann, 2017; Rosmanitz *et al.*, 2019b).

- *O1.1.3 What general interactions can be derived from the site-specific questions to reveal the mutual influence of the paleo-surface and medieval building activity at each study site?*

In Waldaschaff, the development of the relief in *Wahlmich Castle's* surroundings varies greatly from that suggested by the geological map. Before discussing some interactions between the relief development and anthropogenic use, the site's landscape development needs to be described more precisely.

In contrast to the geological map (Weinelt, 1962), the study revealed a stratigraphic anomaly. Remarkably, only borehole W3_BH_c situated further south of the disturbed relief confirmed the geological map; in all other boreholes, the findings considerably diverge from the map (Figure 178).

The spatial relationship between the (grano-)diorite and the Bröckelschiefer requires specific structural and geomorphological preconditions, which can be discussed in two variants of a basic geomorphogenetic model (Figure 179). The first

Figure 178: Comparison of the lithological units after geological map and after the results at Waldaschaff a) location of lithological units by geological map; b) approximate location of lithological units by results – both combined with a selection of the location of surveyed data

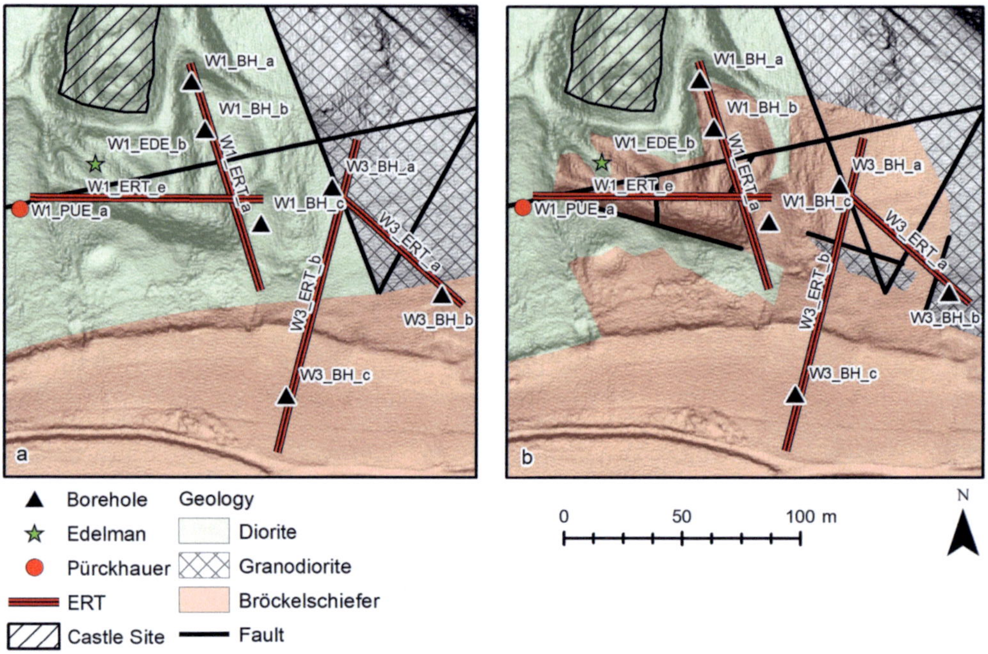

Sources: Weinelt, 1962; Bavarian Agency for Digitalisation, High-Speed Internet and Surveying, 2014; Bavarian Environment Agency, 2014

variant (Figure 179b) is based on a theory suggested by Weinelt (1962). When mapping the Bröckelschiefer formation north of Waldaschaff, he observed the lower boundary rising in a southeast direction. This is the opposite direction to the overall inclination of the Mesozoic formations covering the Bröckelschiefer, and therefore must be explained by the subsidence of local fault blocks (Weinelt, 1962). The activity of the fault also weakened the structure and texture of the bedrock at this position and left it prone to the intense weathering of the Miocene (Jung, 2006; Okrusch *et al.*, 2011). Throughout the Miocene period, the sedimentary cover was eroded, and the Bröckelschiefer has been preserved variably depending on local topography. The bedrock was exposed to a deep chemical dissolution, and the weathering residues prevailed as saprolite. Over the course of the Pliocene and the Pleistocene, the relief was prone to chemical weathering and denudation amplified by fluvial erosion and incision, with a shift to a more accentuated relief and changed morphodynamics, such as solifluction and sliding.

However, the timing of the erosion of the local relief remains unclear due to the lack of dating possibilities, and evidence for sliding or intense solifluction was observed in some cores. Thus, the occurrence of Bröckelschiefer in the local depression of the fault block can also be explained by mass movements during one or more later phases. According to this theory, the Bröckelschiefer deposits came from further up-

Figure 179: Model of the landscape development – Waldaschaff

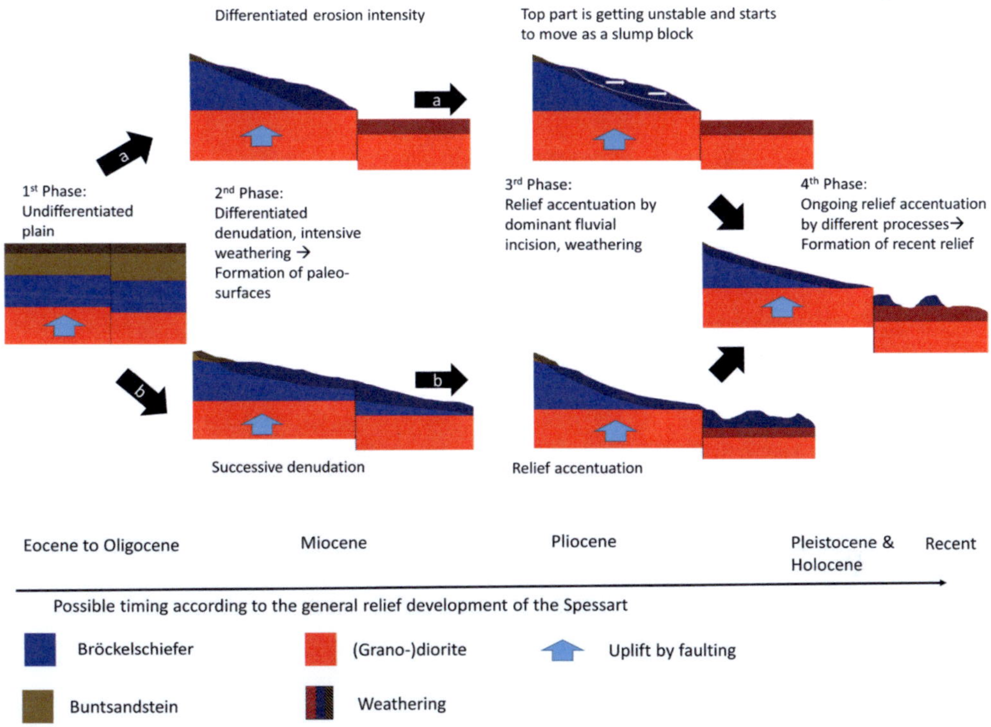

Source: own illustration, based on Jung, 2006; Okrusch et al., 2011

slope and frequently covered the more or less bare saprolites of the (grano-)diorite (Figure 179a). Additionally, Lorenz *et al.* (2010) describe how transported claystone packages were visible in an opening only 500 m away from the area of investigation. There is evidence for both intense solifluction with the development of thick PCBs and for slumping and sliding processes that took place before the latest PCBs developed. In some exposed parts, the weathered surface of the (grano-)diorite's saprolite directly underlies the PCBs that contain debris from the upslope bedrock of Bröckelschiefer and diorite as well, with an inverted stratigraphic succession. Therefore, there is a strong argument for high incision rates during or after the late Pleistocene. The saprolite indicates a tilted surface, following the former relief, and is evenly smoothened by the solifluction processes. Regardless, the observed sliding structures are rooted at depths between 230 and 474 cm, and thus, were not compellingly affected by solifluction. Therefore, mass movement cannot yet be excluded as a secondary process that formed the recent relief and will be studied in future projects.

Combining all the results, it is clear that the study site is or was a geomorphologically and tectonically active area.

This finding indicates that the castle was built in a very special geomorphological and geological situation. This natural relief provides the perfect spot for the creation of a castle overviewing the Aschaff Valley. The castle was placed on top of the isolated hard-rock hill made of (grano-)diorite, whereas the surrounding relief consists of weathered claystone, which was easy to dig to the desired shape to enhance the castle's fortifications. No outer bailey or evidence for other anthropogenic building activity could be found. Thus, the specific geomorphological and geological situation of *Wahlmich Castle* is unique within the entire Spessart and is regarded as an advantageous position. The rough relief and the lithology are prone to erosion, and therefore, the area is considered as unsuitable for arable land, unfavorable for intensive land use.

Chronostratigraphy and critical reflection

An unresolved, major issue is the dating of the displacement of the Bröckelschiefer, which happened before and/or during the Pleistocene solifluction. The dating of cosmogenic nuclides or isotopes seems problematic for this type of question. According to the bottom part of the model (Figure 179), the border between Bröckelschiefer and (grano-)diorite was not necessarily the surface until the movement started, making the dating of cosmogenic nuclides (e.g. Hermanns *et al.*, 2015) ambiguous. Additionally, no dateable sediment trap Lorenz *et al.* (2010) is located in the accumulation area. There is no hint of human activity or relict soil formation phases, charcoal or other material underneath the basal-breccia of any of the boreholes. Therefore, the movement must be older than the human activity within the area and was not caused by intense earthworks or quarrying around the castle site, but instead by broader tectonic or geomorphologic processes. Overall, the suitability of typical mass movement dating methods such as dendrochronology, ^{14}C- and luminescence-dating is highly questionable for this study site (e.g. Stoffel *et al.*, 2013; Bao *et al.*, 2020). Mechanical dating methods such as Schmidt-hammer exposure-age dating (Winkler *et al.*, 2020) require a large number of samples under comparable weathering conditions, a precondition not satisfied by the study site. Hence, the establishment of a proper chronostratigraphy should be a primary aim of any future project incorporating specialized geochronologists.

While dating was not yet appropriate, this study focused on the morphodynamic history of a relief in the vicinity of intense anthropogenic activity.

Even though the resistivity value of the diorite or granodiorite is far higher than the value of the Bröckelschiefer or its weathering products, this value is only around one-tenth of the literature values (Reynolds, 2011). Therefore, the literature values allow for the assumption that the (grano-)diorite in the recorded parts is weathered, fractured and aquiferous, which would fit better the suggested values (Telford *et al.*, 1990). However, the considerable difference between claystone and (grano-)diorite or their weathering products is quite obvious.

The basal-breccia contains a considerable amount of (grano-)diorite debris. This is indicative of sedimentation on top of the (grano-)diorite without Zechstein dolomites or magmatic rocks nested in between (Scheinpflug, 1992). The lack of Zechstein rocks underneath the Bröckelschiefer is typical for the region and explained by its location as part of the accumulation areas of the Zechstein and the intense weathering of the Zechstein dolomites in a hot and wet climate (Okrusch *et al.*, 2011).

Due to the drilling resistance of (grano-)diorite, it was impossible to penetrate more than a couple of centimeters. Comparing this information with the maximum extent of the basal-breccia of about 2 m, close to the area of investigation, which includes large bedrock fragments (from observations from construction works at the nearby motorway; Lorenz *et al.*, 2010), it is possible that the (grano-)diorite within the boreholes could be part of the breccia and not part of the crystalline bedrock.

5.1.2 Hauenstein

Hauenstein, the sedimentation in the talweg was associated with the mill's usage at the castle (H2). The center of the castle hill is made of bedrock (H3), but several collapsed and in-situ wall structures are located on top as well (H4.1). Similar to *Wahlmich Castle*, the outer bailey could not be found yet (H4.2).

- *O1.1.3 What general interactions can be derived from the site-specific questions to reveal the mutual influence of the paleo-surface and medieval building activity at each study site?*

Initially, the position of *Hauenstein Castle* at the bottom of a valley seems an unfavorable spot to build a castle and as well to build a mill. However, after combining all the results at Hauenstein, the topographic position seems to be a suitable place to create a mill at a castle site. The valley could be used to create the necessary mill ponds to use the mill (Bury, 2018), and the hard-rock castle hill made of bedrock improved the representativity and defensibility of the site.

Comparable to Waldaschaff, the study site of Hauenstein displays a variable mutual influence of the relief and the medieval usage of it. Early assumptions of the interactions could be confirmed in parts (e.g. H2) but had to be rejected for others (e.g. H3, H4.2). However, the center of the castle hill and the meadows upslope were proven with geophysical methods and a borehole as being predominantly natural in origin, but anthropogenic usage of the hill massively influenced the entire sedimentation regime at the talweg. By adding information from old maps to the collected data of this study, a high spatial and temporal variability of the fluvial and limnic system is evident.

Visible in the "*Bayerische Uraufnahme 1808–1864*" (Bavarian Ministry of Finance and Regional Identity, 2020), the medieval ponds could survive into the 19th century CE, at least there are some "lakes" mapped at the position of the medieval ones. The bird's-eye view of the different structures (Figure 180) reveals a high temporal and spatial fluctuation compared with the location of fluvial or limnic elements of the old maps. The distribution and size of the sediment bodies seem to be independent of those of the 19th century CE structures. Additionally, the mapped wall structures within the area of investigation have no visible influence on the sediment distribution. The 19th century CE creek only passes the second from north of the present-day wall structures while it would have crossed the others. Therefore, the third and first, from the north, wall structures are probably younger than the 19th century CE map, whereas the second structure could be older. However, this could also be caused by an imprecise map.

Figure 180: Models of the distribution of different topographic features and sediment bodies at Hauenstein a) sediment bodies documented in H2 and present-day creek; b) location of 19th-century creek and "lake"; c) location of wall structures; d) Combined information

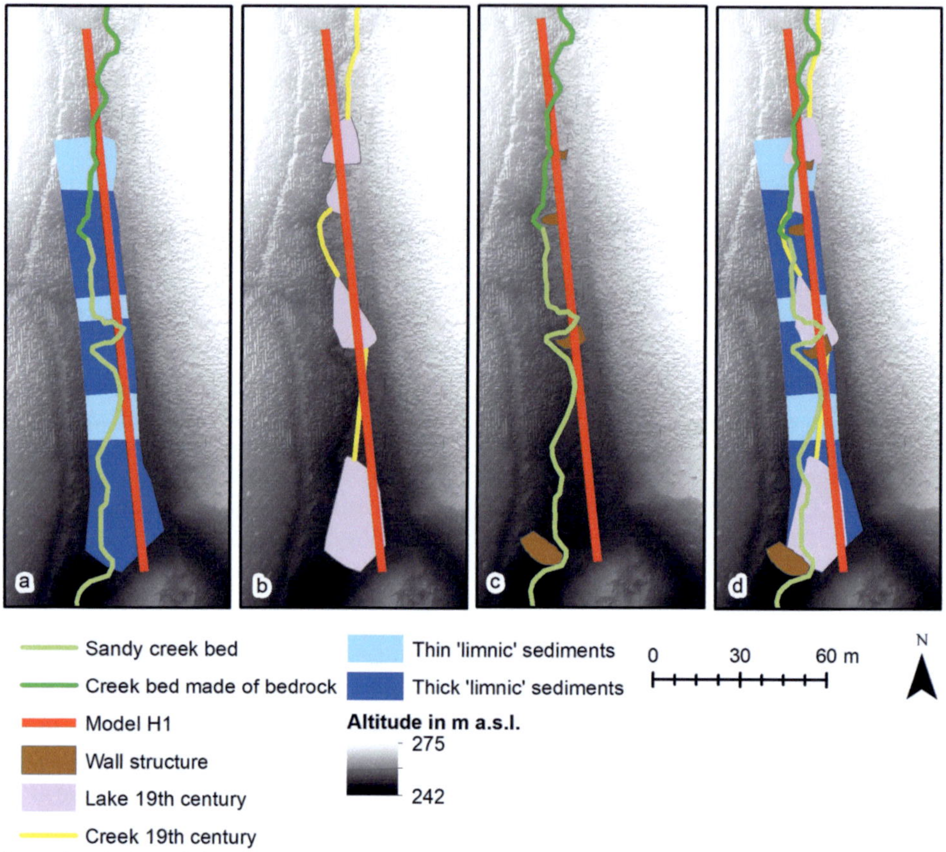

Sources: Bavarian Agency for Digitalisation, High-Speed Internet and Surveying, 2014; Bavarian Environment Agency, 2014; Bavarian Ministry of Finance and Regional Identity, 2020
Map creation: J. Trappe

Analog to *Wahlmich Castle*, all these advantages for creating a castle at the bottom of a valley can be supplemented by disadvantages for other types of usage. The small-sized talweg was most likely not a favorable position for agricultural use or other building activity (Scherer-Hall, 1998) except for the creation of a mill, and accessibility for other usages can also be described as impractical. Even for the usage of a mill, the water flow seems to be very low (Bury, 2018).

Critical reflection

Measuring with single 2D ERT measurements for prospecting building structures seems quite unusual. However, the relief at the castle hill is highly exposed; different building fragments are collapsed and located at the surface. Therefore, a quasi-3D approach could not be pursued further. Additionally, around the time of the

Second World War, different wall structures for undefined use were built on top of the hill, making other methods of measurement difficult to perform (Rosmanitz *et al.*, 2019a). A major issue is the question of whether the vast, high-resistive anomaly in the northeastern area could be the other side of the enclosure wall excavated in the southwestern part. Other anomalies located at the edges of the castle hill are smaller but hint at the possibility of fragments of the wall surrounding the entire castle. However, vast collapsed structures in the northern part indicate that the anomaly consists of such collapsed walls as well.

The partly mixed texture of the sediments in the ponds could be explained by the mixing activity of carps. Carps farming was common in the medieval period in several types of ponds, especially for the about 100 days lasting period of fasting (e.g. Schneider, 2014).

The magnetic distribution within the survey of H4.2 indicates an anthropogenic use of these meadows. However, no evidence for this could be found with the invasive methods, therefore, the high content of magnetite, which was previously recorded in the gneiss (Okrusch & Weinelt, 1965), may cause such magnetic anomalies.

5.1.3 Mömbris

Mömbris, a geophysical prospection enabled a precise excavation of the castle (M4.1).

- *O1.1.3 What general interactions can be derived from the site-specific questions to reveal the mutual influence of the paleo-surface and medieval building activity at each study site?*

As initially suspected, the rectangle leveled area was created by humans as the first 230 cm of the borehole proves. Instead of an expected medieval usage (Gröber & Karlinger, 1916) of this plateau, it could be dated to the 17th century CE (Rosmanitz *et al.*, 2020). Furthermore, the location of the castle was at the northern edge of the plateau and further northward. Even though the plateau covers many medieval influences on the relief, the 18-m-long wall in the center of the castle hill especially indicates a further effect on the relief development via building the castle. On the one hand, the wall was most likely constructed to stabilize the slope (Rosmanitz *et al.*, 2020). Therefore, it can be concluded that the builders were aware of the instability of this relief but tried to handle it. This may sound obvious, but at other sites within the Spessart, such as *Altes Schloss* near Kleinwallstadt (Rosmanitz, 2009), the subsurface made of loess deposits hindered the construction of the castle as intended, and conversion was necessary. On the other hand, the wall consists of two parts belonging to different building phases (Rosmanitz *et al.*, 2020), indicating an underestimation of the relief conditions by the builders or simply an adjustment of the wall to changing relief conditions.

Critical reflection

The position of the anomalies within the ERT and reflectors in the GPR differ slightly from the excavated wall structures in the archaeological findings. This difference could be due to small inaccuracies during the single ERT electrodes' positioning or the differences in measuring the GPR grids. Furthermore, a precise migration and topographic migration of the data are not always possible, and the data had to be interpolated and filtered. Nevertheless, the geophysical prospection allowed a targeted excavation; indeed, the parts north of the plateau belonging to the castle would probably not have been found without this prospection.

Despite the good results that led to the targeted excavation, a further interpretation of the geophysical data would have led to other problems during the interpretation without the knowledge from the archaeological excavation. The relative positions to each other and the meaning and role of the walls would have been especially difficult to identify as belonging to a specific usage. Additionally, the dating of the building structures can generally not be suggested by these methods. Therefore, geophysical prospection helps to target an excavation, but detailed information about the structures is difficult to derive without the ground-truth. In contrast, an archaeological interpretation of the DEM, which has led to good results in many places (e.g. Höfle & Wagener, 2012), would have been absolutely incorrect.

5.1.4 Alzenau

Rannenburg Castle, near Alzenau, is surrounded by loess loam deposits, which were of agricultural use (A1). Many anthropogenic structures and evidence for intensive building activity were found at the castle site itself (A4.1). Nevertheless, an excavation has not yet been conducted, thus, these results need to be proven in the future.

- *O1.1.3 What general interactions can be derived from the site-specific questions to reveal the mutual influence of the paleo-surface and medieval building activity at each study site?*

Rannenburg Castle was located in the westernmost part of the Hahnenkamm, at the edge of the Vorspessart, overlooking the Main Valley. As derived from the depth and size of the neck ditches, the most significant influence on the relief for all the study sites was probably during the medieval period at this castle.

The loess loam-rich northern parts were most likely used for agriculture. At least in some places, colluvial relocation of material allows for such an assumption. However, a precise statement of the mutual influence between medieval building activity and relief development can only be made in the context of future archaeological excavations. Nevertheless, the fertility of loess deposits has been known since the Neolithic, and they were important during the medieval period also (Lang, 2003; Verstraeten *et al.*, 2009). Agricultural usage makes loess deposits prone to erosion (Evrard *et al.*, 2008), which could have rates up to two orders of magnitudes stronger than in non-agricultural landscapes (Montgomery, 2007) and lead to gully erosion (e.g. Bennett & Wells, 2019). Therefore, such an impact can be assumed for the northern area of gully erosion near *Rannenburg Castle*.

Critical reflection

The outcrops collected at the roots of uprooted trees had a depth that depended on the roots' penetrating depth. Thus, some of those outcrops were very shallow. Therefore, the depth of the described soil is variable, and some of the shallow outcrops that do not correlate with the lithology units of the geological map (Okrusch *et al.*, 1967) could be too shallow to describe as loess loam or a substrate with loess loam content.

The data from A4.1 were especially meant as a previous prospection for a planned excavation in the late summer of 2019. Regardless, after the described heavy storm in August 2019, the excavation was not able to be carried out safely and was canceled. Nevertheless, the collected data remains indistinct without ground-truth; being a *Bodendenkmal*, invasive methods such as ground-truth for geophysical prospection, focusing on an archaeological question cannot be performed. As such, the results need further proof and must be considered preliminary. However, they can be used as prospection for future excavations in the area of interest. Therefore, they are published in this dissertation without further proof.

5.1.5 Eisenberg

At the *Eisenberg mining site*, the size, form and depth of collapsed shafts and the position and type of iron deposits were analyzed (E2). Additionally, the existence of wet meadows in the surrounding area was disconnected from anthropogenic use (E3). They occur due to aquiferous layers on top of dense clay-rich layers of Miocene sediments.

- *O1.1.3 What general interactions can be derived from the site-specific questions to reveal the mutual influence of the paleo-surface and medieval building activity at each study site?*

At *Eisenberg's medieval mining site*, many shafts had been dug and several bloomery furnaces constructed (Hasenstein *et al.*, 2019), affecting the relief massively as in other mining sites (Pełka-Gościniak, 2014). However, the results of E3 indicate only a sporadic anthropogenic influence during the period of interest on the relief development of the meadows. They show a mostly natural hydrogeological system on dense and aquiferous layers. Nevertheless, these were later actively drained and used by humans.

Of course, the deposit position has mainly geological reasons in the context of the Vogelsberg vulcanism (Ehrenberg & Hickethier, 1982). Furthermore, in addition to the accessibility of the deposit, other sedimentological and geomorphological processes affected the location of the deposit and, with it, the anthropogenic use of it.

These results, combined with their interpretation, indicate an orebody underneath the basalt. The small-sized iron enrichments within the Miocene sediments can be found in wide areas in the surroundings of the Eisenberg (e.g. E2_OC_b, E3_EDE_a). A random proximate analysis resulted in 19% Fe_2O_3 content in these sediments in which trials determined that about 60% Fe_2O_3 would be necessary to smelt in the excavated bloomery furnaces (Hasenstein *et al.*, 2019; J. Lorenz, 2020, personal communication). However, even if these enrichments exist to a more considerable extent below the basalt. It would probably not justify the effort to dig several-meter-deep shafts through relatively hard basalt if they can be found beneath about one meter of loose sediments in the surroundings, for example, at the wet meadows, as the results of E3 are showing. Additionally, within the shafts, no residues of iron ore were found (Lorenz, 2021), indicating the great importance and value of the ore, as well as contradicting its frequent occurrence in the surroundings. Therefore, the iron enrichments within the Miocene sediments are most likely not the deposits that were mined at the Eisenberg.

Most likely, the basalt becomes weathered and creates "basalt iron ore" (free translated from *"Basalteisenerz"*). These deposits occur through weathering (Pohl, 2005) of the basalt and probably consist of goethite. At least a little of the material was found at the Eisenberg (Figure 181).

The Vogelsberg vulcanism basalts are commonly known for such deposits, and these were fairly economically important in former times and have, therefore, been researched (e.g. Münster, 1905; Köbrich, 1934; Friedrich-Lautz, 1963). In more modern times, they have been economically unimportant and, therefore, have not been investigated in recent studies.

Figure 181: Field picture: iron ore at the Eisenkopf

Photographer: J. Lorenz

The weathering of the basalt needs special conditions for the creation of such minable enrichments. Such conditions can probably only be found at the specific position of the Eisenkopf, even though the basalt extended farther. As far as it was possible, mining was done in an open pit. Subsequently, shafts were constructed in positions of large basalt deposits on top of the ore body (J. Lorenz, 2020, personal communication). Especially at the edges of the basalt body, in other positions of the Eisenkopf, it would be easy to mine in an open pit. Therefore, the position of the orebody at its specific location meant that the Miocene sediments were of major importance. The dense clay layers allow water to accumulate on top and weather the basalt for the creation of this deposit.

Critical reflection

The lithological units' expectable resistivity values overlap each other (Telford *et al.*, 1990; Reynolds, 2011). Therefore, the interpretation of the ERT data at the Eisenberg needs to be considered with caution. The ground-truth could only be performed at a few spots, and along the measurements the distribution of expectable lithological units seems to be variable. Additionally, the value range is narrow, which can cause additional errors. Due to different contents of water, air and iron concretions within the Miocene sediments, as well as possible remnants of the ore body, the groundwater could influence the resistivity distribution within the data more than the conductivity of the lithological units. However, the dataset was compared with the GPR and the prospection pits. In addition, the necessary conditions for the development of the deposit fitted the conceptual model.

At the Neudorfwiesen, the magnetic survey revealed many (mostly dipole) anomalies. Initially, these seemed to belong to some kind of smelting spots or bloomery furnaces. The excavations found that some of these spots in the meadow's western part could be associated with the smelting of the iron (Hasenstein *et al.*, 2019). However, at least the largest of these dipole anomalies has proven to be of natural origin. The iron-rich layers observed in E2_OC_b were most likely spread laterally by the aquiferous layers.

5.2 Transferability of the results

From the important meaning of the site-specific results and interpretations, some transferable assumptions can be derived. However, before focusing on the transferability, the representability of the study sites needs to be discussed.

- *O1.2.1 Is the number and representativity of the chosen site-specific questions and study site high enough?*

Each of the analyzed sites of medieval usage is special from different points of view. Even if comparing similar usage types, their relief position, relief form and lithological and geomorphological features vary significantly. Therefore, the representativity of focusing on precise relief developments or special features is not provided, and the sites and questions must be considered as case studies.

However, taking a broader perspective, considering the micro-relief in total, rather than focusing on specific features and forms, generalized statements can be made.

• *O1.2.2 Which general assumptions can be derived for the interaction of medieval building activity and relief development in the entire Spessart region or even on a larger scale?*

The topographic position is essential for interpretation in castle research (e.g. Scherer-Hall, 1998; Biller, 2016; Pärn, 2019), as for spatial defensive dikes (cf. Küntzel, 2009). Additionally, the geography and geology must be described for each excavation (Bavarian State Office for Monument Preservation, 2020). However, in most cases, the position's meaning is linked to the mesoscale (e.g. Leser, 1995) more than this study's focus on the microscale. For example, a spur castle is known as being an important position for the castle's defensibility and the opportunity of protecting a valley. However, a specific spur's choice from several possible ones is more related to the microscale.

Furthermore, all these investigations were analyzed from an archaeological perspective and not with a geomorphological focus. Therefore, this work supplies a transferable approach to answer remaining objectives at castle sites in the Spessart and beyond.

In addition to the different focus in this study, a landscape development model could be derived for medieval sites. As mentioned above, many of the results, their interpretation and discussions are highly site-specific. However, the broader meaning is transferable to a general, simplified model of landscape development on a microscale in the context of medieval castle sites, at least within the Spessart (Figure 182).

First phase

Different natural processes developed a mostly natural-driven relief comprising several different relief generations (Figure 182a).

Second phase

Medieval earthworks and building structures impacted the processes within the geomorphological and geological setting of the primary landscape. Nevertheless, these building activities depended on the former natural relief (Figure 182b).

Third phase

The impact stopped more or less suddenly and left a surface and subsurface setting that was partly natural, partly anthropogenic in origin (Figure 182).

Fourth phase

This new setting underwent further geomorphological processes to produce the present-day landscape. Additional anthropogenic impacts influence the new development of the relief, creating a landscape of various relief generations from which the paleo-surface and the medieval building activity can be reconstructed (Figure 182d).

Such interaction between humans or human activity with the environment is exemplary and generalized, and is well described in the "human-environment spiral" (*Mensch-Umwelt-Spirale*) of Bork (1998). Bork´s spiral is focussing on soil erosion,

Figure 182: Model of micro-relief development in an area of medieval usage: a) several natural relief generations formed an almost natural surface and subsurface; b) anthropogenic impact created the medieval relief generation; c) the impact stopped suddenly and left a surface and subsurface setting that was partly natural, partly anthropogenic in origin; d) this new setting underwent further geomorphological processes including the possibility of another anthropogenic impact developing the present-day landscape

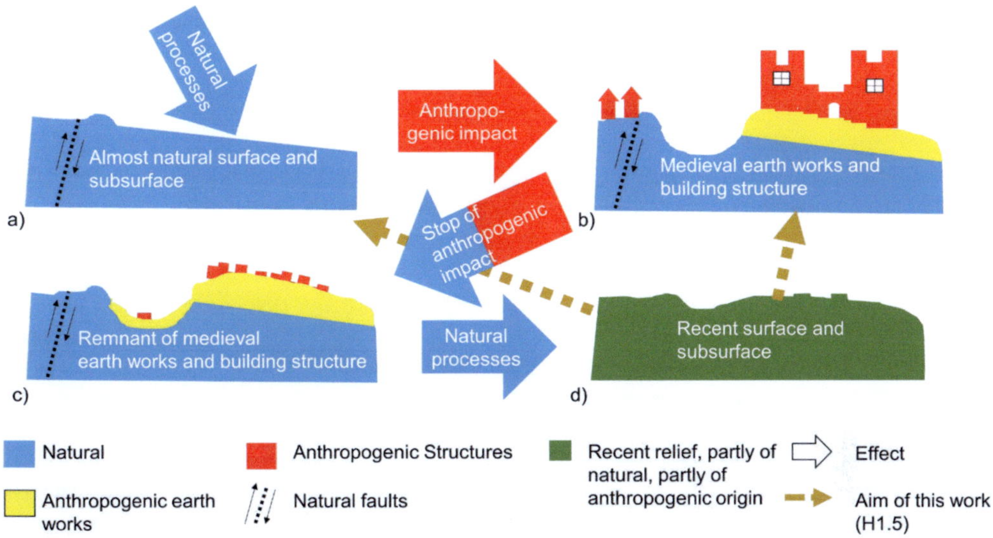

Source: own illustration

the flow of substance or the water balance. Population growth in a stable ecosystem causes soil erosion and the reduction of its fertility. This results in destabilization and population reduction, including reforestation. Due to the reduction in population, a new stable ecosystem can appear in which the population grows again. However, a single loop can be adapted to the medieval castles' geomorphological setting in this investigation area. The phases and particularly the timing of transitions between two of the phases, differ in both loops. Especially the timing between Phases 2 (Figure 182 b) and 3 (Figure 182 c) can be dated exactly to 14 days, which was contractually agreed, and archival documented (Ruf, 1984, 2019). In contrast, this transition can last much longer in the spiral. However, analogies are evident between the spiral and this model, especially when focussing on stable environments being anthropogenically destabilized, which results in a new stable system. Therefore, it must be mentioned in this position.

In addition to this general model of landscape development, additional observations can be made. In Waldaschaff and Hauenstein, the geomorphological and geological disadvantages of a location for alternative uses could often be more important than its advantages for its specific use. As mentioned above, the rough relief and the lithology are prone to erosion, and the area in Waldaschaff is therefore regarded as unsuitable for arable land, nor is it favorable for intensive land use. For *Hauenstein Castle*, the spot was not really perfect for the creation of a mill. Even with the millponds, it was not possible to run the mill most of the time (Bury, 2018). However, for another type of land use, the valley with the ponds and the castle hill was too exposed. For other agricultural use or for the construction of further farm and economic buildings, the position is unsuitable.

Another transferability of the results can be performed: the anthropogenic influence of landscape is known for large scales (e.g. Hooke & Martín-Duque, 2012; Ellis *et al.*, 2013), but also in case studies (Rózsa *et al.*, 2020) or focused on special land-use forms (Pełka-Gościniak, 2014). Such studies focus mainly on the anthropogenic effect on a landscape. Without a doubt, this is the most important aspect and issue today and in the future. However, it seems this aspect results in a subconscious overestimation of the anthropogenic impact by analyzing a cultural landscape, and it likely disguises many ancient, ongoing natural processes and parts of the still existing paleo-surface. Generally, humans tend to connect cultural land more to usage than to paleo-relief. Within this massively affected cultural landscape, many features and processes linked to the paleo-landscape and relief are still ongoing. These features and processes are often associated with the use of the area; they can be the reason for the usage and not caused by it. Nevertheless, this point should not undermine the anthropogenic effect on landscapes and processes. It should only be noted that a special and unique relief in a cultural landscape might have been a special and unique relief prior to usage, which could have affected the anthropogenic influence on it.

The model in Figure 182 is a summary of this study. Based on this model, it needs to be discussed how this work terminologically can be classified: Is it a geoarchaeological approach or a geomorphological approach in an archaeological context?

In the 1970s, the term "geoarchaeology" was developed (e.g. Butzer, 1973), but most geoarchaeological research has been undertaken in the past couple of decades (e.g. Fuchs & Zöller, 2006; Bebermeier & Schütt, 2011). Today, "geoarchaeology" often follows the definition of Rapp and Hill (2006): "[Geoarchaeology is] any earth-science concept, technique, or knowledge base [used for] the study of artifacts and the processes involved in the creation of the archaeological record." This is part of archaeometry, including all the natural scientific methods to answer archaeological questions. There are many more definitions of geoarchaeology that focus on single topics but are also covered by the general definition of Rapp and Hill (e.g. Fuchs & Zöller, 2006). In contrast, Bebermeier and Schütt (2011) pointed out that both geomorphological and archaeological knowledge are involved in geoarchaeology. Therefore, approaches investigating the human effect on the geomorphology on larges scales with a focus on geoscience are often recommended in Germany (e.g. Nicoly *et al.*, 2014) or central Europe (e.g. Bork, 1998).

This study is considered as a geoarchaeological approach according to the definition of Bebermeier and Schütt (2011). The approach combines the knowledge of both disciplines, especially geomorphology, to answer geomorphological questions in the context of archaeological excavations. This study not only uses the methods of "earth science" to answer archaeological questions. However, depending on the definition, this study is considered a geoarchaeological study (Bebermeier & Schütt, 2011) or geomorphological study in an archaeological context (Rapp & Hill, 2006). Therefore, parts of it might be viewed as a geomorphological study in the context of anthropogenic areas, and other parts of it as a geoarchaeological approach.

5.3 Applied methods

Regarding the results at the sites and their transferability, the applied methods and their transferability need to be discussed.

• *O2.1 Is the chosen multi-method approach suitable for answering the main research-question in such variable relief?*

Parts of the applied methods have been used at other important archaeological study sites in Germany. For example, the topographic position of Heuenburg Castle was analyzed using ALS (airborne-laser-scanning) combined with geophysical methods (Bofinger *et al.*, 2014), supplying important information for the interpretation of the castle. At the Limes geophysical prospection (Fassbinder, 2010), a systematic acquisition with ALS (Koch *et al.*, 2017) was performed to identify important areas in the present-day relief. Additionally, various other studies have focused on different aspects of the applied methods (e.g. Siart *et al.*, 2010; Fischer *et al.*, 2016; Lange-Athinodorou *et al.*, 2019).

Therefore, and based on the quality and interpretation of the results, the chosen multi-method approach is considered suitable for answering most of the site-specific questions and objectives for the study sites. However, it might be argued that the chosen approach was only suitable for a descriptive analysis of the landscape in the surroundings of the castle sites, but that the excavations were especially essential to place the results of this work in context.

• *O2.2 What are the most important factors for improving the quality of geophysical prospection in the context of archaeological excavations?*

For each geophysical prospection of an archaeological site, it is necessary to have a difference in the physical parameters of the subsurface created by anthropogenic usage. For example, ditches dug and refilled in loess loam are generally not easy to resolve using ERT and GPR due to a tight value range of the electrical resistivity and dielectric constant between the surrounding material and the infillings. However, due to variable remnant magnetization, such ditches are recognizable using a magnetic survey. In contrast, walls made of Buntsandstein are easy to detect in sand-rich colluvium using ERT and GPR, but not using a magnetic survey (see value ranges in Telford *et al.*, 1990; Reynolds, 2011).

In addition to these basics, most important for the prospection of building structures and other anthropogenic influences on the subsurface is the spatial distribution of the geophysical parameters. It is difficult to define a high-resistive anomaly or a group of reflectors as a wall structure without their relative position. The measured values themselves are not unique, and in some of the study sites, it could not be said that, for example, a value above 1.0 kΩm was such a structure. The results indicate that a high-resistive anomaly in Waldaschaff could either be a less-weathered part of the (grano-)diorite or a wall made of Buntsandstein. With the help of the guidelines for ERT description, especially regarding boundary analysis, it is easier to distinguish these elements via a sharp boundary at the surface, but the data are still not unique. Only their relative position can deliver a clear indication of walls or other manmade structures. In this study, linear subsurface structures with 3D measurements could clearly be distinguished as manmade and in-situ building structures. Therefore, it was refrained from the first tries of an automated interpretation, despite the less-subjective description via the guidelines. However, with the help of the guidelines for ERT description, a generalized, relative resistivity distribution for single study sites is derived. Nevertheless, before the creation of such generalized datasets is attempted, the transferability of the concept needs to be discussed.

- *O2.3 How transferable are the guidelines for ERT description for other study sites?*

The guidelines for ERT description were designed for a more objective description of the data in this study. However, most of the parameters and descriptions can be used universally in various areas of investigations, arrays and systems for measurement and inversion. Nevertheless, a few of the parameters need to be further investigated. For example, the differentiation between r-layer and r-anomaly is especially dependent on the used spacing, and they are not very comparable. Nevertheless, with unique values for the different spacings, it is possible that none or all changes in resistivity need to be classified as r-layers or r-anomalies, respectively. Therefore, the single classified r-anomalies/r-layers need to be regarded as classifiable as r-anomalies/r-layers for the specific spacing rather than in general.

The number of anomalies is challenging to transfer to spacings and especially to electrodes that are far higher or far lower in number than those used. Adding another parameter for the spacing of the measurement made this parameter far too sensitive for it. Even more important is the lack of implementation of the different arrays and the number of data points per model block and due to the amount of data per model block, which varies with depth even in the use of a single array. Another problem is the different sensitivity patterns at different depths that apply with different arrays.

Nevertheless, this trade-off includes a simple formula for more practicality rather than objectivity in the number of anomalies. With the received information regarding the number of anomalies, different datasets of the same array and spacing can be compared. Moreover, different forms, orientations and other parameters are set up for this work's objectives and questions, meaning that other types of approaches need to be added to the guidelines.

The boundary analysis is meant to produce additional standardized and less subjective datasets for the description of ERT data. First, it is necessary to mention that these data are only a different way of displaying the already inverted data. All information obtained using these methods is already within the dataset. The only thing the boundary analysis does is to quantify and standardize the data and the derivation of a different picture of the resistivity distribution.

Nevertheless, there are still parameters that need to be discussed in the following.

Due to geometrical reasons, the inverted data of an inversion program had to be used. Therefore, any inaccuracies forced by the program, the used inversion method and other used parameters remain in the datasets.

The subsequently performed quality management was not specifically designed for deleting outliers generated by mistakes, which contradicts the subsurface. Quality management is a necessary step for keeping the following percentage values in a meaningful range. The empiric value of 3.5 times the 95%-quantile was a compromise between the loss of information and quality management.

The slope tool was used to calculate the percentage slope. This value sets the distance in relation to the resistivity value. This calculation is sensitive to high absolute values. For example, a rise from 100 to 1,000 Ωm has a lower slope than a rise from 64,000 to 65,000 Ωm; therefore, the normalization allows for better transferability and easier interpretation of the data. The new values still depend on the distance between the single measurement points. Therefore, the slope has to be multiplied by the spacing. This can cause an overestimation or underestimation by comparing two datasets of different spacings: otherwise, for example, two model blocks next to each other with a value of 80 and 90 (percent of the maximum value), respectively, would cause a slope of 200% for a 5 m spacing and 2,000% for 0.5 m spacing. In this way, however, the problem is also oversimplified and takes the same transition for all measurements. Therefore, these values need to be considered approximations. These values were then reclassified. After being reclassified, the new values have a linear scale, which was created to display all the important r-layer boundaries of these datasets. For other datasets with a different value range, it may be necessary to customize the new values or extend them.

Nevertheless, other guidelines for different descriptions, such as describing GPR data (Neal, 2004) or for mapping soils (Ad-hoc-Arbeitsgruppe Boden, 2005; Fao, 2015), include similar inaccuracies. For example, changes in the continuity of measured GPR reflectors depend on the intensity and type of the utilized gain or analog field approximations, such as bulk density, the percentage of pores, roots or coarse fragments, depending on who is doing the description. Therefore, the guidelines for ERT description should be customized and checked for other applications but could be helpful for describing the relative distribution of resistivity values in various other studies.

- *O2.4 Is it possible to create a guideline profile with a generalized description of the subsurface's relative resistivity distribution for a single study site?*

Essentially, this objective attempts to link all ERT data of a study to a guideline profile.

With the help of the guidelines for ERT description for single study sites, with comparable results and a high number of profiles, the data can be linked to create a generalized, relative resistivity distribution. For such a distribution, at least two similar r-layers or r-anomalies need to be in each of the profiles, and a large number of profiles is necessary. Therefore, only for Waldaschaff was such a generalization created. The other study sites' results are too variable, especially between the subsites of the different site-specific questions, or the number of measurements can be estimated as too few for linking them. How large the number needs to be is difficult to estimate, but working with statistical parameters, the calculation of a median from a sample size of maximal four was not considered to be purposeful at this point. However, this is not the scope of this study.

Profiles with an inverted relief, or at least an inverted resistivity distribution, need to be deleted. Such profiles are special cases and are important for understanding the topography but would distort the following calculations. In the case of this example, W1_ERT_f was not used.

Due to the quality management of deleting results with the inverted relief, r-layer 1 of each profile is comparable to every other r-layer 1. For r-layers 2 and 3, this is not the rule, but they fit mostly with each other. However, the horizontal changes complicate the comparability.

For anomalies, the process is more complicated. Therefore, the anomaly types of the single profiles were interpreted and classified into three categories: Type 1: PCBs, Type 2: fragments of less-weathered (grano-)diorite and Type 3: wall structures.

When one works mostly with an ordinal scale, the median can be calculated for most of the parameters (homogeneity, r-layer thickness, layering, boundary of r-layers, number of r-anomalies, r-anomaly location, r-anomaly size and r-anomaly boundary). The nominally scaled orientations and r-anomaly forms are presented in their real number. Only the horizontal changes needed to be prepared. Some of the ERT results reveal prominent horizontal changes (2), others none (0); on average, those can be described as possible (1), but no numerical or statistical value justifies it. Additionally, the second r-layer only has values for r-layering, horizontal changes, boundary of r-layers and r-layer orientation whenever a third r-layer exists; and the third layer has values for those parameters when located solely beneath the first and not the second r-layer. Therefore, only a few values create the median. These values refer to the relative position of two layers to each other and are generally noted for the layer above/left of it.

In the guideline profile with a generalized, relative resistivity distribution of Waldaschaff (Table 37), three layers can be distinguished. At the uppermost, a top layer, made of Bröckelschiefer, is mostly homogeneous and small-sized with a medium-distinct layering. Horizontal changes within it can occur but are not always present, and the median boundary is sharp. In contrast, the r-layer orientation is variable. A second r-layer, made of more or less-weathered (grano)diorite with a lower homogeneity but increased thickness, is located underneath it. Horizontal changes are possible, and the layering is highly distinct in this r-layer but the boundary is less developed than within the Bröckelschiefer layer, and the orientation varies. Not easy to distinguish from the second r-layer, a third layer is developed. This layer can be split from the second layer by horizontal changes or is located underneath the first and partly the second layer. The third layer consists of mostly less-weathered (grano-)diorite with a higher homogeneity and a lower resolution to its thickness. The layering is well-developed, and the boundary is sharp whenever it underlies the first layer; its orientation to the first or the second layer is mostly horizontal or vertical.

Table 37: Guideline profile with a generalized, relative resistivity distribution in Waldaschaff: Summary of all ERT data belonging to Waldaschaff; median of homogeneity, r-layer thickness, layering, boundary of r-layers, number of r-anomalies, r-anomaly location, r-anomaly size and r-anomaly boundary; count of values in brackets of r-layer orientation, r-anomaly form and boundary orientation of r-anomalies; adjustment to the entire datasets of horizontal changes

R-layer	*Homogeneity*	*R-layer thickness*	*Layering*	*Horizontal changes*	*Boundary of r-layers*	*R-layer orientation*
R-layer 1 (Bröckelschiefer)	4	1	3	1	5	1, 2 (4x), 3 (2x), 4, 5, 6 (3x), 7 (4x)
R-layer 2 (weathered diorite)	3	3	5	1	4	2, 3, 6
R-layer 3 (less-weathered diorite)	4	2	5	-	5	2
R-anomaly	*Number of r-anomalies*	*R-anomaly location*	*R-anomaly Form*	*R-anomaly size*	*R-anomaly boundary*	*Boundary orientation of r-anomalies*
R-anomaly Type 1 (PCBs)	1	1	2 (5x)	2	4	2 (6x), 4
R-anomaly Type 2 (diorite fragments)	1	1	1 (7x)	2	5	1 (7x)
R-anomaly Type 3 (wall fragments)	3	1	1, 3	1–2	5	2

Source: own research

The bottom two layers cannot always be clearly distinguished from each other, but at least in profiles with three r-layers, this generalization mostly fits.

At the top of the profile, three types of anomalies occur. A quasi-layered anomaly is associated with the remnants of PCBs, as well as two further types that can difficultly be differentiated without further information. The two similar ones are mostly blocky and medium-sized and surrounded by a sharp boundary. Whereas Type 2 is mostly located just beneath the surface and is surrounded entirely by a boundary, Type 3 is located more often in the uppermost position without resolution of the top boundary in the data. Type 3 occurs more frequently in a single profile as soon it appears. Type 2 and 3 can be interpreted as less-weathered fragments of the bedrock and wall structures. The differentiation between regarding a top boundary is error-prone and depends on the used spacing. Therefore, the relative position and ground-truth information are much more important for differentiating these two anomaly types.

The creation of such a guideline profile with a generalized, relative resistivity distribution most likely needs to be modified and tested with a higher number of measurements in different study areas. Additionally, the expressiveness of such data without the database needs to be questioned. However, at this point, it should only be a small outlook regarding the potential of this new type of description.

Critical reflection

In addition to the transferability and used parameters of the newly developed way of describing the data, some other aspects of the used methods need to be discussed as well.

GIS-based terrain analyses

Regarding the calculations of the TPI (Weiss, 2001), the derived landform classification (Deumlich *et al.*, 2010), the TWI and all the used standard surface calculations, such as slope and hillshade, a huge number of further derivations from the DEM were tried. As well as a number of further standard calculations of ArcGIS and SAGA-GIS, such as the aspect, the flow accumulation and other hydrology tools, the Relief Visualization Toolbox (Kokalj *et al.*, 2013) was tested (e.g. Kokalj & Somrak, 2019). The surface and hydrology tools helped identify geomorphological areas of interest and underline some of these results. However, these datasets do not reveal other features and are, thus, not included in this study. The use of the Relief Visualization Toolbox is a suitable tool to highlight archaeological research questions. With the help of different directions for creating a hillshade, the toolbox enables the identification of archaeological areas of interest. However, regarding the sites in this study, no further areas of archaeological interest could be defined, but the already known areas could be sometimes better identified. That said, the archaeological sites have probably been too much anthropogenically overprinted, meaning these methods are not included in this work to retain the comparability between the single study sites. Nevertheless, they could be helpful for defining other areas of interest in comparable studies.

ERT

Most ERT measurements in this study were taken with the Wenner-Schlumberger array. Due to its good horizontal resolution and the possibility of displaying the bottom of larger structures, such as dug ditches or deep wall structures, this method was more often chosen than the others (e.g. Dahlin & Zhou, 2004; Kneisel & Hauck, 2008). The resolution of small obstacles, especially near the surface, is not highly detailed using the Wenner-Schlumberger array (Dahlin & Zhou, 2004; Bery & Saad, 2014). The Dipole-Dipole array, Wenner-β array or Wenner-β-Schlumberger array are more precise in showing such detailed obstacles near the surface (Dahlin & Zhou, 2004; Hauck & Kneisel, 2008; Pebriyanto *et al.*, 2017). Therefore, especially for the prospection of archaeological objects and obstacles, the Dipole-Dipole array could deliver a better resolution, as practiced in the quasi-3D in Waldaschaff (W4.1) and partly in Alzenau (A4.1). Following the theoretical depth resolution of ERT measurement with half the used spacing (Reynolds, 2011) for each quadripole and the recommendation to not use a higher n-value (distance between the electrode pairs; cf. Subsection 3.2.1, Figure 17) than 6 (Dahlin & Zhou, 2004) in Dipole-Dipole measurements, since small profiles such as that used for the quasi-3D grid in Mömbris or H4.1 (both 0.75 m spacing) are prone to measurement errors in depths over 2.25 m. Therefore, the Wenner-Schlumberger array can deliver better resolution if there are archaeological objects expected in deeper areas. Nevertheless, such measurements were additionally performed due to the shorter measurement time using a single channel and lower energy supply in a highly conductive material. Indeed, without using an external battery, the resistivity meter sometimes failed to finish a single 408 quadripole Dipole-Dipole array, especially when using a short spacing.

However, once the collected results were analyzed, the Dipole-Dipole array could deliver a better-fitting image in the other measurement, too, especially for the questions 4.1 in Waldaschaff, Hauenstein, Mömbris and Alzenau.

All measurements were inverted using *RES2DINVx64*. It is expected that different inversion software such as *BERT* (Rücker & Günther, 2018) shows slightly different results (Emmert, 2020). However, due to a usually good fit with the invasive data, this would most likely not cause a different geomorphological or archaeological interpretation of the data. All displayed results were inverted using the robust inversion scheme (L2 norm), which produces more reliable results regarding objects and layers with sharp boundaries (Loke *et al.*, 2003). On the one hand, this reliability had a better fitting with the expected and proven geomorphological and archaeological situations. On the other hand, the least-mean-square inversion scheme (L1 norm) was also used (not shown) and found no situation that allows for a substantially different geomorphological or archaeological interpretation.

Not only does the raw data plays a major role in the single tomogram, but also the used model (normal model/extended model), inversion type and other parameters. Exemplary, different inversion results are compared in Figure 183. The boundary analysis values were added to each other whenever their value was above 1 (the lowest possible value) and then divided by 2 for the comparison, whether boundaries are displayed with both of the compared models or inversions (Figure 183k, l, n, p). To calculate the gradient between the boundaries, the values of the boundary analysis were subtracted from each other (Figure 183i, j, m, o). Therefore, "dark brown" is given as a color whenever the extended model or the robust inversion scheme (L1 norm) delivers a value of 5 and the normal model or least-square inversion scheme (L2 norm) delivers a value of 1 and is upside-down for "violet" values.

Although the gradient orientation of the boundary varies only slightly (Figure 183b, d, f, h), differences within the boundary's intensity are visible in the used models and inversions (Figure 183a, c, e, g). Upon closer inspection of the different inversion results, it is evident that the boundary is spread wider in the least-square inversion scheme (L2 norm). Loke *et al.* (2003) discuss a better fit of the robust inversion scheme (L1 norm) with sharp boundary layers and objects that seems to affect the boundary distribution. Nevertheless, the comparison with the robust inversion scheme (L1 norm; Figure 183i, j) reveals a location deeper of the boundaries of the least-square inversion scheme (L2 norm) results. However, the specific form is displayed in both inversion models (Figure 183k, l).

Figure 183: Comparison of different models and inversion processes for the boundary analysis: a) boundary analysis W1_ERT_a robust inverted, normal model; b) boundary gradient orientation W1_ERT_a robust inverted, normal model; c) boundary analysis W1_ERT_a robust inverted, extended model; d) boundary gradient orientation W1_ERT_a robust inverted, extended model; e) boundary analysis W1_ERT_a least-square inverted, normal model; f) boundary gradient orientation W1_ERT_a least-square inverted, normal model; g) boundary analysis W1_ERT_a least-square inverted, extended model; h) boundary gradient orientation W1_ERT_a least-square inverted, extended model; i) gradient between the single inversions, normal model; j) gradient between the single inversions, blocky model; k) boundaries and their intensity displayed in the results of both inversions of the normal model; l) boundaries and their intensity displayed in the results of both inversions of the extended mode; m) gradient between the different models of the least-square inversion schemes; n) boundaries and their intensity displayed in the results of both models of the least-square inversion schemes; o) gradient between the different models of the robust inversion scheme; p) boundaries and their intensity displayed in the results of both models of the robust inversion scheme

Figure 183:

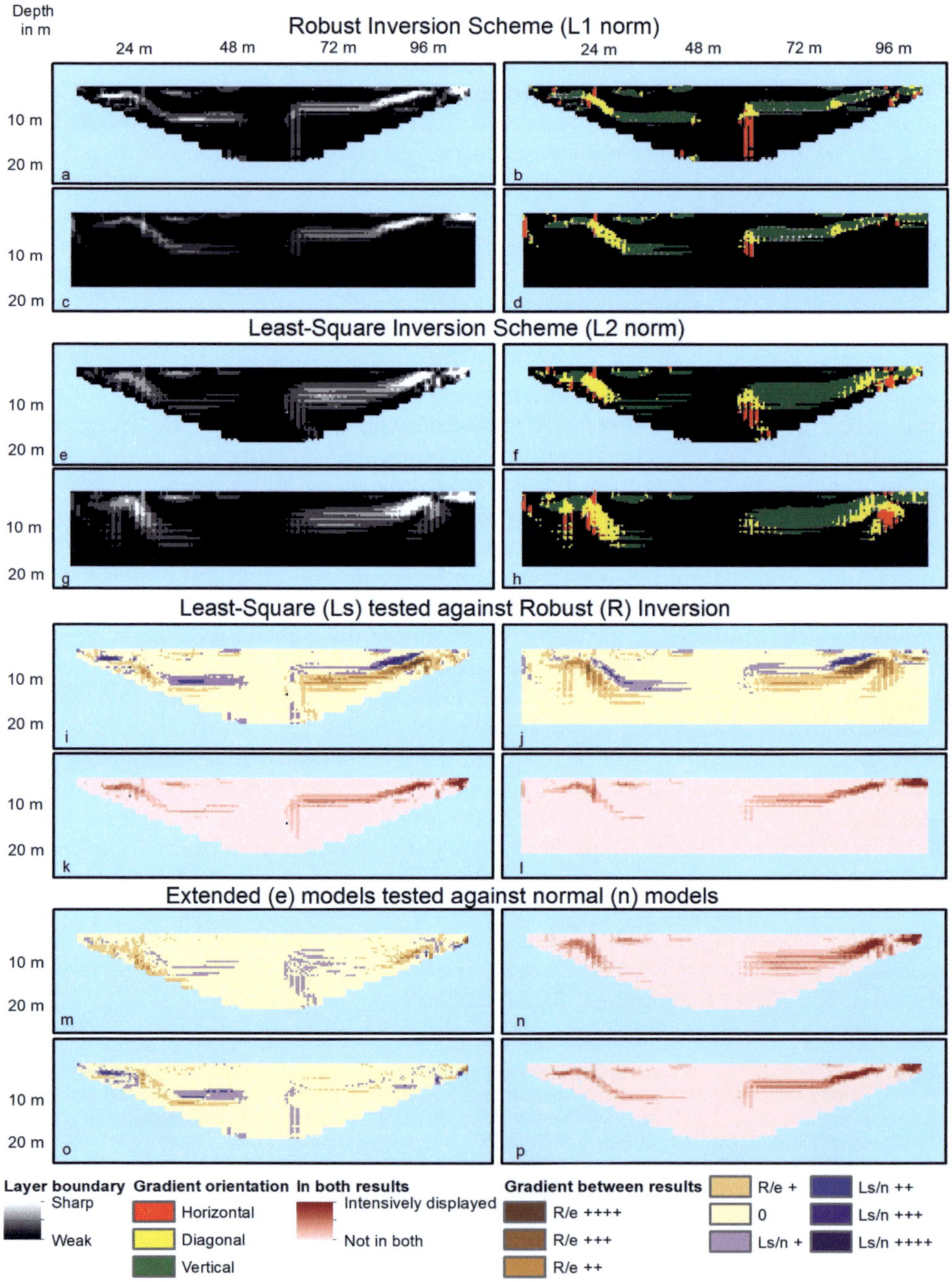

Source: own illustration

The differences between the different models are less distinct than expected. Nevertheless, in various parts, the boundaries of the single models differ. Vertical gradients are especially more intensely displayed within the normal models. Within the other differences, no clear trend is visible (Figure 183m, o). In addition, all major forms and r-layers are displayed within both models (Figure 183n, p).

With the combination of all the comparisons in Figure 183, it is evident that all the different models and inversion types display different boundary distributions. However, all these models and inversion types allow for the same geomorphological interpretation. Nevertheless, the surprisingly different depth positions of the boundaries within the different inversion results indicate, once again, that it is difficult to determine exact and unambiguous depths of different units with these data.

GPR and SRT

The 30-MHz GPR data in Waldaschaff and at the Eisenberg delivered particularly indistinct results. Therefore, the interpretation itself must be considered with caution and could only be made with extra information, such as from other geophysical methods, invasive methods and information from maps (e.g. the position of known faults). Standing alone, the results of these measurements were not able to be interpreted. However, the results support the other data and help extrapolate the developed information. In Waldaschaff, the second interpretation line, most likely the border between saprolite and bedrock of the (grano-)diorite, has no ground-truth at all and is not resolved well in any other geophysical method. However, the other collected data were not meant to resolve this border, but it seems this border is the most suitable possibility for the data provided. It seems plausible that there is a fluent change in dielectricity between weathered and un-weathered (grano-)diorite that reveals the number of reflectors displayed in the data. The maximum gain is reached early in these profiles; thus, the rise in intensity is not caused by an over-gaining of the bottom part of the profile.

The depth within the GPR profile is a function of time, which passed since transmitting the impulse and the expected velocity of radar wave in the specific material. However, only a single expected velocity can be set within the program, as is usual in most GPR-evaluation programs. The depth of the single layers must be taken with caution, and even at a single measurement point, the velocity can vary a great deal (Neal, 2004). For example, examining more closely E2_GPR_a and -b at the Eisenberg (Figure 163 and 164), including four different areas with different materials: The basalt layer on top has expected velocity values of about 0.11 m/ns, whereas the velocity of the Miocene sediments of clay and sand can vary between 0.15 m/ns in dry sand and 0.06 m/ns in wet clay (Blindow, 2005). The Buntsandstein had velocities between 0.11 m/ns and 0.15 m/ns (Reynolds, 2011). Therefore, the velocity between and within the different layers can vary up to a factor of 2.5. Especially at slope positions, the water content differs a great deal and causes a change in velocity. Thus, the depth of single layers needs to be considered with caution. Distinct statements regarding the behavior of single layers can only be made about the uppermost parts. For example, the visibility of a parallel layering of several meters, in a depth of 20 m, is dependent on a continuous substrate and the water content of the layers above. Additionally, the depth of the GPR data can differ a great deal from those of the ERT and the invasive methods. To counter these errors that occur regarding the material's velocity, two methods of generating the velocity of material are possible. On the one hand, small features such as single rocks, artifacts or infrastructure create hyperbolas. Caused by a nonlinear spreading of the radar wavetrains, both hyperbola flanks getting created in more distance to the feature. The hyperbola's width flanks the time, and thus, the average velocity through the material above can be distinguished. On the other hand, a CMP measurement can be made. A known and well-developed reflector is detected using an increasing antenna separation. With this method, a synthetic hyperbola is created and can be used to distinguish the velocity. However, both methods need a single feature or a well-developed layer combined with a comparable site to measure the CMP. In addition, only one velocity for a short position is distinguished, generally still including a significant error for other parts of the profile.

The GPR receiver receives all data within a previously decided time window. The used filters delete all information outside of the used frequency. However, many signals within the time window in the frequency of interest are still not interpretable. Unshielded antennas especially cause many signals that are not of interest. As a result of a hyperbola in the speed of light, surface reflections (e.g. from trees) are easy to distinguish and delete. Other unwanted reflections, such as multiples, mirroring large changes in the dielectric constant on top by refracting the same wavetrain in all directions several times, are sometimes difficult to define and interpret. Nonlinear layering or reflectors, such as concave depressions or single features, such as large rocks or human infrastructure, cause signals that are difficult to interpret and need to be respected. Nevertheless, using migration, these reflections can be filtered in a manner similar to seismic reflectors (McQuillin *et al.*, 1979; Neal, 2004).

The used filters help with the interpretation and help to identify structures in noisy areas. Nevertheless, some variable information can be deleted or transformed using migration or filters. For example, the background subtraction deletes all surface-parallel information, such as air- and ground-waves, but also linear reflectors.

In contrast to the GPR, the velocities of the seismic refraction are more precise due to the measurement setup (Sandmeier & Krummel, 2005; Schrott & Hoffmann, 2008) and traveling along the surface of a material with a changing velocity. The speed is measured at several geophones and can be calculated. Nevertheless, there are other potential errors within such a measurement. For example, the first arrivals of the wave were picked manually (Geometrics, 2009). In this study, due to noise pollution, these were sometimes not easy to identify. Subsequently, an initial model was created manually; for the performed measurements, the default values were used (Geometrics, 2009). Other values would have created results that vary slightly from those displayed. Additionally, some transition layers, such as the basal-breccia, or the bedrock's different weathering status, complicated the interpretation of the results. To be fair, the author admits that he has little experience with this method, complicating a safe interpretation. Therefore, SRT was only used complementarily to underline the results of the other methods.

Magnetic survey

With the help of magnetic surveying in a few cases, easy-to-interpret results were produced. Most important for identifying archaeological and geomorphological features with the help of magnetic surveying is the remnant magnetization of material, which can be correlated to a movement of the material (Reynolds, 2011; Everett, 2013; Fassbinder, 2015). Additionally, for example, post-holes (Becker, 1987) and fireplaces (Le Borge, 1960) can be detected (see in detail Subsection 3.2.4). Even though such anomalies are caused by ferri- and ferromagnetism, its content and the amount of magnetization are quite low in the given examples. Indeed, ferromagnetism can be 10^6 times stronger than diamagnetism or paramagnetism (Telford *et al.*, 1990). Therefore, contamination of such material, most likely artifacts made of iron, especially superimpose such smooth anomalies. Such artifacts caused most of the omnipresent dipole anomalies within the results of the magnetic survey. The sole exception might be the Eisenberg, where the anomalies could be partly associated with iron ore and natural iron concretion. Such artifacts are sometimes of archaeological importance, too, but generally, they are modern scrap. However, measuring an artifact with sensors is the only opportunity to derive in-depth information for single findings (Foerster, 2018). Indeed, this was not the aim of this work and further complications were caused by fences in the surroundings of the measurement, causing strong magnetic signals, visible, for example, in the magnetic survey at Mömbris.

Ground-truth methods

Soundings were conducted for ground-truthing the geophysical results. On the other side, the geophysical methods extrapolated the information from the soundings. Therefore, most of the descriptions of the soundings are generalized. Additionally, several further parameters were noted according to the used soil Charter BKA 5 (Ad-hoc-Arbeitsgruppe Boden, 2005), which was translated into the WRB (Fao, 2015). For example, the color of the material, the detailed texture, redoximophic features, amount of roots and other parameters have been noted. However, only a few of these parameters are displayed in detail when they were necessary for the interpretation of a study site.

Other approaches regarding the sediments, such as micromorphological and soil chemical approaches (e.g. Macphail & Goldberg, 2017) or the analyses of phytolites (e.g. Dietrich *et al.*, 2019) and geophysical parameters (Obrocki *et al.*, 2020), can answer further questions that fit the context. These approaches can be considered future possibilities to supplement the multi-methods approach used in this study.

6 Conclusions and key findings

In the Spessart uplands, numerous medieval archaeological sites are known and have been excavated in recent years by the ASP (Rosmanitz & Bachmann, 2017; Hasenstein *et al.*, 2019; Rosmanitz *et al.*, 2019a, 2019b; Rosmanitz *et al.*, 2020). In the context of these excavations, this study was conducted to identify how relief development and medieval building activity in the Spessart uplands influenced each other.

Regarding the structure of the objectives and discussion, site-specific key findings that focus on the single study sites and site-specific questions are first derived. Subsequently, this local study's comparability with and transferability to a larger scale are considered regarding another bunch of key findings. Furthermore, the key findings according to the methodological approach and the newly developed aspects are discussed. Finally, all findings are summarized in a short synthesis.

6.1 Site-specific key findings (K)

For all five study sites within the Spessart, this study examined which parts are predominantly of natural origin and which were anthropogenically influenced. Such influence was mainly described qualitatively, but slight quantification was also performed for the following derivation of site-specific key findings. At all sites, an interlocking between relief development and medieval building activity can be observed. The site-specific key findings are mostly divided into two parts. The first focuses more on relief development (1), the second more on the area's medieval building activity (2). Of course, these are often inseparable, but the focus can be changed.

6.1.1 Waldaschaff

The castle site of Waldaschaff (W) was located on an isolated hard-rock hill of (grano-)diorite, surrounded by a rough relief made of sedimentary claystone.

WK1 – Relief development

The exposed and rough relief in the surroundings of *Wahlmich Castle* in Waldaschaff is of natural origin. As geophysical and sedimentological data of W1 reveal, the elevated areas and depressions consist of fragmented Bröckelschiefer, which easily weathers. The Bröckelschiefer moved downslope as a gravitative mass movement, triggered most likely by tectonic processes, most visible at the breaking-off edge, consisting of (grano-)diorite. The breaking-off edge belongs to a complex faulting system at the study site, which is described in W3. Therefore, the medieval building activity was performed precisely in this rough relief and did not cause it. However, the region's geological map (Weinelt, 1962) is barely detailed and offers no hint of the Bröckelschiefer deposits in the area of interest.

WK2 – Medieval building activity

Parts of the complex building history can be described in the combination of excavation (Rosmanitz & Bachmann, 2017; Rosmanitz *et al.*, 2019b) and geophysical prospection. The location of the outer bailey remains unclear, however. Nevertheless, at least an anthropogenic usage of the most prominent elevated area is suspected based on magnetic surveying and a prospection pit.

6.1.2 Hauenstein

Hauenstein (H) Castle was located on a hard-rock hill, manually isolated by dug neck ditches, at the bottom of a valley surrounded by two creeks. Archaeological findings (Rosmanitz *et al.*, 2019a) have found evidence for a mill directly at the castle site.

HK1 – Relief development

In contrast to the center of the castle hill, consisting of bedrock, which provides a natural origin, the western valley is greatly affected by humans. Geophysical and sedimentological results reveal several ponds that had dam water for use by the mill. ^{14}C-datings from the bottom of the sediments accumulated in a low-energy fluvial system, directly above the saprolite, are dated to the late usage period. Thus, the deposit of these sediments and the ponds' usage can clearly be linked to the time of the medieval castle in the 14th century CE.

HK2 – Medieval building activity

Different anthropogenic structures found in the valley and the castle site were used to identify areas of archaeological interest and to enable a targeted excavation. Nevertheless, the location of the outer bailey remains unclear and needs further investigation. Incidentally, the origin of round structures visible on the slope and hillshade, combined with a high number of geophysical anomalies in GPR and magnetic measurements at a meadow, could not be identified. An affiliation of these features with modern or ancient agricultural use is expected.

6.1.3 Mömbris

The castle site in Mömbris (M) is located in the town next to a graveyard. A rectangle plateau of about 500 m^2 occurs at the site. It was initially associated with the castle's former extension.

MK2 – Medieval building activity

The present-day plateau was not constructed at the time of the former castle. The previous geophysical prospection using GPR and ERT enabled a targeted archaeological excavation. Some linear anomalies were detected with the help of these methods. These anomalies were proven with the excavation to belong to the castle. Remarkably, the location of the wall structure was beside the suspected areas at the plateau's edges and located partly next to the plateau. Therefore, this geophysical prospection made precise archaeological excavation possible (Rosmanitz *et al.*, 2020).

6.1.4 Alzenau

Rannenburg Castle, near Alzenau (A), is located on a ridge intersected by deep neck ditches into different nearby areas with gully erosion.

AK1 – Relief development

The northwestern exposed slopes are covered with loess loam. The distribution of the loess loam varies only slightly from those on the geological map (Okrusch *et al.*, 1967). Colluvial, relocated layers mixed with charcoal fragments indicate an anthropogenic use of these culmination areas and slopes, probably including deforestation, which would have increased the gully erosion.

AK2 – Medieval building activity

At the castle site, the findings are prohibited from being combined with any type of ground-truth yet. Therefore, the findings necessarily need to be considered preliminary results and could help guide for targeted, future excavations. The area seems to be greatly impacted by humans. Apparent impacts, such as building structures and dug neck ditches, were proven by the results. Additional earthworks, such as the flattening of areas and the massive accumulation of sediments to create plateaus, are expected.

6.1.5 Eisenberg

The *Eisenberg (E) mining site* is marked by an area of collapsed shafts used for mining, surrounded by a few wet meadows.

EK1 – Relief development

The results indicate that dense, clay-rich layers influence the creation of wet meadows in the surroundings of the Eisenberg. Even though these meadows were partly used and are still drained actively in some parts, their existence is most likely caused by natural processes. Beyond these meadows, loess loam and basalt are located at the uppermost layers. In the meadows, aquiferous, sandy layers on top of clay-rich layers of Miocene sediments are located near the surface, which causes stagnic water or groundwater to rise.

EK2 – Medieval earthworks and ore development

The collapsed shafts are more than 8 m in depth, as proved by an outcrop in combination with a Pürckhauer sounding. The geophysical data indicate depths of up to

15 m. The single shafts seem small at their entrance. The shafts are extended over a wider area in deeper parts, most likely at the former orebody.

A weathering deposit of basalt iron ore is suspected based on a combination of archaeological and geological findings. This weathering deposit is located underneath the present-day basalt layer. Most likely, the clay-rich layer is dense, and the aquiferous layers on top allowed a weathering of the basalt to create this ore type. Additionally, small-sized layers with iron-enrichment are located downslope of the mining area and in locations with groundwater outlets.

6.2 Key findings for the Spessart and at a larger scale (LK)

These essential findings link the relief development with medieval building activity and help interpret the archaeological findings and geomorphology. Nevertheless, their significance is spatially very limited. Therefore, the importance of local information for a regional and supra-regional scale are summarized in the following key findings.

LK1 – Landscape development model and medieval castle sites

A generalized model was derived that focuses on several processes and influences that have to be mentioned by analyzing the relief at a castle site. This model displays the relief development in the context of the sites' medieval usages, indicating four phases, including the predominantly natural relief before the medieval usage, the massive anthropogenic impact during the period of interest, a transition phase after usage, and the present-day relief. Due to generalizations, various site-specific influences are necessary to mention as well.

LK2 – Positive and negative location factors

At all the study sites, the relief affected the location choice. Nevertheless, most visible are the often-mentioned military, logistical, resource-based and representative reasons, which are not focused on in this study. However, in addition to these reasons, different choices regarding the locations' exact positions can be inferred. The location of *Wahlmich Castle* in Waldaschaff was precisely chosen in an area of former gravitative mass movement. *Hauenstein Castle*'s location at the bottom of a valley is surrounded by two creeks that created ponds advantageous for using a mill, a choice driven by geomorphological and geological features. The natural development of the ore body in the Eisenberg influenced the anthropogenic use there, whereas the hydrological system in the surroundings of the Eisenberg, especially in several wet meadows, is of natural origin and caused by Miocene sediments.

However, another aspect appears that focuses on the topographic disadvantages of study sites for uses other than the creation of a castle. At *Wahlmich Castle* but also at *Hauenstein Castle*, the advantages of the relief regarding building a castle or mill are likely to be described. Upon closer inspection, the precise location, however, various disadvantages for other uses, such as agriculture or the creation of economic buildings, appear. The location choice for castle creation could be triggered by the advantage of the relief regarding position and its disadvantage for other types of use. Maybe the other usage's negative location factors are as important as the castle's positive location factors, but this aspect should be considered in future castle site research.

LK3 – Possible overestimation of anthropogenic influence

As a cultural landscape, of course, the landscape and relief of the Spessart were massively affected by human activity during the time of interest. Active processes like earthworks, such as dug neck ditches, mining activity and built structures, like walls or ponds, impacted the relief. Passive processes, such as colluvial erosion and accumulation, enhanced gully erosion via agricultural use, and deforestation reinforced this impact.

Nevertheless, it seems that these effects on relief development result in an overestimation of the anthropogenic impact and likely disguise many ancient, ongoing natural processes and parts of the still-existing paleo-surface. Most likely, humans tend to associate a cultural landscape more with the usage of the land than with the paleo-relief. Within this massively affected cultural landscape, many features and processes linked to the paleo-relief continuing. These features and processes are often associated with the use of the area, but they could be the reason for the usage and not caused by it.

6.3 Methodological key findings (MK)

In addition to the regional transferability of single observations and the derived model, the applied multi-method approach and the newly developed methods are transferable to other studies, as demonstrated by the following key findings.

MK1 – Suitability of the methodological approach

The combination of GIS-based terrain analyses, geophysical methods and ground-truth methods seem to be suitable for answering the research questions with proven interpretations of the results. Sometimes these methods need to be extended by additional methods, partly by other specialists, and used flexibly, depending on physical parameters and on-site logistics. Whenever the corresponding methods are missing, the database becomes thin, and the interpretation becomes an approximation.

MK2 – Transferability of the guidelines for ERT description

The developed guidelines were designed for the present study, but transferability to other fields is desirable. Especially when investigating multiple study sites and using many different applications and spacings to answer different questions, a quantitative description seems to be necessary to maintain objectivity. For a broader application in various habitats, including penetration depths between 10 cm and several kilometers, the value ranges most likely need to be extended, or different value ranges for different applications may need to be created. However, an objective description of the data between different investigations of similar topics would increase comparability. Nevertheless, it should always be borne in mind that these guidelines focus only on geometrical information. Therefore, comparability is only guaranteed when using the same measurement and inversion applications and robust data sets.

MK3 – Guideline profile of the generalized relative resistivity distributions

With the help of the guidelines for ERT description, generalized relative resistivity distributions can be derived. Therefore, further measurements at the same study sites could be easier to interpret. However, these generalized datasets are ambiguous. The representativeness must always be questioned, and exceptions confirm the rule, for example, at positions with inverted relief. Nevertheless, such generalized, relative resistivity distribution displays the potential of the added value generated by a more quantitative description of the relative resistivity distribution.

6.4 Synthesis

In the Spessart uplands, the mutual influence between relief development and medieval building activity was analyzed. At four castle sites and one mining site, chosen for archaeological reasons, a multi-method approach was conducted. This approach included GIS-based relief analyses, geophysical measurements and sedimentological investigations.

In retrospect, the site location choices for archaeological reasons ensured that the process understanding was not focused, and archives for geochemical analyses and datings were rarely provided. A better focus on the processes and archives would have improved understanding and quantification from a geomorphological perspective. However, this would have impaired the importance of this study regarding the archaeological meaning of the single sites and their multidisciplinary applicability in the context of historically important sites.

Nevertheless, the key findings' numbers and meanings reveal that the multi-method approach was suitable for achieving the objectives. For all the study sites, it was possible to declare which parts of the relief are predominantly of natural origin and which are anthropogenically influenced (O1.1.1). Therefore, a quantitative categorization was developed, and the intensity of the anthropogenic influence was mapped at the study sites. Furthermore, it was possible to answer 13 of 15 site-specific questions to improve our knowledge of the single study sites, which allowed for precise excavations and put the archaeological findings in the context of the paleo-relief development (O1.1.2). With the help of the remaining two questions, at least an indication of anthropogenic usage of parts of the relief was found, and other parts were excluded from anthropogenic use in the medieval period. At each study site, it was possible to derive a strong interlocking between the paleo-surface development and the site's anthropogenic usage (O.1.1.3). Various examples were found of how relief development influenced the anthropogenic building activity on the microscale and how the building activity affected the relief at the sites.

The representativity of such a few study sites depends on the consideration approach (O1.2.1): upon closer inspection at the different topographic features or building activities, each study site seems unique and special. However, from a broader perspective, general assumptions of the interlocking and mutual influence between paleo-relief development and medieval building activity can be derived (O1.2.2). Four phases of interaction between paleo-relief and medieval building activity can be observed at each of the sites. First, a predominantly natural relief exists. Subsequently, its specific properties were used and integrated into the anthropogenic building activity. This activity ended suddenly, and natural processes and further anthropogenic use influenced the relief at the study sites. Therefore, within the present-day relief exits evidence of the influences of all the phases. Additionally, indications of a stronger influence of negative location factors for other types of usage, in addition to the advantages of the specific usage, were found. Furthermore, a high number of particular relief attributes at the microscale were important for the area's usage.

The chosen multi-method approach was suitable to achieve the questions and to reach the objectives of this study (O2.1). However, for further quantification, the study sites would have had to have been chosen from a more geomorphological perspective and combined with further methods and investigations. Nevertheless, the chosen analysis in combination with the application at specific sites for archaeological reasons seems to be a suitable compromise for achieving the objectives of this study. As the most important factor in improving the quality of geophysical prospection for an archaeological excavation, the relative position of geophysical anomalies, in addition to the subsurface's physical properties, were detected (O2.2). Therefore, a new way of describing the relative resistivity distribution of ERT data was developed. Even though this was specifically established for this study, transferability to other investigations at multiple study sites is desirable (O2.3). However, some adjustments will probably need to be made to adapt the approach to other sites, topics and applications. With the help of this new way of describing the relative resistivity distribution and a high number of measurements, it is possible to create a guideline profile with a generalized relative resistivity distribution (O2.4). These new generalized resistivity distributions can be a blueprint for the interpretation of other measurements at the same study site. Nevertheless, they must be treated with caution, and the number of necessary measurements needs to be tested in future studies and study sites.

In conclusion, this study has improved the understanding of the mutual influence of paleo-relief and medieval building activity on a local scale, but also contributed a little to the regional and supra-regional scale.

7 References

Ad-hoc-Arbeitsgruppe Boden (2005) *Bodenkundliche Kartieranleitung: Mit 103 Tabellen und 31 Listen*, 5th edition. E. Schweizerbart'sche Verlagsbuchhandlung (Nägele und Obermiller): Stuttgart.

Anselmetti, F.S., Hodell, D.A., Ariztegui, D., Brenner, M. & Rosenmeier, M.F. (2007) Quantification of soil erosion rates related to ancient Maya deforestation. *Geology*, 35(10), 915–918. Available from: https://doi.org/10.1130/G23834A.1.

Anthes, G. & Reischmann, T. (2001) Timing of granitoid magmatism in the eastern mid-German crystalline rise. *Journal of Geodynamics*, 31(2), 119–143. Available from: https://doi.org/10.1016/S0264-3707(00)00024-7.

Archie, G.E. (1942) The Electrical Resistivity Log as an Aid in Determining Some Reservoir Characteristics. *Transactions of the AIME*, 146(01), 54–62. Available from: https://doi.org/10.2118/942054-G.

Atya, M.A., Al Khateeb, S.O., Ahmed, S.B., Musa, M.F., Gaballa, M. & Abbas, A.M. et al. (2012) GPR investigation to allocate the archaeological remains in Mut temple, Luxor, Upper Egypt. *NRIAG Journal of Astronomy and Geophysics*, 1(1), 12–22. Available from: https://doi.org/10.1016/j.nrjag.2012.11.002.

Bao, Y., Zhai, S., Chen, J., Xu, P., Sun, X. & Zhan, J. et al. (2020) The evolution of the Samaoding paleolandslide river blocking event at the upstream reaches of the Jinsha River, Tibetan Plateau. *Geomorphology*, 351, 106970. Available from: https://doi.org/10.1016/j.geomorph.2019.106970.

Barker, R.D. (1979) Signal contribution sections and their use in resistivity studies. *Geophysical Journal International*, 59(1), 123–129. Available from: https://doi.org/10.1111/j.1365-246X.1979.tb02555.x.

Bavarian Agency for Digitisation, High-Speed Internet and Surveying (2014) DEM 1 m.

Bavarian Environment Agency (2017a) Digitale Übersichtsbodenkarte TK25-Blatt 5920 Alzenau i.UFr.

Bavarian Environment Agency (2017b) Digitale Übersichtsbodenkarte TK25-Blatt 5921 Schöllkrippen.

Bavarian Environment Agency (2017c) Digitale Übersichtsbodenkarte TK25-Blatt 6021 Haibach.

Bavarian Environment Agency (2020a) Naturraum-Untereinheiten: shapefile.

Bavarian Environment Agency (2020b) UmweltAtlas Bayern.

Bavarian Ministry of Finance and Regional Identity (2020) Geoportal bayern; BayernAtlas.

Bavarian State Office for Monument Preservation (2020) *Vorgaben zur Dokumentation archäologischer Ausgrabungen in Bayern*.

Bebermeier, W. & Schütt, B. (2011) Geoarchaeology - A New Discipline? Editorial to DIE ERDE. *Die Erde - Journal of the Geographical Society of Berlin*, 142(3), 210–212.

Becker, H. (1987) Magnetische Prospektion eines neolithischen Langhauses bei Baldingen, Stadt Nördlingen, Lndkreis Donau-Ries, Schwaben. *Das archäologische Jahr Bayern 1986*, 35–37.

Becker, H. & Fassbinder, J.W.E. (1999) Magnetometry of a Scthian settlement in Siberia near Cicah in the Baraba Steppe. *Archaeological prospection*, 108, 168–172. Available from: https://doi.org/10.11588/monstites.2001.0.22877.

Bennett, S.J. & Wells, R.R. (2019) Gully erosion processes, disciplinary fragmentation, and technological innovation. *Earth Surface Processes and Landforms*, 44(1), 46–53. Available from: https://doi.org/10.1002/esp.4522.

Bery, A.A. & Saad, R. (2014) Enhanced Horizontal and Vertical Resolution in 4-D Electrical Resistivity Tomography for Environmental Slope Study in Penang Island. *Applied Mechanics and Materials*, 661, 73–80. Available from: https://doi.org/10.4028/www.scientific.net/AMM.661.73.

Bhat, G.R., Balaji, S., Bali, B.S., Iqbal, V., Balakrishna & Hussain, H. (2018) Paleoseismological investigations along Joggers Park Fault, Port Blair, South Andaman: Implications towards delineation of blind thrusting and related crustal deformation through ground penetrating radar (GPR) and electrical resistivity techniques (ERT and VES). *Journal of the Geological Society of India*, 91(1), 81–90. Available from: https://doi.org/10.1007/s12594-018-0823-z.

Bianchini Ciampoli, L., Santarelli, R., Loreti, E.M., Ten, A., Benedetto, A. & Atya, M.A. et al. (2020) Structural detailing of buried Roman baths through GPR inspection // GPR investigation to allocate the archaeological remains in Mut temple, Luxor, Upper Egypt. *Archaeological prospection*, 1(1), 12–22.

Biller, T. (2016) *Die mittelalterlichen Stadtbefestigungen im deutschsprachigen Raum: ein Handbuch*. Philipp von Zabern: Darmstadt.

Blecha, V., Fischer, T., Tábořík, P., Vilhem, J., Klanica, R. & Valenta, J. et al. (2018) Geophysical evidence of the Eastern Marginal Fault of the Cheb Basin (Czech Republic). *Studia Geophysica et Geodaetica*, 62(4), 660–680. Available from: https://doi.org/10.1007/s11200-017-0452-9.

Blindow, N. (2005) Bodenradar. In: Knödel, K., Krummel, H. & Lange, G. (Eds.) *Handbuch zur Erkundung des Untergrundes von Deponien und Altlasten: Geophysik.* Springer: Berlin Heidelberg, pp. 389–424.

Bofinger, J., Kurz, S. & Schmidt, S. (2014) Hightech aus der Luft für Bodendenkmale. Airborne Laserscanning (LIDAR) und Archäologie. *Denkmalpflege In Baden-Württemberg – Nachrichtenblatt Der Landesdenkmalpflege*, 153–158. Available from: https://doi.org/10.11588/nbdpfbw.2007.3.12005.

Bork, H.-R. (1998) *Landschaftsentwicklung in Mitteleuropa.* Klett - Perthes: Gotha und Stuttgart.

Boss, M. & Wamser, L. (1984) Eine Waldglashütte des frühen Spätmittelalters bei Schöllkrippen. Landkreis Aschaffenburg, Unterfranken. *Das Archäologische Jahr in Bayern*, 1983, 157–159.

Brinkman, R. (1948) Die mitteldeutsche Schwelle. *Geologische Rundschau*, 36(1), 56–66. Available from: https://doi.org/10.1007/BF01791910.

Bristow, C.S. (2009) Ground penetrating radar in aeolian dune sands. In: Jol, H.M. (Ed.) *Ground Penetrating Radar Theory and Applications.* Elsevier science: Amsterdam, Boston, pp. 273–297.

Büdel, B., Kempf B., Rüppel, J. & Nelle, O. (2021) Der Wald im Spessart: Entwicklung des Spessartwaldes, in press. *Spessart*, 1.

Bury, A. (2018) Hydraulic calculations at the medieval excavation of castle hauenstein. *Bachelorthesis, Geography Würzburg*.

Butzer, K. (1973) Spring Sediments from the Acheulian Site of Amanzi (Uitenhage District, South Africa). *Quaternia*, 17.

Chavez, R.E., Tejero, A., Argote, D.L., Cifuentes, G., Hernández-Quintero, J.E. & García-Serrano, A. (2017) Interior Imaging of El Castillo Pyramid, Chichen Itza, Mexico, Using ERT-3D Methods: Preliminary Results. *Geofísica international*, 56(2), 219–227.

Conyers, L.B. (2012) *Interpreting ground-penetrating radar for archaeology*. Left Coast Press: Walnut Creek, California.

Correia, A. (2019) Archaeological Geophysics in Portugal: Some Survey Examples. *Archaeogeophysics: State of the Art and Case Studies*, 101–124.

Cramer, C. (1952) Landeshoheit und Wildbann im Spessart. *Landeskunde und Kunst des Untermaingebietes*, 1.

Dahlin, T. & Zhou, B. (2004) A numerical comparison of 2D resistivity imaging with 10 electrode arrays. *Geophysical Prospecting*, 52(5), 379–398. Available from: https://doi.org/10.1111/j.1365-2478.2004.00423.x.

Dannath, Y., Larsen, A. & Kirleis, W. (2013) Archäobotanische Untersuchungen einer Grabenverfüllung der mittelalterlichen Burg „Zur Mole" im Spessart. *Hoppea, Denkschrift der Regensburgischen Botanischen Gesellschaft*, 74.

Deiana, R., Bonetto, J. & Mazzariol, A. (2018) Integrated Electrical Resistivity Tomography and Ground Penetrating Radar Measurements Applied to Tomb Detection. *Surveys in Geophysics*, 39(6), 1081–1105. Available from: https://doi.org/10.1007/s10712-018-9495-x.

Deumlich, D., Schmidt, R. & Sommer, M. (2010) A multiscale soil-landform relationship in the glacial-drift area based on digital terrain analysis and soil attributes. *Journal of Plant Nutrition and Soil Science*, 173(6), 843–851. Available from: https://doi.org/10.1002/jpln.200900094.

Diederich, G. & Ehrenberg, K.-H. (1998) *Geologischen Karte von Hessen 1:25 000. Blatt Nr. 5722 Salmünster: mit Erläuterungen*. Hessisches Landesamt für Bodenforschung: Wiesbaden.

Dietrich, L., Meister, J., Dietrich, O., Notroff, J., Kiep, J. & Heeb, J. et al. (2019) Cereal processing at Early Neolithic Göbekli Tepe, southeastern Turkey. *PloS One*, e0215214. Available from: https://doi.org/10.1371/journal.pone.0215214.

Dongus, H. (2000) *Die Oberflächenformen Südwestdeutschlands: geomorphologische Erläuterungen zu topographischen und geologischen Übersichtskarten mit 1 Tabelle*. Borntraeger: Berlin.

Drahor, M., G. (2019) Integrated Geophysical Investigations in Archaeological Sites: Case Studies from Turkey. In: El-Qady, G. & Matwaly, M. (Eds.) *Archaeogeophysics: State of the Art and Case Studies*. Springer International Publishing: Cham, pp. 27–68.

Ehrenberg, K.-H. & Hickethier, H. (1982) *Geologischen Karte von Hessen 1:25 000. Blatt Nr. 5622 Steinau a. d. Straße: mit Erläuterungen*. Hessisches Landesamt für Bodenforschung: Wiesbaden.

Eisen, O., Nixdorf, U., Keck, L. & Wagenbach, D. (2011) Alpine ice cores and ground penetrating radar: combined investigations for glaciological and climatic interpretations of a cold Alpine ice body. *Tellus B: Chemical and Physical Meteorology*, 55(5), 1007–1017. Available from: https://doi.org/10.3402/tellusb.v55i5.16394.

Ellis, E.C., Fuller, D.Q., Kaplan, J.O. & Lutters, W.G. (2013) Dating the Anthropocene: Towards an empirical global history of human transformation of the terrestrial biosphere. *Elementa: Science of the Anthropocene*, 1. Available from: https://doi.org/10.12952/journal.elementa.000018.

El-Qady, G. & Matwaly, M. (Eds.) (2019) *Archaeogeophysics: State of the Art and Case Studies*. Springer International Publishing: Cham.

Emmert, A. (2020) The Internal Structure of Periglacial Landforms - Assessments of Subsurface Variations in Permafrost-related and Frost-related Phenomena by Multidimensional Geophysical Investigations. Available from: https://doi.org/10.25972/WUP-978-3-95826-139-6.

Ermischer, G. (2006) Niederadel im Wandel. In: Rosmanitz, H. (Ed.) *Die Ketzelburg in Haibach: Eine archäologisch-historische Spurensuche*. Heimat und Geschichtsverein Haibach: Haibach, pp. 45–53.

Ermischer, G. (2008) Wege in die Kulturlandschaft: Einige Beispiele aus dem Spessart. *Vermittlung von Kulturlandschaft: Initiative zur Förderung des Kulturlandschaftsbewusstseins*, 89–98.

Ermischer, G. (2019) Der gläserne Wald.: Glasproduktion im Spessart. *Bayerische Archäologie*, 4, 36–41.

Everett, M.E. (2013) *Near-Surface Applied Geophysics*. Cambridge University Press: Cambridge.

Evrard, O., Vandaele, K., Bielders, C. & van Wesemael, B. (2008) Seasonal evolution of runoff generation on agricultural land in the Belgian loess belt and implications for muddy flood triggering. *Earth Surface Processes and Landforms*, 33(8), 1285–1301. Available from: https://doi.org/10.1002/esp.1613.

Fao (2015) *World Reference Base for Soil Resources*. Food and agriculture organization of the United Nations: Rome.

Fassbinder, J.W., Stanjek, H. & Vali, H. (1990) Occurrence of magnetic bacteria in soil. *Nature*, 343(6254), 161–163. Available from: https://doi.org/10.1038/343161a0.

Fassbinder, J.W.E. (2010) Geophysical prospection of the frontiers of the Roman Empire in southern Germany, UNESCO World Heritage Site. *Archaeological prospection*, 17(3), 129–139. Available from: https://doi.org/10.1002/arp.387.

Fassbinder, J.W.E. (2015) Seeing beneath the farmland, steppe and desert soil: magnetic prospecting and soil magnetism. *Journal of Archaeological Science*, 56, 85–95. Available from: https://doi.org/10.1016/j.jas.2015.02.023.

Fassbinder, J.W.E., Asăndulesei, A., Radner, K., Kreppner, J. & Squiteri, A. (2017) Magnetometer prospection of Neo-Assyrian sites in the Peshdar Plain, Iraqi Kurdistan. *12th International Conference of Archaeological Prospection*, 70–76.

Federal Agency for Cartography and Geodesy (2020) DEM 200.

Fertig, J. (2005) Seismik: Prinzip der Methode; Anwendungsmöglichkeiten; Grundlagen. In: Knödel, K., Krummel, H. & Lange, G. (Eds.) *Handbuch zur Erkundung des Untergrundes von Deponien und Altlasten: Geophysik*. Springer: Berlin Heidelberg, pp. 425–466.

Fischer, P., Meurers-Balke, J., Gerlach, R., Bulla, A., Peine, H.-W. & Kalis, A.J. et al. (2016) Geoarchaeological and archaeobotanical investigations in the environs of the Holsterburg lowland castle (North Rhine-Westphalia) – evidence of landscape

changes and saltwater upwelling. *Zeitschrift für Geomorphologie, Supplementary Issues*, 60(1), 79–92. Available from: https://doi.org/10.1127/zfg_suppl/2015/S-00186.
Foerster (2018) Manual: DATA2LINE - Auswertungs- und Projektmanagement software: BASIC/UXO/BM/GEO.
Freeden, W. & Rummel, R. (Eds.) (2017) *Ingenieurgeodäsie: Handbuch der Geodäsie,*. Springer Berlin Heidelberg: Berlin, Heidelberg.
Freymann, K. (1991) *Der Metallerzbergbau im Spessart: Ein Beitrag zur Montangeschichte des Spessarts*. Geschichts- und Kunstverein Aschaffenburg e. V.: Aschaffenburg.
Friedel, H. (2000) Der Standort der Womburg. *Unser Kahlgrund 2001. Heimatjahrbuch*, 46, 21–25.
Friedrich-Lautz, I. (1963) Zusammensetzung und Entstehung der Brauneisenerzlagerstätten in den Tagebauen der Gewerkschaft bei Merlau/Oberhessen. *Notizblatt des Hessischen Landesamtes für Bodenforschung zu Wiesbaden*, 91, 277–317.
Fuchs, M. & Zöller, L. (2006) Geoarchäologie aus geomorphologischer Sicht Eine konzeptionelle Betrachtung (Geoarchaeology from a Geomorphological Perspective — A Conceptual Consideration). *Erdkunde*, 60(2), 139–146. Available from: https://doi.org/10.3112/erdkunde.2006.02.05.
Furman, A., Ferré, T.P.A. & Warrick, A.W. (2003) A Sensitivity Analysis of Electrical Resistivity Tomography Array Types Using Analytical Element Modeling. *Vadose Zone Journal*, 2(3), 416–423. Available from: https://doi.org/10.2136/vzj2003.4160.
Gaffney, C., Gaffney, V., Neubauer, W., Baldwin, E., Chapman, H. & Garwood, P. et al. (2012) The Stonehenge Hidden Landscapes Project. *Archaeological prospection*, 19(2), 147–155. Available from: https://doi.org/10.1002/arp.1422.
Galli, P., Giaccio, B., Peronace, E., Piscitelli, S., Quadrio, B. & Bellanova, J. (2014) Integrated near surface geophysics across the active Mount Marzano Fault System (southern Italy): seismogenic hints. *International Journal of Earth Science*, 103(1), 315–325.
Geometrics (2009) SeisImager/2D Manual: Version 3.3. *avaidable (https://geometrics.com/wp-content/uploads/2019/05/SeisImager2D_Manual_v3.3.pdf)*.
Gerber, R., Felix-Henningsen, P., Behrens, T. & Scholten, T. (2010) Applicability of ground-penetrating radar as a tool for nondestructive soil-depth mapping on Pleistocene periglacial slope deposits. *Journal of Plant Nutrition and Soil Science*, 173(2), 173–184. Available from: https://doi.org/10.1002/jpln.200800163.
Giocoli, A., Galli, P., Giaccio, B., Lapenna, V., Messina, P. & Peronace, E. et al. (2011) Electrical Resistivity Tomography across the Paganica-San Demetrio fault system (L´Aquila 2009 earthquake). *Bollettino di Geofisica Teorica ed Applicata*, 52(3), 457–469. Available from: https://doi.org/10.4430/bgta0029.
Graham, I.D.G. & Scollar, I. (1976) Limitations on magnetic prospection in archaeology imposed by soil properties // Technische und naturwissenschaftliche Beiträge zur Feldarchäologie, 6, 1–124.
Griebel, E. (1982) *Chronik des Marktes Mömbris*. Markt Mömbris.
Gröber, K. & Karlinger, H. (1916) *Die Kunstdenkmäler des Königreichs Bayern: Regierungsbezirk Unterfranken & Aschaffenburg*. Druck und Kommissionsverlag von R. Oldenbourg: München.

Groves, P.D. (2013) *Principles of GNSS, inertial, and multisensor integrated navigation systems*. Artech House: Boston.
Hasenstein, L., Geschwinder, R. & Bergmann, C. (2019) Relikte mittelalterlicher Eisengewinnung und einer Wüstung bei Steinau an der Straße. *Jahrbuch für Archäologie und Paläontologie in Hessen*, 2018, 186–188.
Hauck, C. & Kneisel, C. (Eds.) (2008) *Applied Geophysics in Periglacial Environments*. Cambridge University Press: Cambridge.
Hauck, C. & Mühll, D.V. (2003) Inversion and interpretation of two-dimensional geoelectrical measurements for detecting permafrost in mountainous regions. *Permafrost and Periglacial Processes*, 14(4), 305–318. Available from: https://doi.org/10.1002/ppp.462.
Heine, K. (1994) Bodenzerstörung -ein globales Problem. *Beiträge zu globalen Umweltproblemen*, (Nr. 2), 65–92.
Herget, J. (2006) Die direkten Einflüsse des Menschen auf die Reliefsphäre. *Zeitschrift für Geomorphologie*, 148, 42–47.
Hermanns, R.L., Fauqué, L. & Wilson, C.G.J. (2015) 36 Cl terrestrial cosmogenic nuclide dating suggests Late Pleistocene to Early Holocene mass movements on the south face of Aconcagua mountain and in the Las Cuevas–Horcones valleys, Central Andes, Argentina. *Geological Society, London, Special Publications*, 399(1), 345–368. Available from: https://doi.org/10.1144/SP399.19.
Hessian Administration for Land Management and Geoinformation (2007) DEM 1 m.
Hessian Agency for Nature Conservation, Environment and Geology (2014) *Bodenkarte von Hessen 1 : 25 000, Blatt 5722; Salmünster*. Wiesbaden.
Hessian Agency for Nature Conservation, Environment and Geology (2020) Naturraum-Untereinheiten.
Himmelsbach, G. (2014) Frühneuzeitlicher Viehhandel und Kulturlandschaftsforschung im Spessart. *Bulletin der polnischen historischen Mission*, 9(0), 153–173. Available from: https://doi.org/10.12775/BPMH.2014.006.
Höfle, B. & Wagener, O. (2012) Burgen in der Landschaft - Inszenierung und Entzifferung anhand neuer Methoden. *Symbole der MAcht? Aspekte milltelalterlicher und frühneuzeitlicher Architektur*, 123–153.
Hooke, R.L. & Martín-Duque, J.F. (2012) Land transformation by humans: A review. *GSA Today*, 12(12), 4–10. Available from: https://doi.org/10.1130/GSAT151A.1.
Huggenberger, F. (2015) Niederadel im Spessart. Adelsgeschichte im Spiegel des spätmittelalterlichen Lehnswesens. *Studien zur bayerischen Verfassungs- und Sozialgeschichte*, 31.
Imposa, S., Grassi, S., Di Raimondo, S., Patti, G., Lombardo, G. & Panzera, F. (2018) Seismic refraction tomography surveys as a method for voids detection: an application to the archaeological park of Cava Ispica, Sicily, Italy. *International Journal of Architectural Heritage*, 12(5), 806–815.
Available from: https://doi.org/10.1080/ 15583058.2017.1419311.
Jõeleht, A., Plado, J. & Sibul, I. (2017) Ground-penetrating radar and electrical resistivity tomography for mapping bedrock topography and fracture zones: a case study in Viru-Nigula, NE Estonia; pp. 142–151. *Estonian Journal of Earth Sciences*, 66(3), 142. Available from: https://doi.org/10.3176/earth.2017.11.

Jung, J. (2004) *Tertiärzeitliche Verwitterungsbildungen im Buntsandstein des Südwest-Spessarts und ihre eiszeitliche Aufarbeitung*. Naturwissenschaftlicher Verein Aschaffenburg: Aschaffenburg.

Jung, J. (2006) *GIS-gestützte Rekonstruktion der neogenen Reliefentwicklung tektonisch beeinflusster Mittelgebirgslandschaften am Beispiel des Spessarts (NW-Bayern, SE-Hessen)*. Universität Würzburg.

Käding, K.-C. (1978) Die Grenze Zechstein/Buntsandstein in Hessen, Nordbayern und Baden-Württemberg. *Jahresberichte und MItteilungen des Oberrheinischen Geologischen Vereins*, 60.

Kampfmann, G. & Krimm, S. (1988) *Verkehrsgeographie und Standorttypologie der Glashütten im Spessart*. Geschichts- und Kunstverein Aschaffenburg e. V.: Aschaffenburg.

Kaufmann, G. (2014) Geophysical mapping of solution and collapse sinkholes. *Journal of Applied Geophysics*, 111, 271–288. Available from: https://doi.org/10.1016/j.jappgeo.2014.10.011

Kaufmann, G. & Romanov, D. (2017) The Jettencave, Southern Harz Mountains, Germany: Geophysical observations and a structural model of a shallow cave in gypsum/anhydrite-bearing rocks. *Geomorphology*, 298, 20–30. Available from: https://doi.org/10.1016/j.geomorph.2017.09.027.

Kneisel, C. & Hauck, C. (2008) Electrical methods. In: Hauck, C. & Kneisel, C. (Eds.) *Applied Geophysics in Periglacial Environments.* Cambridge University Press: Cambridge, pp. 3–27.

Knödel, K. (2005) Magnetik. In: Knödel, K., Krummel, H. & Lange, G. (Eds.) *Handbuch zur Erkundung des Untergrundes von Deponien und Altlasten: Geophysik.* Springer: Berlin Heidelberg, pp. 17–42.

Knödel, K., Krummel, H. & Lange, G. (Eds.) (2005) *Handbuch zur Erkundung des Untergrundes von Deponien und Altlasten: Geophysik*. Springer: Berlin Heidelberg.

Köbrich, C. (1934) Geologisches und bergwirtschaftliches von den oberhessischen Eisenerzen. *Zeitschrift der Deutschen Geologischen Gesellschaft*, 86, 329–333.

Koch, J., Griessel, F., Hüssen, C.-H., Kothieringer, K. & Rahrig, M. (2017) Archäologische Forschung am Raetischen Limes bei Zandt im Köschinger Forst. *Nachrichtenblatt der deutschen Limeskommission*, 2016(2), 32–37.

Koch, K., Wenninger, J., Uhlenbrook, S. & Bonell, M. (2009) Join interpretation of hydrological and geophysical data: Electrical resistivity tomography results from a process hydrological research site in the Black Forest Mountains, Germany. *Hydrological Processes*, 23, 1501–1513. Available from: https://doi.org/10.1002/hyp.7275.

Kokalj, Ž. & Somrak, M. (2019) Why Not a Single Image? Combining Visualizations to Facilitate Fieldwork and On-Screen Mapping. *Remote Sensing*, 11(7), 747. Available from: https://doi.org/10.3390/rs11070747.

Kokalj, Ž., Zakšek, K., Oštir, K., Pehani, P., Čotar, K. & Somrak, M. (2013) Relief Visualization Toolbox (RVT): program.

Kolb, P. & Krenig, E.-G. (1989) *Unterfränkische Geschichte: Von der germanischen Landnahme bis zum hohem Mittelalter*. Echter: Würzburg.

Kolb, P. & Krenig, E.-G. (1992) *Unterfränkische Geschichte: Vom hohen Mittelalter bis zum Beginn des konfessionellen Zeitalters*. Echter: Würzburg.

Křivánek, R. (2020) The contribution of new geophysical measurements at the previously excavated neolithic rondel area near B ylany, c entral B ohemia. *Archaeological prospection*, 27(1), 39–52. Available from: https://doi.org/10.1002/arp.1755.

Kunz, J. & Kneisel, C. (2020) Glacier–Permafrost Interaction at a Thrust Moraine Complex in the Glacier Forefield Muragl, Swiss Alps. *Geosciences*, 10(6), 205. Available from: https://doi.org/10.3390/geosciences10060205.

Lagies, M. (2005) Neue pollenanalytische Forschungen in Spessart und Odenwald – eine Zusammenfassung. *Carolinea*, 63, 113–134.

Landau, G. (1958) *Beiträge zur Geschichte der alten Heer- und Handelsstraßen in Deutschland*. Bärenreiter-Verlag: Kassel und Basel.

Lang, A. (2003) Phases of soil erosion-derived colluviation in the loess hills of South Germany. *CATENA*, 51(3-4), 209–221. Available from: https://doi.org/10.1016/S0341-8162(02)00166-2.

Lange, G. (2005) Gleichstromgeoelektrik. In: Knödel, K., Krummel, H. & Lange, G. (Eds.) *Handbuch zur Erkundung des Untergrundes von Deponien und Altlasten: Geophysik*. Springer: Berlin Heidelberg, pp. 128–173.

Lange-Athinodorou, E., El-Raouf, A.A., Ullmann, T., Trappe, J., Meister, J. & Baumhauer, R. (2019) The sacred canals of the Temple of Bastet at Bubastis (Egypt): New findings from geomorphological investigations and Electrical Resistivity Tomography (ERT). *Journal of Archaeological Science: Reports*, 26, 101910. Available from: https://doi.org/10.1016/j.jasrep.2019.101910.

Larsen, A., Bork, H.-R., Fuelling, A., Fuchs, M. & Larsen, J.R. (2013) The processes and timing of sediment delivery from headwaters to the trunk stream of a Central European mountain gully catchment. *Geomorphology*, 201, 215–226. Available from: https://doi.org/10.1016/j.geomorph.2013.06.022.

Larsen, A., Heckmann, T., Larsen, J.R. & Bork, H.-R. (2016) Gully catchments as a sediment sink, not just a source: Results from a long-term (~12 500 year) sediment budget. *Earth Surface Processes and Landforms*, 41(4), 486–498. Available from: https://doi.org/10.1002/esp.3839.

Le Borge, E. (1960) Influence du feu sur les propertietes magnétiques du sol et sur celles du schiste et du granite. *Ann. Geophys.*, 16, 159–195.

Leser, H. (1995) *Geomorphologie*, 7th edition. Westermann: Braunschweig.

Li, M., Zhang, Z., Nobes, D.C. & Yang, J. (2019) Integrated Geophysical Methods for Detecting Archaeological Han Dynasty Tombs. In: El-Qady, G. & Matwaly, M. (Eds.) *Archaeogeophysics: State of the Art and Case Studies*. Springer International Publishing: Cham, pp. 125–136.

Linck, R. & Fassbinder, J.W.E. (2014) Determination of the influence of soil parameters and sample density on ground-penetrating radar: a case study of a Roman picket in Lower Bavaria. *Archaeological and Anthropological Sciences*, 6(1), 93–106. Available from: https://doi.org/10.1007/s12520-013-0145-4.

Loke, M.H. (2015) RES2DINVx64 ver. 4.05 with multi-core and64-bit support: RES2DINVx64 Manual. *www.geotomosoft.com*.

Loke, M.H. (2020) Tutorial: 2-D and 3-D electrical imaging surveys (Revision date: 17th August 2020). *(Avaidable http://geotomosoft.com/coursenotes.zip)*.

Loke, M.H., Acworth, I. & Dahlin, T. (2003) A comparison of smooth and blocky inversion methods in 2D electrical imaging surveys. *Exploration Geophysics*, 34(3), 182–187. Available from: https://doi.org/10.1071/EG03182.

Lorenz, J. (Personal communication 4 December 2020).

Lorenz, J. (2021) *http://spessartit.de/Eisen: website (accessed 18.01.2021)*.

Lorenz, J., Okrusch, M., Geyer, G., Jung, J., Himmelsbach, G. & Dietl, C. (2010) *Spessartsteine: Spessartin, Spessartit und Buntsandstein - eine umfassende Geologie und Mineralogie des Spessarts ; geographische, geologische, petrographische, mineralogische und bergbaukundliche Einsichten in ein deutsches Mittelgebirge*. Lorenz: Karlstein a. Main.

Lowrie, W. (2011) *Fundamentals of geophysics*, 2nd edition. Cambridge Univ. Press: Cambridge.

Macphail, R.I. & Goldberg, P. (2017) *Applied Soils and Micromorphology in Archaeology*. Cambridge University Press.

Marquart, M. (2017) *Beiträge zu Vorgeschichte des Aschaffenburger Landes im Spiegel der Sammlungen des Aschaffenburger Stiftsmuseums*. Geschichts- und Kunstverein Aschaffenburg: Aschaffenburg.

Masini, N., Capozzoli, L., Chen, P., Chen, F., Romano, G. & Lu, P. et al. (2017) Towards an Operational Use of Geophysics for Archaeology in Henan (China): Methodological Approach and Results in Kaifeng. *Remote Sensing*, 9(8), 809. Available from: https://doi.org/10.3390/rs9080809.

Mayoral, A., Granai, S., Develle, A.-L., Peiry, J.-L., Miras, Y. & Couderc, F. et al. (2020) Early human impact on soils and hydro-sedimentary systems: Multi-proxy geoarchaeological analyses from La Narse de la Sauvetat (France). *The Holocene*, 30(12), 1780–1800. Available from: https://doi.org/10.1177/0959683620950390.

McQuillin, R., Bacon, M. & Barclay, W. (1979) *An Introduction to Seismic Interpretation: Reflections Seismics in Petroleum Exploration*. Gulf Publishing Company: Houston.

Meschede, M. (2018) *Geologie Deutschland: Ein prozessorientierter Ansatz*, 2nd edition. Springer Spektrum: Berlin.

Metwaly, M., Green, A.G., Horstmeyer, H. & Maurer, H. Abbas, A. M. (2019) Combined Seismic Tomographic and Ulta-Shallow Seismic Reflection Study of an early Dynastic Mastaba, Saqqara, Egypt. In: El-Qady, G. & Matwaly, M. (Eds.) *Archaeogeophysics: State of the Art and Case Studies*. Springer International Publishing: Cham.

Meynen, E. & Schmithüsen, J. (Eds.) (1955) *Handbuch der naturräumlichen Gliederung Deutschlands: 2. Lieferung*. Selbstverlag der Bundesanstalt für Landeskunde Remagen: Remagen.

Miller, B. (2014) Relief Analysis Toolbox: Toolbox for ArcGIS.

Mol & Preston (2010) The writing's in the wall : a of new prelimnary application of electrical resistivity tomography within archaeology. *Archaeometry*, 52(6), 1079–1095. Available from: https://doi.org/10.1111/j.1475-4754.2010.00516.x.

Montgomery, D.R. (2007) Soil erosion and agricultural sustainability. *Proceedings of the National Academy of Sciences of the United States of America*, 13268–13272. Available from: https://doi.org/10.1073/pnas.0611508104.

Mueller, S. (2011) *New insights about Pleistocene periglacial slope deposits and pedogenesis in the Hessian Spessart Mountains*. Univ. Frankfurt am Main, Dissertation, 2011: Frankfurt am Main.

Mueller, S. & Thiemeyer, H. (2014) Formation and transformation of Pleistocene periglacial slope deposits in the Spessart Mountains (Hesse, Germany). *Zeitschrift für Geomorphologie*, 58(3), 91–113. Available from: https://doi.org/10.1127/0372-8854/2014/S-00157.

Münster, H. (1905) Die Brauneisenerzlagerstätten des Seen- und Ohmtals am Nordrand des Vogelsgebirges.: Ein Beitrag zur Kenntnis der Basalteisensteine. *Zeitschrift für practische Geologie und Lagerstättenkunde*, 13, 242–268.

Murawski, H. (1965) Der Spessart als Teilgebiet der Mitteldeutschen Schwelle. *Geologische Rundschau*, 54(2), 835–852. Available from: https://doi.org/10.1007/BF01820758.

Neal, A. (2004) Ground-penetrating radar and its use in sedimentology: principles, problems and progress. *Earth-Science Reviews*, 66(3-4), 261–330. Available from: https://doi.org/10.1016/j.earscirev.2004.01.004.

Nicoly, A., Raab, A., Raab, T., Rösler, H., Bönisch, E. & Murray, A.S. (2014) Evidence of (pre-) historic to modern landscape and land use history near Jänschwalde (Brandenburg, Germany). *Zeitschrift für Geomorphologie*, 58(2). Available from: https://doi.org/10.1127/0372-8854/2014/S-00162.

Nisengard, J.E., Isaacson, J.S., Ferguson, J.F., Hinz, E. & Gauthier, R. (2008) Seismic Refraction at Ancestral Puebloan Sites on the Pajarito Plateau, New Mexico. *Journal of Field Archaeology*, 33(1), 85–100. Available from: https://doi.org/10.1179/009346908791071349.

Nuzzo, L., Leucci, G. & Negri, S. (2009) GPR, ERT and magnetic investigations inside the Martyrium of St Philip, Hierapolis, Turkey. *Archaeological prospection*, 16(3), 177–192. Available from: https://doi.org/10.1002/arp.364.

Obrocki, L., Vött, A., Wilken, D., Fischer, P., Willershäuser, T. & Koster, B. et al. (2020) Tracing tsunami signatures of the ad 551 and ad 1303 tsunamis at the Gulf of Kyparissia (Peloponnese, Greece) using direct push in situ sensing techniques combined with geophysical studies. *Sedimentology*, 67(3), 1274–1308. Available from: https://doi.org/10.1111/sed.12555.

Okrusch, M., Geyer, G. & Lorenz, J. (2011) *Sammlung geologischer Führer Band 106: Spessart: Geologische Entwicklung und Struktur, Gesteine und Minerale*. Gebrüder Bornträger: Stuttgart.

Okrusch, M., Streit, R. & Weinelt, W. (1967) *Geologischen Karte von Bayern 1:25 000. Blatt Nr. 5920 Alzenau i. UFr.: mit Erläuterungen*. Bayerisches Geologisches Landesamt: München.

Okrusch, M. & Weinelt, W. (1965) *Geologischen Karte von Bayern 1:25 000. Blatt Nr. 5921 Schöllkrippen: mit Erläuterungen*. Bayerisches Geologisches Landesamt: München.

Pärn, A. (2019) Topographie als „Machtfaktor“, 85–94. Available from: https://doi.org/10.11588/dgamn.2019.0.69671.

Pebriyanto, Y., Dahlan, K. & Sari, Y.W. (2017) Electrical Resistivity Tomography Using Wenner β - Schlumberger Configuration for Anomaly Detection in The Soil. *IOP Conference Series: Earth and Environmental Science*, 58, 12012. Available from: https://doi.org/10.1088/1755-1315/58/1/012012.

Pełka-Gościniak, J. (2014) Environmental aspects of relief transformation (Silesian Upland, Southern Poland). *Environmental & Socio-economic Studies*, 2(4), 13–20. Available from: https://doi.org/10.1515/environ-2015-0045.

Pohl, W.L. (2005) *Mineralische und Energie-Rohstoffe: Eine Einführung zur Entstehung und nachhaltigen Nutzung von Lagerstätten*, 5th edition. E. Schweizerbart'sche Verlagsbuchhandlung (Nägele und Obermiller): Stuttgart.

Posselt, M., Zickgraf, B. & Dobiat, C. (Eds.) (2007) *Geophysik und Ausgrabung: Einsatz und Auswertung zerstörungsfreier Prospektion in der Archäologie*. Marie Leidorf GmbH: Rahden/Westf.

Preuschoff, M. (1996) Der Spessart - ein geologisch-petrologischer Überblick. *Wissenschaftliche Landeskunde und Landnutzung im Spessart Bestandsaufnahme und Perpektiven*, 194.

Ramsey, C.B. (2013) OxCal 4.2: Program.

Rapp, G. & Hill, C.L. (2006) *Geoarchaeology: The Earth-Science Approach to Archaeological Interpretation*, 2nd edition. Yale University Press: New Haven and London.

Reimer, H. (1897) *Urkundenbuch zur Geschichte der Herren von Hanau*. Hessisches Urkundenbuch: Leipzig.

Reimer, P.J., Bard, E., Bayliss, A., Beck, J.W., Blackwell, P.G. & Ramsey, C.B. et al. (2013) IntCal13 and Marine13 Radiocarbon Age Calibration Curves 0–50,000 Years cal BP. *Radiocarbon*, 55(4), 1869–1887. Available from: https://doi.org/10.2458/azu_js_rc.55.16947.

Reynolds, J.M. (2011) *An introduction to applied and environmental geophysics*, 2nd edition. Wiley-Blackwell: Chichester.

Ristić, A., Govedarica, M., Pajewski, L., Vrtunski, M. & Bugarinović, Ž. (2020) Using Ground Penetrating Radar to Reveal Hidden Archaeology: The Case Study of the Württemberg-Stambol Gate in Belgrade (Serbia). *Sensors (Basel, Switzerland)*, 20(3), 607. Available from: https://doi.org/10.3390/s20030607.

Rosmanitz, H. (2009) Burgenforschung im Spessart. Das „Alte Schloss" in Kleinwallstadt. *Beiträge zur Archäologie in Unterfranken*, 243–286.

Rosmanitz, H. (2012) *Die Altenburg zwischen Sulzbach und Leidersbach: die prähistorische Wallanlage zwischen Main und Spessart*. Ph.C.W. Schmidt: Neustadt an der Aisch.

Rosmanitz, H. & Bachmann, S. (2017) Waldaschaff, Lkr. Aschaffenburg, Burg Wahlmich, Maßnahmen-Nr. M-2016-799-1_0: Archäologische Untersuchung, Mai - August 2016.

Rosmanitz, H., Bachmann, S. & Galina, I. (2020) Mömbris, Lkr. Aschaffenburg, Burg Mömbris, Maßnahmen-Nr. M-2019-743-1_0.

Rosmanitz, H., Bachmann, S. & Geißlinger, M. (2019a) Krombach, Lkr. Aschaffenburg, Burg Hauenstein, Maßnahmen-Nr. M-2017-1232-1_0: Archäologische Untersuchung, August bis November 2017.

Rosmanitz, H., Bachmann, S. & Geißlinger, M. (2019b) Waldaschaff, Lkr. Aschaffenburg, Burg Wahlmich, Maßnahmen-Nr. M-2018-871-1_0.

Rosmanitz, H. & Gresse, S. (2011) The castle project in the Spessart – scientists and volunteers explore a cultural landscape: Papers by International conference "Lordship and Castle landscapes - Franconian and international studies in comparison". *Concilium medii aevi*, 14, 101–120. Available from: https://doi.org/10.11588/cma.2011.0.77401.

Rózsa, P., Incze, J., Balogh, S. & Novák, T.J. (2020) A novel approach to quantifying the degree of anthropogenic surface transformation – the concept of 'hemeromorphy'. *Erdkunde*, 74(1), 45–57. Available from: https://doi.org/10.3112/erdkunde.2020.01.03.

Rücker, C. & Günther, T. (2018) Boundless Electrical Resistivity Tomography: BERT software package.

Ruf, T. (1984) *Die Grafen von Rieneck Genealogie und Territorienbildung: I. Genealogie 1085 bis 1559 und Epochen der Territorienbildung*. Freunde Mainfränkischer Kunst und Geschichte e.V., Würzburg: Würzburg.

Ruf, T. (2019) Zur Geschichte einiger Spessartburgen im 12. und 13. Jahrhundert. *Aschaffenburger Jahrbuch für Geschichte, Landeskunde und Kunst im Untermaingebiet*, 33, 9–93.

Sabel, K.-J. (1996) Gestein, Relief und Böden im Spessart. *Wissenschaftliche Landeskunde und Landnutzung im Spessart Bestandsaufnahme und Perpektiven*, 194.

Sandmeier, K.-J. & Krummel, H. (2005) Refraktionsseismik: Prinzip der Methods. In: Knödel, K., Krummel, H. & Lange, G. (Eds.) *Handbuch zur Erkundung des Untergrundes von Deponien und Altlasten: Geophysik*. Springer: Berlin Heidelberg, pp. 530–534.

Sass, O., Bell, R. & Glade, T. (2008) Comparison of GPR, 2D-resistivity and traditional techniques for the subsurface exploration of the Öschingen landslide, Swabian Alb (Germany). *Geomorphology*, 93(1-2), 89–103. Available from: https://doi.org/10.1016/j.geomorph.2006.12.019.

Schäfer, R. (1991) *Glashütten Glasindustrie Glashandel im Spessart*. Oskar Rummel.

Schäfer, R. (2000) *Die Herren von Eppstein*. Historische Kommision Nassau: Wiesbaden.

Schecher, O. (1969) *Die Grafen von Rieneck: Zur Geschichte eines mittelalterlichen Hochadelsgeschlechts in Franken*. Schriften des Geschichtsverains Lohr a. Main: Lohr am Main.

Scheinpflug, R. (1992) *Main-Spessart-Geologie: mit 7 Tabellen*. Mineralogisch-Paläontologische Privatsammlung Scheinpflug: Lohr am Main.

Scherer-Hall, R. (1998) Zur Typisierung von Burgen und Schlösser und deren Erscheinungsformen. *Burgen und Schlösser*, 2, 87–93.

Schneider, A. (2014) Die drei ehemaligen herzoglich-württembergischen Fischweiher bei Nabern (Kreis Esslingen). Ein gefährdetes Zeugnis der herrschaftlichen Teichwirtschaft des späten Mittelalters und der frühen Neuzeit, 192–197. Available from: https://doi.org/10.11588/nbdpfbw.1989.4.13749.

Schrott, L. & Hoffmann, T. (2008) Refraction seismics. In: Hauck, C. & Kneisel, C. (Eds.) *Applied Geophysics in Periglacial Environments*. Cambridge University Press: Cambridge, pp. 57–80.

Schrott, L. & Sass, O. (2008) Application of field geophysics in geomorphology: Advances and limitations exemplified by case studies. *Geomorphology*, 93(1-2), 55–73. Available from: https://doi.org/10.1016/j.geomorph.2006.12.024.

Schütt, B. (2006a) Aktuelle Mensch-Umwelt-Interaktion. *Zeitschrift für Geomorphologie*, 148, 41–42.

Schütt, B. (2006b) Die indirekten Einflüsse des Menschen auf die Reliefsphäre. *Zeitschrift für Geomorphologie*, 148, 47–55.

Semmel, A. (1964) Junge Schuttdecken in hessischen Mittelgebirgen. *Notizbl. hess. L.-Amt Bodenforsch*, 275–285.

Sensors & Software Inc. (2016) EKKO_Project: Users Guide.

Siart, C., Hecht, S., Holzhauer, I., Altherr, R., Meyer, H.P. & Schuhkraft, G. et al. (2010) Karst depressions as geoarchaeological archives: The palaeoenvironmental reconstruction of Zominthos (Central Crete), based on geophysical prospection, sedimentological investigations and GIS. *Quarternary International*, 216(1-2), 75–92. Available from: https://doi.org/10.1016/j.quaint.2009.06.020.

Sørensen, R., Zinko, U. & Seibert, J. (2006) On the calculation of the topographic wetness index: evaluation of different methods based on field observations. *Hydrology and Earth System Sciences*, 10(1), 101–112. Available from: https://doi.org/10.5194/hess-10-101-2006.

STD (2016) *Stratigraphische Tabelle von Deutschland*. GeoForschungszentrum: Potsdam.

Stegemann, W. (1964) Paläontologie und Stratigarphie einiger Sedimentvorkommen im Vogelsberg. *Bericht der Oberhessischen Gesellschaft für Natur- und Heilkunde zu Gießen*, 33(4), 251–279.

Stoffel, M., Butler, D.R. & Corona, C. (2013) Mass movements and tree rings: A guide to dendrogeomorphic field sampling and dating. *Geomorphology*, 200, 106–120. Available from: https://doi.org/10.1016/j.geomorph.2012.12.017.

Strabo (1836) *Strabo´s Geographie*. Metzler: Stuttgart.

Telford, W.M., Geldart, L.P. & Sheriff, R.E. (1990) *Applied geophysics*, 2nd edition. Cambridge Univ. Press: Cambridge.

Torge, W. (2002) *Geodäsie*. Walter de Gruyter: Berlin, New York.

Trappe, J. & Kneisel, C. (2019) Geophysical and Sedimentological Investigations of Peatlands for the Assessment of Lithology and Subsurface Water Pathways. *Geosciences*, 9(3), 118. Available from: https://doi.org/10.3390/geosciences9030118.

Ullrich, B., Meyer, C. & Weller, A. (2007) Geoelektrik und Georadar in der archäologischen Forschung: geophysikalische 3D-Untersuchungen in Munigua (Spanien). In: Wagner, G.A. (Ed.) *Einführung in die Archäometrie*. Springer: Berlin, Heidelberg, pp. 76–95.

Verstraeten, G., Rommens, T., Peeters, I., Poesen, J., Govers, G. & Lang, A. (2009) A temporarily changing Holocene sediment budget for a loess-covered catchment (central Belgium). *Geomorphology*, 108(1-2), 24–34. Available from: https://doi.org/10.1016/j.geomorph.2007.03.022.

Wagner, G.A. (Ed.) (2007) *Einführung in die Archäometrie*. Springer: Berlin, Heidelberg.

Wamser, L. (1979) *Glashütten im Spessart: Gewinne und Verluste*. Ausgrabungsnotizen aus Bayern.

Wamser, L. (1982) Neue Ausgrabungen mittelalterlicher Spessart-Glashütten bei Schöllkrippen, Landkreis Aschaffenburg, Unterfranken. *Das Archäologische Jahr in Bayern*, 188–189.

Wedepohl, K.H. (2003) *Glas in Antike und Mittelalter: Geschichte eines Werkstoffs*. E. Schweizerbart´sche Verlagsbuchhandlung: Stuttgart.

Weinelt, W. (1962) *Geologischen Karte von Bayern 1:25 000. Blatt Nr. 6021 Haibach: mit Erläuterungen*. Bayerisches Geologisches Landesamt: München.

Weiss, A. (2001) *Topographic Position and Landforms Analysis, ESRI user conference*, 2001.

Welc, F., Mieszkowski, R., Lipovac-Vrkljan, G. & Konestra, A. (2017) An attempt to integration of different geophysical methods (magnetic, GPR and ERT); a case study from the late Roman settlement on the Island of Rab in Croatia. *Studia Quaternaria*, 34(1), 47–59. Available from: https://doi.org/10.1515/squa-2017-0004.

Winkler, S., Matthews, J.A., Haselberger, S., Hill, J.L., Mourne, R.W. & Owen, G. et al. (2020) Schmidt-hammer exposure-age dating (SHD) of sorted stripes on Juvflye, Jotunheimen (central South Norway): Morphodynamic and palaeoclimatic implications. *Geomorphology*, 353, 107014. Available from: https://doi.org/10.1016/j. geomorph.2019.107014.

Zerbe, S. (1997) Stellt die potentielle natürliche Vegetation (PNV) eine sinnvolle Zielvorstellung für den naturnahen Waldbau dar? *Forstwissenschaftliches Centralblatt*, 116(1-6), 1–15. Available from: https://doi.org/10.1007/BF02766877.

Zheng, Y., Wang, S., Feng, J., Ouyang, Z. & Li, X. (2005) Measurement of the complex permittivity of dry rocks and minerals: application of polythene dilution method and Lichtenecker's mixture formulae. *Geophysical Journal International*, 163(3), 1195–1202.
Available from: https://doi.org/10.1111/j.1365-246X.2005.02718.x.

Appendix

Overview

Appendix 1: Technical overview Waldaschaff: ERT
Appendix 2: Technical overview Waldaschaff: GPR
Appendix 3: Technical overview Waldaschaff: MAG
Appendix 4: Technical overview Hauenstein: ERT
Appendix 5: Technical overview Hauenstein: GPR
Appendix 6: Technical overview Hauenstein: MAG
Appendix 7: Laboratory report Pornan Radiocarbon Laboratory; 12/06/2018
Appendix 8: Technical overview Mömbris: ERT
Appendix 9: Technical overview Mömbris: GPR
Appendix 10: Technical overview Mömbris: MAG
Appendix 11: Technical overview Alzenau: ERT
Appendix 12: Technical overview Alzenau: GPR
Appendix 13: Technical overview Alzenau: MAG
Appendix 14: Technical overview Eisenberg: ERT
Appendix 15: Technical overview Eisenberg: GPR
Appendix 16: Technical overview Eisenberg: MAG

Waldaschaff

Appendix 1: Technical overview Waldaschaff: ERT

Name	*Spacing*	*Array*	*Electrodes*	*Equipment type*
W1_ERT_a	1.5 m, 72 elec.	WenSl	1*a; 32 lvl.; 1216 qdp.	Syscal Junior Switch 72
W1_ERT_b	3 m, 36 elec.	WenSl	1*a; 16 lvl.; 288 qdp.	Syscal Junior Switch 72
W1_ERT_c	3 m, 36 elec.	WenSl	1*a; 16 lvl.; 288 qdp.	Syscal Junior Switch 72
W1_ERT_d	3 m, 36 elec.	WenSl	1*a; 16 lvl.; 288 qdp.	Syscal Junior Switch 72
W1_ERT_e	3 m, 36 elec.	WenSl	1*a; 16 lvl.; 288 qdp.	Syscal Junior Switch 72
W1_ERT_f	1 m, 36 elec.	WenSl	1*a; 16 lvl.; 288 qdp.	Syscal Junior Switch 72
W1_ERT_g	1 m, 36 elec.	WenSl	1*a; 16 lvl.; 288 qdp.	Syscal Junior Switch 72
W2_ERT_a	1 m, 36 elec.	WenSl	1*a; 16 lvl.; 288 qdp.	Syscal Junior Switch 72
W2_ERT_b	1 m, 36 elec.	WenSl	1*a; 16 lvl.; 288 qdp.	Syscal Junior Switch 72
W2_ERT_c	1 m, 36 elec.	WenSl	1*a; 16 lvl.; 288 qdp.	Syscal Junior Switch 72
W3_ERT_a	3 m, 2*36 elec.	WenSl	1*a; 16 lvl.; je 288 qdp.	Syscal Junior Switch 72 [1]
W3_ERT_b	3 m, 36 elec.	WenSl	1*a; 16 lvl.; 288 qdp.	Syscal Junior Switch 72
W4.1_ERT_a	2 m, 36 elec.	WenSl	1*a; 16 lvl.; 288 qdp.	Syscal Junior Switch 72
W4.1_ERT_b	2 m, 72 elec.	WenSl	1*a; 32 lv.l; 1216 qdp.	Syscal Junior Switch 72
W4.1_3DERT_a	0.5 m, 7* 36 elec.	DipDip	1*a; 16 lvl.; 408 qdp.	Syscal/Pro [2]

Specials: [1] Roll-along 18 elec. overlap, [2] Quasi 3D 1m interline spacing

Appendix 2: Technical overview Waldaschaff: GPR

Name	*Frequency*	*Shield*	*Gain*	*Processing, Grid*	*Equipment type*
W1_GPR_a	30MHz	No	Start: 120 Attenuations: 10 dB/m MaxGain: 1500 Filter width: 5m	Bandpass-filter (% Freq): Fc1: 40; Fp1: 80; Fp2: 120; Fc2: 160; Topographic Migration; Background subtraction; Vel. 0.07 m/ns	PulsEKKO PRO System
W3_GPR_a	30MHz	No	Start: 120 Attenuations: 10 dB/m MaxGain: 1500 Filter width: 5m	Bandpass-filter (% Freq): Fc1: 40; Fp1: 80; Fp2: 120; Fc2: 160; Topographic Migration; Background subtraction; Vel. 0.07 m/ns	PulsEKKO PRO System
W3_GPR_b	30MHz	No	Start: 120 Attenuations: 10 dB/m MaxGain: 1500 Filter width: 5m	Bandpass-filter (% Freq): Fc1: 40; Fp1: 80; Fp2: 120; Fc2: 160; Topographic Migration; Background subtraction; Vel. 0.07 m/ns	PulsEKKO PRO System
W3_GPR_c	30MHz	No	Start: 120 Attenuations: 10 dB/m MaxGain: 1500 Filter width: 5m	Bandpass-filter (% Freq): Fc1: 40; Fp1: 80; Fp2: 120; Fc2: 160; Topographic Migration; Background subtraction; Vel. 0.07 m/ns	PulsEKKO PRO System
W3_GPR_d	30MHz	No	Start: 120 Attenuations: 10 dB/m MaxGain: 1500 Filter width: 5m	Bandpass-filter (% Freq): Fc1: 40; Fp1: 80; Fp2: 120; Fc2: 160; Topographic Migration; Background subtraction; Vel. 0.07 m/ns	PulsEKKO PRO System
W3_GPR_e	30MHz	No	Start: 120 Attenuations: 10 dB/m MaxGain: 1500 Filter width: 5m	Bandpass-filter (% Freq): Fc1: 40; Fp1: 80; Fp2: 120; Fc2: 160; Topographic Migration; Background subtraction; Vel. 0.07 m/ns	PulsEKKO PRO System
W3_GPR_f	30MHz	No	Start: 120 Attenuations: 10 dB/m MaxGain: 1500 Filter width: 5m	Bandpass-filter (% Freq): Fc1: 40; Fp1: 80; Fp2: 120; Fc2: 160; Topographic Migration; Background subtraction; Vel. 0.07 m/ns	PulsEKKO PRO System
W3_GPR_g	30MHz	No	Start: 120 Attenuations: 10 dB/m MaxGain: 1500 Filter width: 5m	Bandpass-filter (% Freq): Fc1: 40; Fp1: 80; Fp2: 120; Fc2: 160; Topographic Migration; Background subtraction; Vel. 0.07 m/ns	PulsEKKO PRO System
W4.1_3DGPR_a	100MHz	No	Start: 4 Attenuations: 1.5 dB/m MaxGain: 20	3D in x and y direction, 1m interline spacing	PulsEKKO PRO System

Appendix 3: Technical overview Waldaschaff: MAG

Name	Filter	Value range
W4.2_MAG_a	Zero mean filter min.: 10; max.: 10	10– -10 nT
W4.2_MAG_b	Zero mean filter min.: 10; max.: 10	10– -10 nT
W4.2_MAG_c	Zero mean filter min.: 10; max.: 10	10– -10 nT

Hauenstein

Appendix 4: Technical overview Hauenstein: ERT

Name	Spacing	Array	Electrodes	Equiment type
H2_ERT_a	1 m, 2* 72 elec.	WenSl	1*a; 18 lvl.; 936 qdp.	Syscal Junior Switch 72 [1]
H2_ERT_b	0.5 m, 36 elec.	WenSl	1*a; 16 lvl.; 288 qdp.	Syscal Junior Switch 72
H3_ERT_a	3 m, 36 elec.	WenSl	1*a; 16 lvl.; 288 qdp.	Syscal Junior Switch 72
H4.1_ERT_a	0.5 m, 36 elec.	WenSl	1*a; 16 lvl.; 288 qdp.	Syscal Junior Switch 72
H4.1_ERT_b	0.5 m, 36 elec.	WenSl	1*a; 16 lvl.; 288 qdp.	Syscal Junior Switch 72
H4.1_ERT_c	0.5 m, 36 elec.	WenSl	1*a; 16 lvl.; 288 qdp.	Syscal Junior Switch 72

Specials: [1] Roll-along 18 elec. overlap

Appendix 5: Technical overview Hauenstein: GPR

Name	Frequency	Shield	Gain	Processing, Grid	Equiment type
W4.2_3DGPR_a	100MHz	No	Start: 10 Attenuations: 10 B/m MaxGain: 500	3D in y direction, 1m interline spacing Velocity: 0.08m/ns	PulsEKKO PRO System

Appendix 6: Technical overview Hauenstein: MAG

Name	Filter	Value range
H4.2_MAG_a	Zero mean filter min.: 10; max.: 10	50– -50 nT

AMS-^{14}C-datings

Appendix 7: Laboratory report Pornan Radiocarbon Laboratory; 12/06/2018

"Results of calibration of 14C dates – order 14281/18

Given are intervals of calendar age, where the true ages of the samples encompass with the probability of ca. 68% and ca. 95%. The calibration was made with the OxCal software.

OxCal v4.2.3 Bronk Ramsey (2013); r:5
IntCal13 atmospheric curve (Reimer et al 2013)
Haue1 C2 R_Date(500,30) → *sample ID: H2_AGE_a*
68.2% probability
1414AD (68.2%) 1437AD
95.4% probability
1333AD (0.7%) 1337AD
1398AD (94.7%) 1449AD
Haue1 C3 R_Date(670,30) → *sample ID: H2_AGE_b*
68.2% probability

1281AD (38.6%) 1305AD
1364AD (29.6%) 1385AD
95.4% probability
1274AD (53.1%) 1320AD
1351AD (42.3%) 1391AD
Haue2 C5 R_Date(300,30) → *sample ID: H2_AGE_c*
68.2% probability
1522AD (47.2%) 1575AD
1585AD (3.1%) 1590AD
1625AD (17.9%) 1646AD
95.4% probability
1489AD (69.6%) 1604AD
1611AD (25.8%) 1654AD

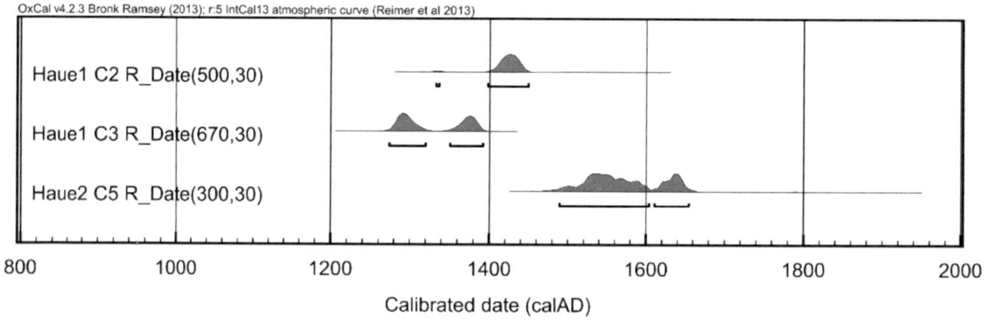

Mömbris

Appendix 8: Technical overview Mömbris: ERT

Name	*Spacing*	*Array*	*Electrodes*	*Equiment type*
M4.1_3DERT_a	0.5 m, 17* 36 elec.	WenSl	1*a; 16 lvl.; 288 qdp.	Syscal Junior Switch 72 [1]

Specials: [1] Quasi 3D 0.5m interline spacing

Appendix 9: Technical overview Mömbris: GPR

Name	*Frequence*	*Shield*	*Gain*	*Processing, Grid*	*Equiment type*
M4.1_3DGPR_a, M4.1_3DGPR_b, M4.1_3DGPR_c,	250MHz	Yes	Start: 2 Attenuations: 20 dB/m MaxGain: 2000	3D in x direction, 1m interline spacing	NOGGIN System

Appendix 10: Technical overview Mömbris: MAG

Name	*Filter*	*Value range*
M4.1_MAG_a	Zero mean filter min.: 10; max.: 10	10– -10 nT

Alzenau

Appendix 11: Technical overview Alzenau: ERT

Name	*Spacing*	*Array*	*Electrodes*	*Equiment type*
A1_ERT_a	3 m, 36 elec.	WenSl	1*a; 16 lvl.; 288 qdp.	Syscal Junior Switch 72
A1_ERT_b	3 m, 36 elec.	WenSl	1*a; 16 lvl.; 288 qdp.	Syscal Junior Switch 72
A1_ERT_c	1.5 m, 36 elec.	WenSl	1*a; 16 lvl.; 288 qdp.	Syscal Junior Switch 72
A1_ERT_d	3 m, 36 elec.	WenSl	1*a; 16 lvl.; 288 qdp.	Syscal Junior Switch 72
A4.1_ERT_a	1 m, 36 elec.	WenSl	1*a; 16 lvl.; 288 qdp.	Syscal Junior Switch 72
A4.1_ERT_b	1.5 m, 36 elec.	DipDip	1*a; 16 lvl.; 408 qdp.	Syscal Junior Switch 72
A4.1_ERT_c	1 m, 36 elec.	WenSl	1*a; 16 lvl.; 288 qdp.	Syscal Junior Switch 72

Appendix 12: Technical overview Alzenau: GPR

Name	*Frequency*	*Shield*	*Gain*	*Processing, Grid*	*Equiment type*
A4.1_GPR_a	250MHz	Yes	Start: 15 Attenuations: 15 dB/m MaxGain: 2000 Filter width: 5m	Bandpass-filter (% Freq): Fc1: 40; Fp1: 80; Fp2: 120; Fc2: 160; Topographic Migration; Background subtraction; Vel. 0.07 m/ns	*NOGGIN System*
A4.1_GPR_b	250MHz	Yes	Start: 15 Attenuations: 15 dB/m MaxGain: 2000 Filter width: 5m	Bandpass-filter (% Freq): Fc1: 40; Fp1: 80; Fp2: 120; Fc2: 160; Topographic Migration; Background subtraction; Vel. 0.1 m/ns	*NOGGIN System*
A4.1_GPR_c	250MHz	Yes	Start: 15 Attenuations: 15 dB/m MaxGain: 2000 Filter width: 5m	Bandpass-filter (% Freq): Fc1: 40; Fp1: 80; Fp2: 120; Fc2: 160; Topographic Migration; Background subtraction; Vel. 0.1 m/ns	*NOGGIN System*
A4.1_GPR_d	250MHz	Yes	Start: 15 Attenuations: 15 dB/m MaxGain: 2000 Filter width: 5m	Bandpass-filter (% Freq): Fc1: 40; Fp1: 80; Fp2: 120; Fc2: 160; Topographic Migration; Background subtraction; Vel. 0.1 m/ns	*NOGGIN System*
A4.1_3DGPR_a	250MHz	Yes	Start: 5 Attenuations: 15 dB/m MaxGain: 2000	3D in x and y direction, 1m interline spacing; Velocity: 0.1 m/ns	*NOGGIN System*
A4.1_3DGPR_b	250MHz	Yes	Start: 5 Attenuations: 15 dB/m MaxGain: 2000	3D in x and y direction, 1m interline spacing; Velocity: 0.082 m/ns	*NOGGIN System*
A4.1_3DGPR_c	250MHz	Yes	Start: 5 Attenuations: 15 dB/m MaxGain: 2000	3D in x and y direction, 1m interline spacing; Velocity: 0.1m/ns	*NOGGIN System*

Appendix 13: Technical overview Alzenau: MAG

Name	*Filter*	*Value range*
H4.2_MAG_a	Zero mean filter min.: 10; max.: 10	10– -10 nT
H4.2_MAG_c	Zero mean filter min.: 10; max.: 10	10– -10 nT
H4.2_MAG_d	Zero mean filter min.: 10; max.: 10	10– -10 nT
H4.2_MAG_e	Zero mean filter min.: 10; max.: 10	10– -10 nT

Eisenberg

Appendix 14: Technical overview Eisenberg: ERT

Name	*Spacing*	*Array*	*Electrodes*	*Equipment type*
E2_ERT_a	5 m, 36 elec.	WenSl	1*a; 18 lvl.; 288 qdp.	Syscal Junior Switch 72
E2_ERT_b	5 m, 36 elec.	WenSl	1*a; 16 lvl.; 288 qdp.	Syscal Junior Switch 72

Appendix 15: Technical overview Eisenberg: GPR

Name	*Frequency*	*Shield*	*Gain*	*Processing, Grid*	*Equipment type*
E2_GPR_a	30MHz	No	Start: 120 Attenuations: 10 dB/m MaxGain: 1500 Filter width: 5m	Bandpass-filter (% Freq): Fc1: 40; Fp1: 80; Fp2: 120; Fc2: 160; Migration; Background subtraction; Vel. 0.11 m/ns	PulsEKKO PRO System
E2_GPR_b	30MHz	No	Start: 120 Attenuations: 10 dB/m MaxGain: 1500 Filter width: 5m	Bandpass-filter (% Freq): Fc1: 40; Fp1: 80; Fp2: 120; Fc2: 160; Migration; Background subtraction; Vel. 0.11 m/ns	PulsEKKO PRO System
E3_GPR_a	100MHz	No	Start: 20 Attenuations: 10 dB/m MaxGain: 1500 Filter width: 5m	Bandpass-filter (% Freq): Fc1: 40; Fp1: 80; Fp2: 120; Fc2: 160; Migration; Background subtraction; Vel. 0.11 m/ns	PulsEKKO PRO System
E3_GPR_b	100MHz	No	Start: 20 Attenuations: 10 dB/m MaxGain: 1500 Filter width: 5m	Bandpass-filter (% Freq): Fc1: 40; Fp1: 80; Fp2: 120; Fc2: 160; Migration; Background subtraction; Vel. 0.07 m/ns	PulsEKKO PRO System
E3_GPR_c	100MHz	No	Start: 20 Attenuations: 10 dB/m MaxGain: 1500 Filter width: 5m	Bandpass-filter (% Freq): Fc1: 40; Fp1: 80; Fp2: 120; Fc2: 160; Migration; Background subtraction; Vel. 0.07 m/ns	PulsEKKO PRO System
E3_GPR_d	100MHz	No	Start: 20 Attenuations: 10 dB/m MaxGain: 1500 Filter width: 5m	Bandpass-filter (% Freq): Fc1: 40; Fp1: 80; Fp2: 120; Fc2: 160; Topographic Migration; Background subtraction; Vel. 0.07 m/ns	PulsEKKO PRO System
E3_GPR_e	30MHz	No	Start: 20 Attenuations: 10 dB/m MaxGain: 1500 Filter width: 5m	Bandpass-filter (% Freq): Fc1: 40; Fp1: 80; Fp2: 120; Fc2: 160; Topographic Migration; Background subtraction; Vel. 0.07 m/ns	PulsEKKO PRO System

Appendix 16: Technical overview Eisenberg: MAG

Name	*Filter*	*Value range*
E3_MAG_a	Zero mean filter min.: 10; max.: 10	10– -10 nT